Securing the Private Sector

Securing the Private Sector

Protecting US Industry in Pursuit of National Security

Darren E. Tromblay

LYNNE RIENNER PUBLISHERS

BOULDER
LONDON

The views expressed in this book are solely those of the author and do not represent the perspectives of any US government agency.

Published in the United States of America in 2022 by
Lynne Rienner Publishers, Inc.
1800 30th Street, Suite 314, Boulder, Colorado 80301
www.rienner.com

and in the United Kingdom by
Lynne Rienner Publishers, Inc.
Gray's Inn House, 127 Clerkenwell Road, London EC1 5DB
www.eurospanbookstore.com/rienner

Library of Congress Cataloging-in-Publication Data
Names: Tromblay, Darren E., author.
Title: Securing the private sector : protecting US industry in pursuit of national security / Darren E. Tromblay.
Description: Boulder, Colorado : Lynne Rienner Publishers, Inc., 2022. | Includes bibliographical references and index. | Summary: "Explores how the complex web of intelligence agencies has struggled to protect private economic and industrial interests that are vital to US national security"— Provided by publisher.
Identifiers: LCCN 2021051338 (print) | LCCN 2021051339 (ebook) | ISBN 9781955055123 (hardback) | ISBN 9781955055383 (ebook)
Subjects: LCSH: Computer security—United States. | Technology and state—United States. | National security—United States. | Public-private sector cooperation—United States.
Classification: LCC QA76.9.A25 T7355 2022 (print) | LCC QA76.9.A25 (ebook) | DDC 005.8—dc23/eng/20220118
LC record available at https://lccn.loc.gov/2021051338
LC ebook record available at https://lccn.loc.gov/2021051339

British Cataloguing in Publication Data
A Cataloguing in Publication record for this book is available from the British Library.

Printed and bound in the United States of America

♾ The paper used in this publication meets the requirements of the American National Standard for Permanence of Paper for Printed Library Materials Z39.48-1992.

5 4 3 2 1

For Kate

Contents

Acknowledgments

I AM INDEBTED TO THE MANY PEOPLE WHO HAVE IMPARTED THEIR professional wisdom over the years and who have inspired and encouraged my writing on national security—John, Fred, Molly, Rob, Deb, Dan, and many others.

I am also grateful to those dear people who kept me sane as I wrote this—especially Joe, Jim, KT and Dr. Pat, Connie and Joe, Sylvia and Garry, Dame Judith, BZ and George, Kay Frances, Suzanne, the totality of the "Trinity Tones," and Dr. Tom.

Finally, many thanks to my editor, Marie-Claire Antoine.

1

Private Industry and National Security

THE US GOVERNMENT HAS NEVER BEEN THE SOLE DE FACTO GUARantor of the country's national security. As a concept, national security has continually evolved in its meaning in order to accommodate new geopolitical and technological realities. The private sector plays a significant role in both of these areas by virtue of what it produces, where it does business, and with whom. Through these decisions, the private sector enhances or denies the government's capabilities with which to maintain power.

A particular challenge on which this book focuses is defining the security relationship between the government and elements of the private sector that do not rely on the government for their livelihood. Whereas certain business sectors (e.g., cleared defense contractors) function as extensions of the government and are attuned to national security considerations, many of the most innovative entities, the work of which has significant implications for US national security, do not naturally view their operations in the context of the national interest. This leaves them vulnerable to exploitation or disruption by threat actors, both state and nonstate, who view these entities as soft targets. The challenge for the US government is to bridge the gap in understanding between governmental and industry awareness of threats.

The Private Sector and US Elements of National Power

The flavor of each US National Security Strategy changes with new geopolitical and technological developments. Identification of new threats means that the United States must be able to pivot toward emerging challenges without having to completely retool its approach and develop bespoke solutions to new concerns. It must instead ensure that it has access to the fundamental tools of geopolitics that it can deploy against new challenges—wherever and from whomever those might emerge. These tools are known as elements of national power and, broadly speaking, consist of diplomacy, information, military, and economics.[1] Ensuring that the US government has access to them allows Washington to address multiple contingencies.

Industry has long been an essential partner in developing elements of national power. For instance, the Lockheed company's "Skunk Works" in Burbank, California, was integral to the development of the U-2 reconnaissance aircraft.[2] The U-2's facilitation of intelligence collection helped to develop the informational advantage of the United States during the Cold War. In this paradigm, industry relies on government patronage and follows its cues on national security. It essentially functions as an extension of the government. However, curation and enhancement of elements of national power are no longer solely the purview of the US government. Many of the capabilities that contribute to elements of national power are increasingly the domain of the private sector.

The relationship between the private sector, the US government, and the acquisition and advancement of capabilities that will support elements of national power has evolved, especially since the end of the Cold War. Increasingly, the private sector innovates and produces new technologies absent government patronage (and therefore absent responsibility to government sponsors). Even in those instances when government has taken a venture capital approach through bodies such as In-Q-Tel and the Defense Innovation Unit, it has been playing catch-up by buying into technologies that are already in development. Furthermore, the private sector is responsible for the bulk of US critical infrastructure, which is essential to elements of national power, particularly economics and information. The mechanisms for ensuring that government and industry find a common understanding of national security, despite responding to different incentives, is the subject of this book.

Dynamics of the Relationship Between the US Government and the Private Sector

Absent its role in directly commissioning technology, the US government and the private sector have developed rules and norms that define their relationship. The most clear-cut regulations are the statutes that define what goods and services industry can provide to whom, and under what conditions transactions can take place. Additionally, there are a number of self-imposed factors, including preservation of market share and ideological pandering, that inform the private sector's willingness, or lack thereof, to work with the US government (even as certain companies test the limits of cooperation with adversarial regimes).

Even though it often acts in its own interests, decoupled from concerns about US national security, the private sector has the ability to develop, or degrade, elements of national power. The field of economics is the element of national power that most people would associate with industry. Thanks to defense contracts, the private sector is also an inextricable participant in developing the military element of national power. Additionally, it is a key player in determining the status of the information element of national power, thanks to its role in developing and deploying means of communication. The ability to instantaneously unleash information on a global scale has been disruptive in both positive and negative ways. It increases "transparency" (although this term has sometimes been hijacked by malignant actors such as WikiLeaks), but also makes deception easier to commit (gullible people consume disinformation and act on it, for example).

Both transparency and deception have real-world implications for US statecraft. They can validate or undercut the narratives that the US government promotes globally. Additionally, private sector facilitation of information flows can strengthen the grip of US adversaries over their countries as well as weaken US allies' ability to support Washington, and, absent gatekeeping, it can allow foreign actors to interfere with US society and politics in a variety of nefarious ways.

Finally, the private sector's decisions have implications for diplomacy through their impacts on the circumstances that US policymakers must navigate. Decisions to sell or not sell certain capabilities to foreign governments change the carrots and sticks that Washington can wield. Furthermore, social media have influenced political outcomes and thereby have the potential to elevate a regime, with which the United States must contend, or unseat an allied government.

Other Vulnerabilities and Profound Consequences

Foreign powers—overtly and clandestinely—can benefit from targets that are not readily linked to, but nevertheless have implications for, elements of US national power. Political scientist Ashley Tellis identified that understanding national power not only is an accounting of visible assets, but also entails unpacking capabilities such as the aptitude for innovation and the quality of the knowledge base.[3] Identifying the linkages between non-obvious targets and elements of national power not only protects the capabilities on which the US government relies, but also helps the private sector to safeguard assets that it might not immediately think of as targets until it is too late to prevent harm to the bottom line.

Consistent with Tellis's assessment, innovation not immediately associated with elements of national power nevertheless has eventual implications for protecting and promoting them. At the time of this writing, the world was struggling through the Covid-19 pandemic. Health has been a long-standing concern for the United States. The 2010 Department of Homeland Security's *Quadrennial Homeland Security Review* explicitly cited the potential catastrophic impact, equal or greater than deliberate malicious attacks, that a pandemic could cause for the United States.[4] In 2014, the department similarly noted that "a devastating pandemic remains the highest homeland security risk."[5]

Foreign actors have long targeted the ability of the United States to effectively innovate toward solutions to wide-ranging health problems. During the 1940s, the Soviet Union attempted to acquire knowledge that would help the country to mass-produce penicillin, going so far as to approach a US company about purchasing a penicillin plant for erection in the Soviet Union. In the early 1950s, a Soviet agent attempted to gather information regarding details about a new process for synthesizing cortisone out of cheap and abundant raw materials that would enable mass production of the substance.[6] Jump ahead to 2020 and the Russians were still trying to siphon off Western research. The United Kingdom's National Cyber Security Centre announced that hackers, who almost certainly were working on behalf of Russian intelligence, targeted vaccine research in the United States, United Kingdom, and Canada.[7] In July 2020, the US Department of Justice indicted Chinese hackers, working on behalf of China's Ministry of State Security, for targeting Covid-19 research.[8]

The private sector also knowingly provides knowledge to hostile actors. For instance, McKinsey, the global consulting company, has helped China's regime to strengthen its grip over the country.[9] With fewer internal challenges, an authoritarian regime such as China's can focus its efforts outward to challenge the United States, forcing the United States to devote military resources (and by extension economic resources in order to develop effective defense technology) as well as diplomatic resources to countering the China threat.

Harm to the United States via Attacks on the Private Sector

Because the private sector is positioned to influence US elements of national power, it is also a direct, kinetic target of threat actors who are seeking to disrupt the ability of the United States to pursue desired policy outcomes. Attacks—especially sabotage and acts of terrorism—on industry, including private sector–owned infrastructure, have the potential to deny the US government tools it needs to achieve strategic objectives. Informational and economic elements of national power are the most immediate casualties in the case of such attacks. However, by focusing US resources inward, attacks on critical infrastructure have the potential to distract from diplomatic and military objectives.

Foreign powers (and domestic actors) have historically targeted US infrastructure. Attacks such as the bombing of the Black Tom railroad yard in 1916 by German agents have on occasion been kinetic in nature.[10] More recent threats have had the potential to turn a cyber attack into physical destruction. In 2013, an Iranian hacker obtained unauthorized access to the supervisory control and data acquisition systems of a dam in Rye, New York.[11] Iran and other entities have also historically probed the US electrical grid.[12]

Additionally, foreign actors have threatened to disrupt elements of US national power through activities that have the potential to affect less tangible, but equally essential, functions. For example, in 2016 the United States indicted several Iranian entities associated with the Iranian Revolutionary Guard for attacks on multiple companies in the US financial sector. These attacks disabled websites, prevented customers from accessing accounts, and incurred tens of millions of dollars in remediation costs.[13]

Closing the Loop: A Necessary Relationship

In order to protect its elements of national power, the US government has had—and will continue—to engage in activities directed at securing the private sector from state and nonstate threats. Among the many challenges in this area is the ability to reach a consensus with US industry about what constitutes security and what are industry's responsibilities, both as an entity regulated by the government and as a corporate citizen, in upholding security.

Even if the US government and US companies had a completely congruent understanding of security, which, due to differing incentives, they do not, there would be additional challenges to securing the private sector. Chief among these is the infrastructure for sharing information. Threats are multifaceted, and mitigation of those threats requires a wide range of expertise. Historically, the US government has struggled to address these issues. The challenge has toggled between a single agency being required to handle too many functions (e.g., the National Infrastructure Protection Center of the Federal Bureau of Investigation [FBI]) or too many agencies handling one function (e.g., aspects of cyber-related challenges divided between the FBI and the Department of Homeland Security).

There is not currently, nor has there ever been, an effective mechanism for establishing coherent and meaningful relationships between the government and private sector entities. In the late 1990s, the United States edged toward this by encouraging the creation of an information sharing and analysis center (ISAC), which was supposed to gather, analyze, sanitize, and disseminate private sector information to industry. The National

Infrastructure Protection Center (NIPC) would then disseminate information to the private sector.[14] However, the NIPC's functions were subsequently scattered across government and the ISAC concept became stovepiped, with individual industries each establishing an ISAC. It is time to revisit this concept and develop a clearinghouse for threat information, on foreign entities' intelligence collection and terrorism activities, that has implications for private sector targets. This body would also help to broker relationships between industries and the appropriate government agencies in order to deploy resources—such as the Department of Homeland Security's Cybersecurity Advisers—in furtherance of disrupting threats and mitigating vulnerabilities.

Structure of the Book

This book examines the history and complexity of the relationship between the US government and private industry in seeking to protect industry's contributions to elements of national power. The intent is to develop, through an examination of how these relationships have evolved, a better understanding of how best to engage the private sector in areas of shared security concerns.

Chapter 2 covers the rules of the road for the relationship between government and the private sector. It begins with a discussion of the laws that govern to whom the private sector can provide what, and when the "what" can go to the "whom." Then it specifically addresses the two types of laws—those that govern the "what" (e.g., the Arms Export Control Act) and those that govern the "whom" (e.g., the Trading with the Enemy Act)—that regulate the private sector's relationships with foreign entities. Additionally, it discusses the laws that govern what a foreign entity can and cannot do vis-à-vis aspects of the private sector (i.e., foreign investment and economic espionage). The chapter also provides an in-depth discussion of deemed exports (the transmission of knowledge rather than tangible technology). This will be a continuing problem in an increasingly globalized research and development ecosystem. It is also an area where foreign governments have pushed the boundaries of US laws in order to siphon knowledge through engagement with US companies and experts. Finally, Chapter 2 identifies the informal dynamics of the interaction between government and the private sector (e.g., US government investment, politics, and foreign relations) that complicate the relationship.

Chapter 3 examines the problems of counterintelligence in the private sector. It discusses the long-standing reality that foreign governments directly target the US private sector. (The private sector, because of this vulnerability, also provides a first line of defense for identifying what capabilities foreign governments are attempting to acquire as well as the methodologies and tactics foreign governments and other threat actors employ.) The chapter discusses the ways in which the US government has countered the threat to the private sector both through coordination across government agencies and by enlisting the American public. It also covers the various initiatives that the government has developed to increase industry's awareness of intelligence threats.

Chapter 4's topic is counterproliferation and counterterrorism. Unlike counterintelligence, both of these functions focus on preventing items from reaching dangerous end-users, rather than on the protection of an informational advantage. Although counterproliferation involves the exfiltration of technology and technological knowledge from the United States, the chapter focuses more on those items that could go boom in the night (or any other time of day). It then addresses how state and nonstate terrorist actors may deploy illicitly acquired technology or knowledge against the US private sector, including critical infrastructure, and discusses the steps taken by the US government to harden private sector targets against attacks from state and nonstate actors. It concludes with how geospatial intelligence (GEOINT) and imagery intelligence (IMINT) could contribute to the protection of critical infrastructure.

Chapter 5 covers the growth of the US government's cybersecurity activities, specifically as they pertain to protecting the private sector. It traces cybersecurity from the foundations that the FBI established, especially in its protection of networks through the National Infrastructure Protection Center, to the Department of Homeland Security's multiple, successive cybersecurity organizations. The reader should view the cyber milieu not as its own threat but rather as an environment that facilitates intelligence and terrorist threat actors.

Chapter 6 tackles some crosscutting considerations created by the changing nature of technology and threat scenarios (i.e., the intersection of actors, implements, and vulnerabilities). The increasingly complex threat environment has prompted the United States to engage in activities beyond its borders in order to protect US interests through the establishment of norms and the collection of information. A skilled, knowledgeable work force is essential to addressing the factors that have made securing the private sector a global challenge. The chapter juxtaposes the ever-expanding challenge with the government's perpetual struggle to hire and retain expertise capable of implementing the government's initiatives vis-à-vis the private sector.

Finally, Chapter 7 revisits the relationship between the public and private sectors. It discusses how government and industry can make common cause around the concept of human security, structure engagement to mitigate the fragmentation of information-sharing, and create opportunities for private sector expertise to inform government.

Notes

1. Congressional Research Service, *Defense Primer: Information Operations* (Washington, DC, 2020), https://crsreports.congress.gov/product/pdf./IF/IF10771.

2. Gregory W. Pedlow and Donald E. Welzenbach, *The CIA and the U-2 Program, 1954–1974* (Langley: Central Intelligence Agency, 1998), https://www.cia.gov/library/center-for-the-study-of-intelligence/csi-publications/books-and-monographs/the-cia-and-the-u-2-program-1954-1974/u2.pdf.

3. Gregory F. Treverton and Seth G. Jones, *Measuring National Power* (Santa Monica: RAND, 2005).

4. Department of Homeland Security, *Quadrennial Homeland Security Review* (Washington, DC, 2010), https://www.dhs.gov/sites/default/files/publications/2010-qhsr-report.pdf.

5. Department of Homeland Security, *Quadrennial Homeland Security Review* (Washington, DC, 2014), https://www.dhs.gov/sites/default/files/publications/2014-qhsr-final-508.pdf.

6. Federal Bureau of Investigation, *Soviet Intelligence Targets in the United States, 1946–1953* (Washington, DC, 1953), https://www.governmentattic.org/2docs/FBI_Monograph_Soviet-Targets-US_1953.pdf.

7. Chris Fox and Leo Kelion, "Coronavirus: Russian Spies Target Covid-19 Vaccine Research," *BBC News,* July 16, 2020, https://www.bbc.com/news/technology-53429506.

8. Department of Justice, "Two Chinese Hackers Working with the Ministry of State Security Charged with Global Computer Intrusion Campaign Targeting Intellectual Property and Confidential Business Information Including COVID-19 Research," https://www.justice.gov/opa/pr/two-chinese-hackers-working-ministry-state-security-charged-global-computer-intrusion.

9. Walt Bogdanich and Michael Forsythe, "How McKinsey Has Helped Raise the Stature of Authoritarian Governments," *New York Times,* December 15, 2018, https://www.nytimes.com/2018/12/15/world/asia/mckinsey-china-russia.html.

10. Federal Bureau of Investigation, "Black Tom 1916 Bombing," https://www.fbi.gov/history/famous-cases/black-tom-1916-bombing.

11. Department of Justice, "Seven Iranians Working for Islamic Revolutionary Guard-Corps Affiliated Entities Charged for Conducting Coordinated Campaign of Cyber Attacks Against U.S. Financial Sector," March 24, 2016, https://www.justice.gov/opa/pr/seven-iranians-working-islamic-revolutionary-guard-corps-affiliated-entities-charged.

12. Center for Strategic and International Studies, *Significant Cyber Incidents Since 2006,* https://csis-website-prod.s3.amazonaws.com/s3fs-public/201106_Significant_Cyber_Events_List.pdf.

13. Department of Justice, "Manhattan U.S. Attorney Announces Charges Against Seven Iranians for Conducting Coordinated Campaign of Cyber Attacks Against U.S. Financial Sector on Behalf of Islamic Revolutionary Guard Corps–Sponsored Entities," March 24, 2016, https://www.justice.gov/usao-sdny/pr/manhattan-us-attorney-announces-charges-against-seven-iranians-conducting-coordinated.

14. Presidential Decision Directive/NSC-63, May 22, 1998, https://fas.org/irp/offdocs/pdd/pdd-63.htm.

2

Regulating the Transfer of Technology and Knowledge

THE UNITED STATES HAS, FROM ITS EARLIEST YEARS TO THE PRESENT day, implemented a variety of legislative and regulatory measures meant to prevent the exfiltration of sensitive technology and knowledge to US adversaries and competitors. Two types of legislation form the regime governing the export of technology and knowledge. The first category is the legislation that focuses on disrupting transactions and seizing assets. This allows the United States to deny foreign threats from making purchases that could harm US interests. The second type of legislation focuses on preventing hostile foreign actors from acquiring certain types of technology and knowledge, by regulating specific items (see Table 2.1). In addition to statutes, the United States regulates foreign acquisitions of US private sector entities through the Committee on Foreign Investment in the United States (CFIUS). In addition to the formal regulations, informal regimens—and tensions—have further governed the relationship between the US government and US commercial interests.

US policymakers have been concerned about the exportation of items that could aid its adversaries since the country's founding. In 1775, the First Continental Congress outlawed exports to Great Britain. Several decades—and the denouement of one revolution—later, President Thomas Jefferson implemented the Embargo Act, in 1807, which terminated US exportation to any overseas destination. (This measure was in place for slightly more than a year before Congress walked back the act.) In response to the eruption of the American Civil War, the US Congress implemented measures to halt exports to Confederate states.[1]

World War I forced the largely isolationist United States to contend with the realities of its geopolitical position—it was a significant power and was consequently in the crosshairs of enemies who wanted it sidelined before it could intervene. Its response was the Trading with the Enemy Act

Table 2.1 Key US Legislative and Regulatory Measures Since 1917

Preventing the Movement of Technology to Threat Actors Abroad		Preventing Threat Actors from Acquiring Knowledge Through Investments in the US	
1917	Trading with the Enemy Act		
1949	Export Control Act		
1969	Export Administration Act (rewritten in 1979)		
		1975	Creation of CFIUS (Executive Order 11858)
1976	Arms Export Control Act	1976	International Investment Survey Act
1977	Emergency Economic Powers Act		
		1988	Exon-Florio Amendment to the Omnibus Trade and Competitiveness Act
		1993	Byrd Amendment to the National Defense Authorization Act
1996	Economic Espionage Act		
		2007	Foreign Investment and National Security Act
2012	Theft of Trade Secrets Clarification Act		
2018	Export Control Reform Act	2018	Foreign Investment Risk Review Modernization Act

of 1917.[2] According to the act, an "enemy" included any individuals, partnerships, and corporations of any nationality, or resident with the territory, of countries with which the United States was at war, or any entity doing business with such a territory or country. According to the Congressional Research Service, this act functioned as the primary means to impose sanctions, which would help the US fight the Cold War, from the post–World War II 1940s through the early 1970s.[3]

Under the act, the president, in time of war or other national emergency, could exercise four primary levers of power. These were: regulatory powers with respect to financial transactions; regulatory power with respect to any property in which a foreign country or its nationals had an interest; power to vest any property or interest of any foreign country or national; and the powers to hold, use, administer, liquidate, sell, or otherwise deal with property for the benefit of the United States.[4]

As it assessed its role vis-à-vis the world, in the run-up to World War II, the United States again took steps to prohibit the exportation of goods that might help hostile countries. Public Law 703, enacted in July 1940, to expedite the strengthening of the national defense, included provisions to halt the provision of goods to certain foreign parties. According to the law, the US president could—when national defense made it necessary—"prohibit or curtail the exportation of any military equipment or munitions, or component parts thereof, or machinery, tools, or material, or supplies necessary for the manufacture, servicing, or operation thereof."[5]

This law provided the basis for the creation of a concept that would be a fundamental aspect of counterproliferation. Expanding on the law, President Franklin D. Roosevelt issued a succession of proclamations, during the following year, that required exporters to acquire a license from the Department of State prior to sending a wide variety of items—including arms, ammunition, and implements of war, as well as a wide variety of basic materials and commodities—to purchasers abroad.[6] Congress clarified the law in 1941, 1942, 1944, and 1945 through additional legislation. This process of clarification included the inclusion of technical data (in contrast to physical materials).[7] The movement of knowledge, through the globalization of business and academia, has become an ever-thornier problem.

The Export Control Act of 1949 was the first significant post–World War II export legislation. It was originally passed as an emergency measure in the face of domestic shortages and the threat of inflation produced by abnormal foreign demand.[8] According to this act, there were three reasons for imposing controls: prevent the export of scarce goods, the absence of which would have a deleterious impact on US industry and national economic performance; promote the foreign policy of the United States; and ensure national security by restricting the exportation of goods and technology that would contribute to the military capabilities of hostile countries.[9]

Under the Export Control Act, the US Department of Commerce became the agency with primary responsibility for administering and enforcing export controls. Commerce's acquisition of export control responsibilities provided an early indication of the need for interagency expertise. Its International Enforcement Branch (a predecessor to the current Bureau of Industry and Security), in Commerce's Office of International Trade, recruited personnel from multiple US government agencies including the Federal Bureau of Investigation (FBI), US Customs (a Department of Treasury agency), and military agencies.[10]

For approximately two decades, despite the act's initial emergency origins, Congress routinely renewed the Export Control Act. In 1969, Congress supplanted it with the Export Administration Act, which, according to the Congressional Research Service, replaced the Export Control Act's "near-embargo characteristic." Implementation of the Export Administration Act was through the Export Administration Regulations, which the Department of Commerce's Bureau of Industry and Security administers. In 1979, Congress comprehensively rewrote the act.[11]

In the mid-1970s the United States introduced a new measure that had significant implications for private industry's transfer of technology. The Arms Export Control Act of 1976 provided the US executive branch with statutory authority over the export of defense articles and services.[12]

Congress, in promulgating this act, specifically linked the sales of military technology to the support of US foreign policy.[13] According to the act, exporters could sell defense articles and services only to friendly countries and only for internal security, legitimate self-defense, to enable the recipient to participate in collective security arrangements, and—in less developed countries—to construct public works and facilitate activities that would enhance economic and social development.[14]

The International Traffic in Arms Regulations constitute the vehicle for implementing Arms Export Control Act. The regulations describe the requirements for obtaining approval to promote major defense equipment and to export defense articles.[15] According to the Arms Export Control Act, the articles and services that the president designates constitute the US Munitions List, which is published as part of the regulations.[16] The Neutrality Act of 1935 provided the impetus to the initial publication, in 1936, of the State Department–issued International Traffic in Arms Regulations.[17] Under the Neutrality Act, entities within the United States could not export arms, ammunition, or implements of war to foreign nations at war, and arms manufacturers were required to apply for export licenses.[18]

In the 1970s, Congress revisited the Trading with the Enemy Act. During the 1930s, Congress had expanded the act in order to allow the president to declare a national emergency in peacetime, as well as wartime. Congressional inquiries in the 1970s discovered that—under the act—the United States had technically been in a state of emergency since 1933. No less than four different states of emergency were in effect simultaneously. The mid-1970s were a bad time for presidential power—as the country looked askance at the White House in the aftermath of Watergate and the revelations of the House and Senate intelligence inquiries (one of which was, coincidentally, led by Senator Frank Church, who was also one of the chairs over the bipartisan special committee that assessed the implications of the Trading with the Enemy Act). Congress, in 1977, sought to assert its power vis-à-vis the executive branch by passing the International Emergency Economics Powers Act, with the intent of reining in presidential power.[19]

Unlike the Export Control Act, the Export Administration Act, and the Arms Export Control Act, International Emergency Economics Powers Act focuses on disrupting transactions with, rather than the transfer of specific technologies to, hostile entities. It permits—in response to "any unusual and extraordinary threats, which has its source in whole or substantial part outside the United States, to the national security, foreign policy, or economy of the United States"—the investigation and curtailment of financial and property transactions with foreign countries and nationals. Additionally, the act permits—in time of armed hostilities or when attacked by a foreign country or foreign national—the confiscation of any property of foreign persons, foreign organizations, or foreign countries that planned, authorized, aided, or engaged in the hostilities or the attack.[20] The United States first used the International Emergency Economics Powers Act in response to the Iranian hostage crisis of 1979.[21]

The act has intersected with other export-related regulations. Presidents have used it to maintain dual-use export controls when the Export Administration Act lapsed. For instance, in 1983, President Ronald Reagan determined that "unrestricted access of foreign parties to United States commercial goods, technology, and technical data and the existence of certain boycott practices of foreign nations" constituted—in conjunction with expiration of the Export Administration Act—an "unusual and extraordinary threat to the national security."[22]

Additionally, the International Emergency Economics Powers Act has, like the Export Control Act of 1949, been explicitly used to advance specific foreign policy objectives. While the International Emergency Economics Powers Act was originally directed at foreign governments, it has also proved to be a tool for addressing nonstate actors and transnational threats. Presidents have used the act to target nonstate entities such as terrorist organizations, corporations, and narcotics traffickers. In 1990, President George H. W. Bush established the precedent of using the act in response to dangerous transnational trends when he took aim at the proliferation of chemical and biological weapons. Other transnational problems at which the United States has leveled the act include human rights abuses, slavery, denial of religious freedom, political repression, public corruption, and the degradation of democracy. Targets of the act have evolved as threat actors and have turned emerging technologies against the United States. President Barack Obama, for instance, applied the powers against "persons engaging in malicious cyber-enabled activities." Acknowledging that private industry's contributions to technology are essential to national security, the United States has used the act to combat the theft of trade secrets.[23]

The Export Control Reform Act of 2018 is the most recent significant piece of export regulation legislation. This act replaced the Export Administration Act as the mechanism for governing the export of dual-use goods and services.[24] It is informed by the understanding that the United States must "maintain its leadership in the science, technology, engineering, and manufacturing sectors, including foundational technology that is essential to innovation."[25] The specific role of private industry in US national security is clearly indicated by the legislation's statement that US leadership "requires that United States persons are competitive in global markets."[26]

Committee on Foreign Investment in the United States

Foreign powers attempt to create new vulnerabilities, as well as exploit existing ones. Through foreign direct investment they can get closer to both the widgets and the work force possessing valuable knowledge. The US government has established a mechanism—known as the Committee on Foreign Investment in the United States (CFIUS)—that is meant to prevent foreign entities from making inroads that get them closer to US intellectual property. Unfortunately, while it is a worthwhile concept, the CFIUS process has historically focused narrowly on a small subset of transactions and has done little to disrupt foreign siphoning of knowledge from US industry.

The CFIUS is a relatively obscure interagency body that has a significant role in governing the acquisition of assets by entities abroad. Its mandate is to assess the national security implications of foreign powers' acquisitions in the United States, via foreign direct investment, of existing US entities. President Gerald Ford implemented the CFIUS in 1975 via an executive order.[27] The committee currently operates under the auspices of the Department of the Treasury and draws its authority from the Defense Production Act of 1950, amended by the Foreign Investment and National Security Act of 2007, and by the Foreign Investment Risk Review Modernization Act of 2018.[28]

However, the CFIUS does not address several key concepts. First, it operates on an insufficient definition of national security. Additionally, it does not assess "greenfield"—locations not previously developed (e.g., not the acquisition of a US company)—outposts of foreign enterprises, or the implications of US companies' expansion abroad (which are significant, from a counterintelligence perspective, as they may create new vulnerabilities). Finally, the CFIUS is largely (though not entirely) reliant on notification from companies about a pending transaction.

As with subsequent evolutions of the body, the creation of the CFIUS was the result of political pressure: an attempt to placate a Congress that had grown increasingly concerned about the rapid increase in investments by Organization of the Petroleum Exporting Countries (OPEC) in US portfolio assets and a belief that some of the OPEC members were making investments for political, rather than economic, reasons. By creating the CFIUS, the White House hoped to dissuade Congress from enacting new restrictions on foreign investment.[29] The International Investment Survey Act of 1976 strengthened the executive branch's hand in reviewing foreign transactions, by providing the president clear and unambiguous authority to collect information on international investment.[30]

The CFIUS was a relatively underpowered and uncertain body during its early years. According to the executive order that established it, the CFIUS was to have the primary continuing responsibility within the executive branch for monitoring the impact of foreign investment in the United States. The body was, among its other functions, supposed to furnish the US government with an informational advantage by preparing analyses of trends and significant developments pertaining to foreign investment in the United States as well as review investment in the United States that, in the judgment of the CFIUS, might have significant implications for US national interests. However, this ambitious agenda was not matched by a reality that saw the CFIUS meet only ten times in its first five years of existence. Furthermore, according to an assessment by the Congressional Research Service, the committee seemed unable to decide whether it should focus on the political or the economic aspects of foreign direct investment in the United States.[31]

Exon-Florio

The next significant development in the evolution of the CFIUS occurred in 1988 with the passage of the Exon-Florio Amendment to the Omnibus Trade and Competitiveness Act. Again, congressional concerns provided the impetus for legislation, with the issue in question being the proposed sale, in 1987, of Fairchild Semiconductor, by its owner, the French firm Schlumberger Limited, to the Japanese firm Fujitsu.[32] This contributed to a broader concern that foreign takeovers of US firms could not be stopped by anything less than the president's declaration of a national emergency, or regulators' invocation of federal antitrust, environmental, or securities laws.[33] Exon-Florio gave the president the authority to block proposed or pending foreign mergers, acquisitions, or takeovers of persons engaged in interstate commerce in the United States that threatened to impact national security.[34] (Before the president can invoke this authority, they must determine that other US laws are inadequate or inappropriate for protection of national security. Additionally, the president must have credible evidence that the foreign investment will impact national security. This represented an effort to distinguish legitimate concerns from political considerations.) President Ronald Reagan implemented provisions of the Omnibus Trade Act, via an executive order, and, as part of this, delegated to the CFIUS the president's authority to conduct reviews, undertake investigations, and make recommendations. This was despite the fact that the legislation did not specifically mention the CFIUS. As a Congressional Research Service report noted, "this transformed CFIUS from a purely administrative body to one with a broad mandate and significant authority."[35] With this authority, companies were increasingly inclined to negotiate with, or shy away from, the CFIUS, in advance of a potential denial of a transaction.

What Exon-Florio failed to do was update the conceptualization of what constituted national security. The original legislation would have broadened the definition beyond the traditional concept of military and defense issues to also include a strong economic component.[36] (Other measures do prohibit foreign investment in certain critical sectors such as maritime, aircraft, banking, resources, and power in order to prevent foreign control over public services and public interest activities.)[37] However, the Reagan administration objected to this change.[38] Limiting the scope of its work created a revised identity for the CFIUS. According to the Department of Commerce, the passage of Exon-Florio shifted the role of the CFIUS from monitoring overall foreign investment in the United States to determining how foreign mergers, acquisitions, and takeovers of US companies would impact national security (in the limited sense of national security).[39] Defining what meets the criteria of national security will continue to be an evolving issue reflecting changes in technology, business practices, and geopolitics.

Under Exon-Florio the president must consider twelve factors when deciding to block a foreign acquisition. (The CFIUS is the body to which these considerations have been delegated.) The Exon-Florio provisions are useful lenses through which to filter the relationship between private industry and national security. These cover three areas: technological capabilities that can enhance elements of national power; the preservation of critical infrastructure; and the impact that the private sector can have on US foreign policy.

First, the provisions highlight the role that private industry plays in giving the United States a technological edge. Specifically, provisions require consideration of the potential effects inherent to a transaction on US technological leadership in areas affecting US national security as well as the impact of a foreign acquisition on US critical technologies. Additionally, Exon-Florio requires consideration of how foreign control of domestic industries and commercial activity could affect the capability and capacity of the United States to meet the requirements of national security.[40] These provisions suggest that governments of multiple countries (including the United States) rely on the private sector to enhance their respective elements of national power.

The second theme that emerges from a reading of Exon-Florio provisions is the role of private industry in protecting the United States against terrorism and coercion (both physical and economic) from foreign powers that might attempt to disrupt functions essential to Americans' well-being. Specifically, the provisions address private industry's control of critical infrastructure. Among these provisions are considerations of whether the transaction has a security-related impact on critical infrastructure in the United States and the potential effects on US critical infrastructure, including major energy assets.[41]

Finally, Exon-Florio prompts considerations of how private industry—by introducing or providing capabilities to foreign powers—can have an impact on US foreign policy. This is especially apparent in the need for assessing the potential effect of the transactions on the sales of military goods, equipment, or technology to a country that supports terrorism or proliferates missile technology or chemical and biological weapons. Relatedly, Exon-Florio requires US policymakers to consider whether specific transactions—identified by the secretary of defense—pose a regional military threat to the interests of the United States.[42]

The private sector can also impact foreign policy by withholding capabilities that the United States needs in order to implement policies of its choice. Exon-Florio acknowledges this in its considerations of domestic production needed for projected national defense requirements. Its provisions also require an assessment of the capability and capacity of domestic industries to meet national defense requirements including the availability of human resources, products, technology, and materials.[43]

Byrd Amendment

The Byrd Amendment to the National Defense Authorization Act for fiscal year 1993 gave the CFIUS additional responsibilities. Until this point, reviews happened either at the voluntary submission of a proposal to the CFIUS process, or in response to a CFIUS member agency's request for the review of a specific transaction. Although the CFIUS relied, significantly, on voluntary notifications, the committee believed that it had been notified of most foreign investment in key areas with implications for national security, since investors had a strong incentive to seek CFIUS approval prior to a transaction, as the president could order divestitures of transactions. This was not the case. According to a 1995 US General Accounting Office (GAO) report, many foreign investments in high-technology and defense-related industries went unreported to the CFIUS. The GAO determined that "the CFIUS process alone cannot be relied on to surface transactions posing potential national security concerns."[44]

In an effort to heighten oversight of foreign activity, the Byrd Amendment made it mandatory for the CFIUS to investigate proposed mergers, acquisitions, or takeovers in cases where two criteria are met: the acquirer is controlled by or acting on behalf of a foreign government, and the acquisition results in control of a person engaged in interstate commerce in the United States that could affect national security. The mandatory review of "covered" transactions has evolved to identify whether a transaction threatens to impair national security, or the foreign entity is controlled by a foreign government, or would result in control of any critical infrastructure that could impair national security.[45]

The US government has made efforts to enhance its scrutiny of foreign investments. The CFIUS can do its own research to identify significant mergers in cases where the parties did not submit notifications. However, there have been cases where the CFIUS's own research failed to detect these incidents, such as when a Hong Kong company that traded with China purchased a US company that produced ball bearings for US military aircraft. When the company self-disclosed the merger—approximately a year after the merger had happened—national security and export control concerns were raised.[46] Such transactions are concerning since technology and knowledge are eminently transferable and may be in the hands of a foreign actor by the time the CFIUS and the president determine that a divestiture is necessary.

Other members of the CFIUS can bring transactions of concern to the committee's attention.[47] For instance, the Department of Commerce—through the Bureau of Industry and Security—assists the CFIUS with ferreting out non-notified transactions. The bureau—in its work on non-notified transactions—focuses on determining whether the US company produces or trades in "critical technologies" that can be indicators of national security risk inherent to the transaction.[48] The identification of foreign investments that pose national security concerns continues to be a challenge for the United States, as indicated by the National Counterintelligence Strategy of 2020–2022, which discusses the need to identify and counter foreign investments in the United States that pose a national security threat.[49] The Department of Homeland Security (DHS), which became a member of the CFIUS in 2003, has also taken on similar tasks. It established an early warning program in order to identify transactions of potential concern before companies made formal CFIUS filings. According to 2005 Senate testimony, this program focused on US critical infrastructure and industrial-based technology companies. Part of this effort included outreach by the DHS to the parties involved and requests for technical and financial briefings from those entities.[50]

Foreign Investment and National Security Act

In 2007, Congress passed the Foreign Investment and National Security Act. As with previous legislation, the political climate provided context for the legislation. In the years leading up to the act, the Chinese had attempted to acquire the oil company Unocal, and more recently, Dubai Ports World pursued purchase of a British firm that operated several key US ports. (In both cases, the potential buyers abdicated the playing field after encountering hostile public opinion.) The Foreign Investment and National Security Act addressed six problems identified by Congress: the principal members of the CFIUS at times seemed to be ill-informed concerning the outcomes of reviews and investigations regarding proposed or pending investment transactions; the CFIUS had interpreted incorrectly the requirements under the statutes for investigation of transaction that involved firms that were owned or controlled by a foreign government; reporting requirements had not provided Congress with enough information about the operations and actions of the CFIUS for members to fulfill their oversight responsibilities; the CFIUS had exercised too much discretion in choosing which transaction it investigated; the definition of "national security" used by the CFIUS was no longer adequate in a post–September 11 world; and deadlines placed on the CFIUS to complete reviews and investigations of investment transactions did not always provide adequate time for the committee to complete its reviews and investigations.[51]

The Foreign Investment and National Security Act introduced several significant changes to the CFIUS process. It gave statutory recognition to the CFIUS and established the secretary of the Treasury as its chair. It also added the Department of Energy as a member and granted the president permission to add any other members deemed necessary.[52] The executive order that implemented the act added five members of the CFIUS to observe and, as appropriate, participate in and report to the president: director of the Office of Management and Budget; chairman of the Council of Economic Advisers; the national security adviser; the assistant to the president for economic policy; and the assistant to the president for homeland security and counterterrorism. The Foreign Investment and National Security Act built upon Exon-Florio's consideration of critical infrastructure by specifically modifying the definition of entities relevant to national security to include critical infrastructure such as energy deposits and power plants. It also included "homeland security" and "critical industries" as broad categories of economic activity that could be subject to a CFIUS review.[53] Consistent with the physical concerns created by critical infrastructure, the CFIUS would consider the location of acquired businesses and the proximity to certain government facilities such as military outposts.[54]

Additionally, the Foreign Investment and National Security Act introduced the director of national intelligence (DNI) as a participant in the process. Although legislation did not make the DNI a member of the CFIUS, it did require the DNI to carry out an analysis of any threat to US national security resulting from any merger, acquisition, or takeover.[55] Analyses incorporate the views from all affected or appropriate intelligence agencies with respect to the transaction. Furthermore, the DNI must ensure that the intelligence community remains engaged in the collection, analysis, and dissemination to the CFIUS of any additional relevant information that may become available during the course of any investigation conducted.[56] Presumably, this requires ongoing awareness of what the foreign presence is doing, since a 2006 change to the CFIUS process, by the administration, allowed the CFIUS to reopen a review of a deal and to overturn its approval at any time if the CFIUS believed that the companies materially failed to comply with the terms of the arrangement.[57]

In accordance with the Foreign Investment and National Security Act, the CFIUS must now provide confidential briefings to Congress as requested and produce an annual report. These measures make investment regulation a subject of national policy debate, as opposed to a technocratic decision. The divergence of the two is demonstrated by the proposed acquisition of P&O (which operated several major US ports) by Dubai Ports World in 2006, which the CFIUS approved but which Congress opposed, as well as by concerns about the Chinese-owned CNOOC's attempt to purchase Unocal, an oil company. Political decisionmaking also introduces a broader rubric than the formal CFIUS process. For instance, according to *The Economist* in 2007, the United States made it clear that it could consider the openness of other countries' markets when their governments are trying to buy US companies.[58]

Foreign Investment Risk Review Modernization Act

US policymakers attempted to close some of the remaining gaps in the CFIUS process with the Foreign Investment Risk Review Modernization Act of 2018. According to the Congressional Research Service, the act grew from concerns that "the national security landscape [had] shifted in recent years and so [had] the nature of the investments that pose the greatest potential risk to national security."[59] The act added several factors to those that Exon-Florio introduced. These new considerations align with the themes that emerge from the Exon-Florio provisions. First, they seek to ward off foreign policy challenges by considering transactions that involve a country of "special concern" that has a strategic goal of acquiring critical technology or critical infrastructure that could affect US leadership in areas of national security. Additionally, they acknowledge the role of the private sector in introducing game-changing technologies that could enhance governments' elements of national power by exacerbating or creating new cybersecurity vulnerabilities or by giving a foreign government a significant new capability to engage in malicious cyber-enabled activities. Finally, the act has a corollary with Exon-Florio in its attempts to ward off physical or economic coercion by assessing the potential effect of the cumulative control of, or pattern of transactions involving, any one type of critical infrastructure, energy asset, critical material, or critical technology by a foreign government or person.[60]

The act also includes a twenty-first-century consideration: the concept of personally identifiable, genetic, or other sensitive data. "Big data" in the hands of US adversaries has been an increasing concern—and one that probably seemed like nothing more than a distant Orwellian nightmare when the original Exon-Florio provisions were introduced—for a number of years. The 2014 and 2015 hacks with a nexus to China that compromised the US Office of Personnel Management, which maintains the records of all federal employees, brought attention to US adversaries' interest in data that through which they could sift to build dossiers on Americans.[61] Chinese hackers were also responsible for a breach of the Anthem insurance company, which, by early 2015, netted the hackers approximately 80 million records on Anthem customers and employees.[62] In addition to acquisition via illicit means, Chinese actors have attempted to obtain identifying information—such as genomic data—through investments in the health sector.[63] Chinese entities have even attempted to acquire data about the sexual preferences of Americans. In 2019, the Chinese company Beijing Kulnun Tech completed its purchase of Grindr, a gay dating app, based in West Hollywood, California.[64] The Grindr database includes information about the user's location, messages, and HIV status. In mid-2019, the CFIUS, operating under the Foreign Investment Risk Review Modernization Act, directed Beijing Kunlun Tech to divest Grindr by June 2020.[65]

Additionally, the Foreign Investment Risk Review Modernization Act changed the CFIUS by making several additional categories of transactions subject to review. Two of these—investments in certain US businesses that afford a foreign person access to material nonpublic technical information in possession, or direction, of the US business, and any change in a foreign investor's rights resulting in foreign control of a US business—are natural extensions of activities regulated by the CFIUS. A third—regulating a transaction based on purchase, lease, or concession by or to a foreign person of real estate located in proximity to sensitive government facilities—acknowledges how a foreign threat may leverage its footprint to engage in activity harmful to the United States.[66]

China's effort to obtain a sensitive site provides context for the latter provision. In 2012, China's nationally owned Ralls Corporation acquired multiple US companies that existed for the purpose of developing wind farms in north-central Oregon.[67] The property that Ralls acquired in this transaction included land near a naval weapons training facility in Boardman, Oregon, that was a test site for unmanned aerial vehicles (UAVs).[68] It is difficult to believe that Ralls's acquisition was entirely coincidental. UAV technology was, contemporaneously, a hot item for Chinese intelligence collection, as indicated by the case of Hui Sheng Shen and Huan Ling Chang, who—as of late 2011—began conspiring to export UAV technology to China.[69] Ralls, in 2015, ultimately agreed to divest of the wind farms.[70] (China, however, has relentlessly continued to collect against unmanned vehicle technology, as indicated by the 2016 case of Wenxia Man, whom the United States convicted for conspiring to export a General Atomics UAV to China.)[71]

Even after passage of the Foreign Investment Risk Review Modernization Act, foreign governments seeking to acquire US technological expertise—at the expense of US industry—can arguably continue to do so via "greenfield" investments. The CFIUS governs acquisitions of US assets, but China has avoided having to make acquisitions by simply setting up US branches of its own technology firms. According to the Department of Defense's Defense Industrial Unit, Chinese firms have done this to access US talent and technology. For instance, in 2013, Baidu established an Institute for Deep Learning in Silicon Valley in order to compete with US technology firms, such as Google, Apple, and Facebook, for talent in the artificial intelligence field. Several years later, the Zhong Guan Cun Innovation Center opened in Silicon Valley.[72] The CFIUS, however, remains underpowered in its ability to call these developments into question, since it can only currently review property acquisitions when the sites are in proximity to US government facilities such as military installations or property with national security sensitivities.[73] This allows China as well as other US adversaries and competitors to leverage US human capital, by hiring talent on US soil, to build the capabilities that they can use to challenge US interests.

Deemed Exports

Tangible transactions are difficult to disrupt. Intangible transactions—specifically the transfer of knowledge—are even more problematic. Historically, multiple foreign governments have engaged in activities that were directed at siphoning off knowledge from experts in the United States. The Export Administration Regulations note that there are multiple routes via which information can conceivably be released to a foreign national who is on US soil: visual inspection of US-origin equipment and facilities; oral exchanges of information in the United States or abroad; and the application to situations abroad of personal knowledge or technical experience acquired in the United States.[74] Foreign governments have tried to acquire information through all of these methods.

Visual Inspection

The Soviet Union provided multiple Cold War examples of how a foreign government could manipulate US contacts to provide opportunities for visual inspection of information. For instance, the Soviet Union's government-run Amtorg Trading Company would advertise for employees. (At one point approximately half of Amtorg's Soviet personnel were either known or suspected to have intelligence affiliations.)[75] When interviewees arrived at Amtorg's offices, they would receive instructions to bring blueprints of former projects in order to demonstrate their aptitude. The Soviets would photograph the blueprints and forward them to Russia.[76] Skip ahead to 2020. In July, a Chinese agent pleaded guilty in the United States after creating a fake consulting company that solicited résumés, which the agent passed to Chinese intelligence operatives. After he identified individuals of interest, the Chinese agent would solicit the individuals to write reports and provide non-public information, which he claimed were for clients in Asia. He wasn't completely lying; the government of China was indeed an Asian client.[77]

In addition to direct contacts with employees or former employees of US facilities in which foreign governments are interested, foreign governments may use the lure of trade deals to extract valuable information from companies attempting to prove their market mettle. For instance, during the Cold War, Soviet-bloc officials used correspondence with firms to obtain blueprints, photographs, production statistics, and other data about

US equipment and production methods.[78] In 1965, the FBI advised Congress that Soviet-bloc personnel could simply review advertisements that appeared in scientific and industrial publications to identify items of interest. They would then initiate direct correspondence with the US industrial concerns and, through this method, frequently succeeded in obtaining photographs, blueprints, and detailed specifications about the most recent industrial developments in the United States.[79]

Business-to-business transactions also provide opportunities for ill-intentioned inspection. The FBI identified this problem as early as 1941, when it advised that managers of US industrial plants should use caution when communicating with companies outside the United States so as not to disclose confidential plans or information that may have been in their possession.[80] Firms chartered in the United States can legally purchase controlled US technology and study it without violating US export controls.[81] In 1962, Congress learned that an Amtorg representative purchased approximately $30,000 worth of electronic equipment from a US firm, claiming that the equipment would be used in the United States.[82] Other Eastern-bloc countries established similar operations. During the latter part of the Cold War, there were approximately thirty communist-owned, US-chartered firms in the United States (five Soviet, seventeen Polish, five Czechoslovakian, and three Hungarian).[83] Although these firms were technically prohibited from removing the technical data from the United States, the potential for knowledge, derived from inspections on US soil, to disappear behind the Iron Curtain was real. The Central Intelligence Agency (CIA) confirmed in a 1982 assessment that Poland, Czechoslovakia, and Hungary were among the principal East European intelligence services leveraged by the Soviets for acquisition of Western technology.[84]

Clearly foreign governments have long relied on the power of purchasing to acquire US industrial information. According to a 1953 FBI monograph, Amtorg's publication division produced a monthly magazine during the late 1940s titled *American Engineering and Industry,* for distribution in the Soviet Union. This Russian-language publication included pictures, drawings, and descriptive data regarding engineering and industrial products that the Soviet Union could purchase.[85] Amtorg also annually prepared the publication *Catalogue of American Engineering,* which in 1946 numbered 5,000 pages.[86]

Illicit photography of items in which foreign governments have an interest has long been a way to facilitate visual inspection. In 1965, for instance, the FBI advised Congress that Soviet-bloc officials had attended various conventions and symposia throughout the United States and had used these visits as opportunities to photograph material on exhibit.[87] More recently, in 2004, the National Counterintelligence Executive noted that clandestine filming of equipment was a standard collection technique.[88] One relatively recent instance involved a Chinese national who, during an international arms exhibit, videotaped every static display and took accompanying notes.[89]

Furthermore, foreign powers have historically taken the concept of visual inspection, through commercial interactions, to the next level by engineering personal visits to US industrial interests. For instance, in 1960, the FBI disclosed that Soviets working for Amtorg would request permission for Soviet officials to visit industrial facilities—premised on the prospect of purchases—throughout the United States. The Soviets would string US companies along with the suggestion that orders for products would be forthcoming if products were deemed satisfactory. Soviet contracts consistently included provisions that afforded Soviet inspectors the privilege of reviewing all of the merchandise before it was shipped to Russia. Along the same lines, the Soviets used the inducement of promised purchases to demand blueprints of the product that they claimed was under consideration for purchase.[90] This was consistent with the duplicitousness demonstrated by Amtorg's demands for blueprints and other information from prospective employees.

Access to US technology, under the auspices of potential purchases, provided the Eastern bloc with opportunities to expand its collection at US facilities beyond the items available for sale. For instance, a Soviet guest visiting a Boeing facility applied adhesive to his shoes in order to collect metal samples.[91] In addition to demanding information associated with products that were available for purchase, the Soviets would, at times, attempt to solicit information that had nothing to do with the product under contract.[92] One step beyond this was the use of visits to industrial facilities to recruit insiders. According to a 1960 FBI assessment, Amtorg officials would attempt to gain the confidence of employees in plants that had contracts with the Russian government and—through these employees—obtain blueprints, which Amtorg would copy.[93]

Opportunities for deemed export also facilitated intelligence collection by formal foreign intelligence services. Line X—the KGB entity responsible for collection of science and technology—seeded Soviet delegations to US firms and laboratories with its officers.[94] Similarly, the East-West Exchange Program, which was an outgrowth of a 1955 US-Soviet summit, provided for reciprocal visits in multiple fields including industry.[95] In 1965 the FBI advised Congress that it was established Soviet policy to pepper delegations such as those that participated in the East-West Exchange Program with one or more full-time KGB officers, as well as for the KGB to task participating scientists. Upon returning from the United States, the scientists would provide the KGB with comprehensive reports about the technical information—including descriptions of installations visited, research being conducted, and the status of particular projects.[96] The United States carried out similar exchanges with Eastern-bloc countries—including Czechoslovakia, Hungary, and Bulgaria—which the US intelligence community later identified as proxies for Soviet collection of technological information.[97]

While the FBI was worrying about the opportunities that the East-West Exchange Program provided to foreign governments, other elements of the US intelligence community were looking to the program as an opportunity to advance US interests. According to a CIA document from 1969, the US government planned to exploit this program for intelligence purposes. Furthermore, the CIA's East-West Exchange Program staff—the intelligence advisers to the Department of State—vetted proposals to identify exchanges of intelligence interest as well as evaluated the net intelligence gains that any proposal would produce.[98]

Oral Exchanges

Foreign powers have frequently taken advantage of their presence in the United States to elicit information through oral exchanges. For instance, in 1954, a Soviet assistant air attaché struck up a relationship, during a reception at a Soviet satellite embassy, with the representative of an aeronautical manufacturer. The Soviet baldly suggested that he wanted information the next time the two met.[99]

Technology-related conferences and industry association events are other venues for exchanges that can veer into deemed export territory. In 1986, the Senate Select Committee on Intelligence assessed that the Soviets and allied intelligence services had, for a number of years, regularly attended scientific, technical, and industrial conferences, including functions held in the United States.[100] As an example of the enduring nature of this problem, the FBI noted—in 1961—that between 1959 and 1960, Soviet officials attended approximately 141 technical, scientific, and general business conventions and expositions.[101] The hunting must have been good, since more than two decades later the Senate Select Committee on Intelligence, in its 1986 report, stated that the Soviets considered the information obtained from these types of events to be extremely significant in the development of their military projects.[102] There was certainly the possibility that the Soviets might glean protected wisdom through oral transmission at these functions, as, according to a 1982 CIA assessment, Soviet officials sent written reports—that recounted lectures and briefings they had attended—back to their headquarters.[103] (They also, on occasion, used those briefings as a starting point for identifying individuals of interest. A Soviet attaché, attending a convention for radio engineers in 1953, approached an individual who had presented a paper at the convention and inquired about the individual's willingness to talk with the attaché.)[104]

Eastern-bloc visits to such events provided opportunities to engage individuals knowledgeable about technology of interest to US adversaries.

For instance, in 1963 the FBI advised Congress that during 1962, ninety-five Soviet-bloc officials had attended sixty-five technical, scientific, and military conferences, exhibits, and symposia throughout the United States. These visits afforded the Soviet officials with opportunities to develop initial contacts with persons who had potential for intelligence recruitment.[105] The FBI continued to observe Soviet-bloc activities at technology-oriented events. In 1965, then–FBI director J. Edgar Hoover advised Congress that during 1964, eighty-six Soviet-bloc officials had visited sixty-one conventions, symposia, and exhibits in various parts of the United States.[106] The FBI had previously seen specific examples of this type of behavior. In 1953, two Soviet assistant military attachés had attended a convention for radio engineers and, after learning that an exhibitor had a Russian name, asked if the exhibitor would be interested in attending social functions at the Soviet embassy in Washington, DC.[107]

In addition to visits, foreign intelligence apparatuses have—over multiple decades—identified the value of networking to identify individuals of interest. As early as 1953, according to an FBI monograph, a Soviet official sought to join the American Rocket Society.[108] Other countries have followed suit. Chi Mak, who was convicted of providing US defense information to China, received instructions from his handler to engage in networking through professional associations and conferences.[109] In an era of social-networking sites, such as LinkedIn, it is easier than ever for hostile and adversarial countries to identify and digitally approach individuals in specific fields of interest.

Application of Personal Knowledge or Technical Expertise Abroad

The application of American know-how on behalf of US adversaries and competitors is a long-standing problem that has arguably become increasingly problematic as global collaboration becomes a business reality. In 1942, Hoover disclosed that a German national in the United States who had been working on behalf of the German consul recruited between 300 and 500 skilled mechanics to return to Germany and assist that country with overcoming a shortage of skilled labor.[110]

Again, the opportunity for deemed export violations is created by commercial dealings. In 1951, a Soviet representative demonstrated interest in engaging a US firm to train Soviet personnel in electronics.[111] Although not a directly commercial example, it is worth noting that one East European intelligence entity routinely invited US professors to teach in the country's universities. This, of course, provided an opportunity for a hostile country to gain expertise through the academic transmission of knowledge. Additionally, the intelligence service saw this arrangement as an opportunity to select individuals to recruit as sources. However, the service was surprised, since it rarely had to bother with recruitments because, according to a defector, the academics were so talkative.[112]

China has historically pursued—and continues to pursue—an aggressive strategy of facilitating interactions on its soil that could lead to deemed export violations. These activities—like the Soviet shenanigans—pre-date the formulation of the Export Administration Regulations. Since 1949, the year in which the Chinese Communist Party seized control of mainland China, China has elicited—and received—support from a sympathetic Chinese community in the United States who have been willing to lend their expertise to the establishment of the communist regime's capabilities. The Chinese in US Science Association (which later became the Association of Chinese Scientific Workers in the USA) formed in 1949. There was no doubt about what this group's intention was. According to its official handbook, it sought to unite and cooperate with scientific workers in China in an effort to advance the work of scientific development in that country. The association adopted a resolution asserting that the core of its work was the study of scientific techniques in order to prepare for their return to China. A 1950 letter to the association indicated that the Chinese government had established a special committee to handle members' return to that country.[113] Although the association disbanded in 1951, its marshaling of the Chinese diaspora on behalf of a US adversary's advancement was only the beginning of a persistent effort, which continues unabated, to transfer knowledge from the United States to a hostile power.

During the intervening decades, China continued to use the application of US-gained expertise to establish its military capabilities. In 1955, Kim Xuesen, who had worked at the California Institute of Technology, returned to China and ultimately helped to establish China's space and missile programs.[114] In addition to tapping its US diaspora in order to effect the transfer of knowledge, China also induced US companies—using the lure of lucre—to provide technological know-how on Chinese soil. In 1979, the McDonnell aircraft company reached a deal to build parts in that country. In return, China demanded the technology and the right to produce increasingly large pieces of the planes that it purchased, from McDonnell, in Chinese factories. As Chinese factories absorbed the manufacturing know-how from production of nose cones and fuselages for airliners, emerging versions of Chinese fighter aircraft began featuring better-produced fuselages and smoother aluminum skins.[115] This represented not simply an opportunity for reverse engineering but rather an opportunity to master technological processes upon which China could innovate and enhance its military capabilities.

China has institutionalized this process of extracting knowledge from the United States through what it calls "talent plans." These plans are conceptually similar to the playbook that China has used since 1949. Zhao Ziyang—who served as general secretary of the Communist Party of China from 1987 to 1989—very openly indicated China's view of its population abroad when he stated that China was storing its talent overseas, as opposed to suffering a brain-drain, which it could eventually tap. The Chinese government incorporated this mentality into its 2006 National Medium and Long-Term Program for Science and Technology, which explicitly assessed that it needed to attract "high caliber talents from overseas" with an emphasis in areas where China was deficient. To achieve this objective, the program called for the government to attract individuals with desired expertise to return to China for patriotic purposes and establish talent recruitment organizations.[116]

There are multiple talent plans, but the one with the greatest implications for the US commercial sector is the Thousand Talents plan (also known as the Recruitment Program of Global Experts).[117] China launched this in December 2008 and it has brought more than 4,000 foreigners to a number of Chinese institutions including the country's companies.[118] (One cannot really use the shorthand of "private sector" about companies under an authoritarian regime.) This program specifically targeted several categories of overseas experts including technical managerial professionals in senior positions at internationally known companies or financial institutions and entrepreneurs holding intellectual property rights or key technologies who possess overseas experience.[119]

China's efforts to siphon talent from the United States have had results, as indicated by US companies' loss of executives (and their knowledge) to Chinese firms. Reporting in 2020, the *New York Times* stated that 600 talent program recruits had worked for US companies.[120] Artificial intelligence is a significant area of technological competition between China and the US, and the former has sought to acquire expertise from the latter. For instance, a former Microsoft vice president and artificial intelligence researcher, Qi Lu, left that company to become the chief operating officer of the Chinese technology firm Baidu.[121]

When employees of US firms leave their US employers, they—on occasion—take not only general knowledge but also proprietary information. For instance, You Xiaorong, a researcher who worked on coatings for Coca-Cola beverage cans, received a Thousand Talents offer based on the secrets she stole.[122] Another talent plan recruit, Hongjin Tan, applied to the Thousand Talents program and stole approximately a million dollars' worth of trade secrets from his former employer, an Oklahoma-based petroleum company. Talent program applicant Shan Shi specifically stated that he would "digest" and "absorb" relevant technology in the United States.[123] Yet another talent program applicant, Long Yu—who worked for a US defense contractor—was arrested in 2014 after he attempted to take hundreds of gigabytes of export-controlled proprietary data, including design information for military jets, to China. Yu explicitly told the Chinese government that he intended to help China to "mature its own aircraft engines."[124]

Talent programs may facilitate longer-term transfer of technology and expertise in addition to the one-off incidents of theft. For instance, Chinese institutions equip researchers whom they have recruited—with generous stipends in the medium to high six figures—with cutting-edge research facilities in China. In certain cases, foreign recruits are allowed to maintain overseas affiliations and establish labs that mirror their US facilities.[125] This creates a continuing conduit for the migration of US research and development to China.

Interestingly, China has made efforts to obfuscate the talent programs. The *New York Times* reported in 2020 that China had renamed the Thousand Talents program (which by this point was associated with the outflow of intellectual property from the United States) to the National High-End Foreign Experts Recruitment Plan.[126] The Chinese government instructed talent recruitment organizations not to put the term "Thousand Talents" in written promotional materials. Furthermore, in the way that authoritarian governments are expert, China began erasing references to the talent plan. Chinese universities stopped promoting the program on their websites, and the official site of the Thousand Talents program deleted the names of participating scientists.[127]

The program's quasi-underground nature is indicated by episodes in the United States involving participants' illegal obfuscation of their participation. In June 2020, Charles Lieber, the former chair of the Chemistry and Chemical Biology Department at Harvard, was charged with lying, on multiple occasions, to federal authorities about his involvement with the Thousand Talents program.[128] Another Thousand Talents participant, Xiao-Jiang Li—formerly of Emory University—obfuscated his involvement by filing a false tax return that did not reflect income earned while working for multiple Chinese universities.[129]

Economic Espionage

The ways in which foreign governments can manipulate commercial concerns in furtherance of acquiring a deemed export also provide opportunities for economic espionage. However, unlike a deemed export violation, economic espionage is theft from an unwitting victim rather than a mutually beneficial, albeit illicit, transaction. By the mid-1990s, the US government recognized that the systematic theft of US proprietary information was a grave problem for the economy and for national security. Senator Herb Kohl, a sponsor of the Economic Espionage Act, put the need for such legislation in stark terms when he stated that "it would not be unfair to say that America has become a full-service shopping mall for foreign governments and companies who want to jump start their businesses with stolen trade secrets."[130]

Until the mid-1990s, federal prosecutors lacked an effective tool to prosecute the theft of proprietary—but unclassified—private sector information. Although the Espionage Act of 1917 facilitated the prosecution of threats to national security, there was not a similar statute to protect the work of the private sector if it was not engaged in sensitive work for the US government. Instead, the FBI predicated its investigations of such activity on a hodge-podge of statutes including those covering the interstate transportation of stolen property as well as wire and mail fraud.[131]

In 1996, Kohl and Senator Arlen Specter introduced a bill to specifically prohibit economic espionage activity. The Economic Espionage Act of 1996 advanced two primary objectives: the protection and promotion of national and economic security. The act made the theft or misappropriation of a trade secret a federal crime. Unlike the Espionage Act of 1917, the offense involves business information rather than classified or national defense information. This is of particular national security interest as some of the country's most cutting-edge technologies are now solely the domain of private industry.

The Economic Espionage Act provides two distinct but related offenses. The first violation outlaws theft of trade secrets for the benefit of a foreign entity (economic espionage). Economic espionage involves the intent to benefit a foreign entity or at least the knowledge the theft could have that result. It does not require intent to injure the owner of the trade secret. The second violation criminalizes the misappropriation of trade secrets with the intent to convert the trade secret to the economic benefit of anyone other than the owner and to injure the owner of the trade secret. According to the *US Attorney's Manual,* some of the factors assessed in determining whether to initiate an economic espionage or trade secret case involving these offenses include the scope of the criminal activity, including evidence of involvement by a foreign government, foreign agent, or foreign instrumentality; the degree of economic injury to the trade secret owner; the type of trade secret misappropriated; the effectiveness of available civil remedies; and the potential deterrent value of the prosecution.[132]

Global developments have continued to outpace US national security in a variety of fields, and countering economic espionage is one of them. Dispersion of multinational corporations has limited the applicability of the Economic Espionage Act. The Department of Justice can only prosecute trade secret thefts that occur abroad when the defendants are US persons or organizations, or an act in furtherance of the offense is committed in the United States.[133]

Although economic and industrial espionage involve different intent elements, both require proof that the intellectual property is a trade secret. A trade secret must have three elements. First, the trade secret must involve information. This may include "all forms and types of financial, business, scientific, technical, economic, or engineering formation . . . whether tangible or intangible, and whether or how stored, compiled, or memorialized physically, electronically, graphically, photographically, or in writing."[134] Second, it must have independent economic value from not being generally known to the public and not being readily ascertainable through proper means by the public. The value of the trade secret can include the competitive advantage of the owner by using the trade secret, the costs for an outsider to duplicate the trade secret, and the lost advantages to the trade secret owner resulting from the disclosure to a competitor. Third, the owner of the trade secret must have taken reasonable measures to keep the information a secret. These steps can involve physical security like locks, passwords, limiting access, use of nondisclosure agreements, labeling, and adequate training for employees. It is not necessary for the trade secret owner to have taken every conceivable step to protect the property; the owner must only have taken "reasonable measures."[135]

The Economic Espionage Act provides for significant prison terms and fines for violations, but its effectiveness in deterring trade secret theft remains in question.[136] For a foreign company, the penalty for a violation may be considered merely a cost of doing business and well worth the risk. The number of prosecutions under the act is relatively few considering the magnitude of the problem. As of 2013 there had been approximately a hundred indictments under the Economic Espionage Act. However, it has been more common for prosecutors to charge defendants under the theft of trade secrets, rather than the espionage, element of the act, since the former charge does not require the prosecution to prove that the defendant acted with the intent to benefit the foreign power.[137]

In 2012, the Theft of Trade Secrets Clarification Act updated the Economic Espionage Act in a significant way. With the passage and enactment of the clarifying legislation, the definition of trade secret was expanded beyond goods to include services. It also broadened the scope of secrets protected by the law from those placed in interstate or foreign commerce to those that were used in or intended for use in interstate or foreign commerce. The government recognized the need for this modification, following the trial of Goldman Sachs employee Sergey Aleynikov, who allegedly stole the code for high-speed computerized trading operations but was not convicted under the Economic Espionage Act because the code was in-house proprietary information that was never intended as an item for commercialization (unlike most trade secrets, which provide the foundation for an eventually lucrative product).[138]

Informal Dynamics of Public Sector–Private Sector Interaction

The framework of laws and regulatory regimes covering economic espionage, technology transfer, and foreign direct investment are the formal rules of the road for government–private sector interactions pertaining to national security–related business practices. However, realities established by government–private sector investments and negotiations provide an equally influential, albeit de facto, set of expectations about both sides' behavior. The US government is now largely just one more customer with which the private sector can choose to do—or not do—business. Potentially problematic is the reality that industry may also choose to do business—and provide capabilities to—foreign adversaries and competitors.

Additionally, the private sector's decisions have significant implications for US national security capabilities. Technology and information developed and maintained by the private sector can prove invaluable to the US government in disrupting state and nonstate threat actors (ranging from hostile foreign governments to domestic criminal entities). This paradigm, which has been evolving since the end of the Cold War, is illustrated by Apple's intransigent refusal to assist the FBI with accessing the iPhone that the perpetrator of a 2015 attack had used. Furthermore, the majority of US critical infrastructure is controlled by the private sector. Failure of these facilities (e.g., the electrical grid) can cause havoc not only for the company involved but also for significant portions of the country.

Investment

Government research priorities have shifted drastically since the end of the Cold War, opening space for private industry to develop significant technologies—that have ramifications for US national security—unbeholden to government clients. In 1964, the US government funded approximately 67 percent of US research and development. By 2016 the government provided only 24 percent of US research and development funding. The government and private sector passed each other like ships in the night. By 2016, industry accounted for 67 percent of US research and development funding.[139]

However, the United States has not entirely abdicated technological development to the private sector. As of 2017, the US government was the largest funder of basic research, at 42 percent, in comparison to the private sector's 30 percent. Basic research is the experimental or theoretical work primarily directed at acquiring new knowledge about the underlying foundations of phenomena and observable facts. In other words, without the US government's investment, industry would have to direct more resources toward establishing the scientific underpinnings of the technology that it can instead focus on commercializing. Consistent with this, business was the primary funder of applied research and development, at 55 percent in 2017. Applied research is directed toward a specific practical objective.[140] By comparison, in 2017 the federal government provided only 13 percent of the funding.[141]

Adopting and Adapting

It is unlikely that the US federal government will catch up with the private sector's funding of research and development. The US government did not cut its constant dollars—in fact it doubled them—in research and development funding, and still was surpassed by a growth in the private sector's research and development investment.[142] Consequently, the United States must learn to adopt and adapt technologies that private industry has already developed into practical applications. According to William Carter, a fellow at the Center for Strategic and International Studies, the United States needs to continue to foster private sector innovation, as this field has become "the source for most new strategic technologies." The challenge for the United States, according to Carter, will be to leverage "commercial advances more effectively than opponents" of the United States.[143]

Multiple US agencies have, in response to this new relationship, resorted to another private sector innovation, venture capital, to access the private sector's work on behalf of US national security. In-Q-Tel was one of the earliest examples of this approach. The CIA established In-Q-Tel in 1999 as a private, nonprofit firm that provides venture capital to small companies developing technologies of interest to the US intelligence community.[144] In-Q-Tel's mission is consistent with the new reality that the United States will increasingly rely on the work of the private sector. According to a 2000 CIA document, the "continued ability of the Agency to add value will be largely a function of its ability to harness the technological advances being made in the private sector."[145] Rather than working through the standard government procurement process (which can dissuade companies from doing business with the government), In-Q-Tel uses private industry contractual terms. Its interactions with private sector companies are directed at adapting the technologies for the government's needs.[146]

In 2000 the CIA was not completely convinced that In-Q-Tel would succeed, but despite this doubt also believed that success was essential. As its then–inspector general noted, "with the creation of In-Q-Tel, the first significant step was taken to involve knowledgeable outsiders in the Agency's work but this effort [was] still in its infancy and the probability of success uncertain."[147] However, the inspector general continued, a "way must be found to identify and harness the capabilities of this world to the Agency's purposes. . . . That is why [the inspector general believed] In-Q-Tel simply has to succeed" and "Agency managers and overseers must find a way to make it work."[148]

In-Q-Tel does appear to have succeeded. As of 2012, it had developed programs with more than 160 startups, which had yielded 297 pilot initiatives. Although it is a project of the CIA, In-Q-Tel serves a variety of US government customers. As of 2012, the Defense Intelligence Agency ranked with the CIA as In-Q-Tel's two largest customers.[149] The DHS has worked successfully with the company to identify innovative startups and invest in them to adapt their emerging commercial technologies.[150] The FBI has also used In-Q-Tel to review new technologies that have yet to be fully developed for the commercial market. The intelligence community writ large has adopted and funded more than a hundred of the technologies funded by In-Q-Tel. Additionally, In-Q-Tel contributes to the growth of the tech-oriented private sector through sales of its shares in ventures to companies, which have included Google and IBM.[151]

The Department of Defense (DoD) established a similar enterprise in the form of its Defense Innovation Unit. In 2015 the DoD created the Defense Innovation Unit–Experimental with the objectives of finding private sector technologies that may prove to be of interest to the DoD and identifying defense technologies in which the tech sector might have an interest.[152] Success was hardly guaranteed. In 2016 the secretary of defense felt it necessary to completely overhaul the office's leadership.[153] The experimental division became permanent in 2018. As of that year, the Defense Innovation Unit had offices in the multiple tech sector–centric regions in the United States: Silicon Valley; Austin, Texas; Boston, Massachusetts; and Washington, DC.[154]

As if In-Q-Tel and the DoD were not enough of a crowd in the field of US government venture capital, the Department of Homeland Security also entered the fray. In 2015, then–DHS secretary Jeh Johnson announced that the DHS planned to open an office in Silicon Valley in order to build relationships with industry and recruit talent from the technology sector.[155] This became the DHS Silicon Valley Innovation Program, under the department's Science and Technology Directorate.[156] The program has provided funds to a variety of projects including securing video feeds on the Internet of Things; various uses of blockchain; and drone technology capable of enhancing border security.[157]

Politics

Relying on the private sector introduces a degree of volatility in the form of political pressure from external actors. Many companies clearly no longer need government contracts and may actually view them as problematic to companies' reputation (and market share). Companies have taken a battering from an external, hyperactive activist culture. For instance, in 2019, protesters attempted to disrupt the Amazon Web Services Summit with demands that Amazon cut its ties with the DHS's Immigration and Customs Enforcement.[158] Such smoke and noise can have real impacts. In 2019, the open-source software company Chef indicated that it would allow its contracts with Immigration and Customs Enforcement as well as Customs and Border Protection to expire, following protests.[159]

Companies also face demands from their own employees regarding whom the companies take (or do not take) as customers. In 2018, Google decided not to renew its contract with the Pentagon for artificial intelligence work. Google took this position after approximately 4,000 of its employees signed a petition demanding a clear policy stating that neither

Google nor its contractors would ever build "warfare technology."[160] In 2019, a number of Google employees signed a petition demanding that the company cease working with Immigration and Customs Enforcement, and Customs and Border Protection.[161]

Additionally, the private sector may take an activist stance—or perhaps shroud financial interests within a cloak of altruism—against US government policies. This dates to at least the early 1990s. In 1993, the US government proposed requiring the installation of a decryption key (a "clipper chip") designed by the National Institute for Standards and Technology and the National Security Agency, in US government–purchased telephones.[162] While the government did not demand that the manufacturers include this chip in devices sold to the private sector, officials hoped that this would become an industry standard due to the federal government's significant purchasing power.[163] (If companies had to make a significant quantity of products with higher specifications, they might make all of their products at that specification in order to avoid having to stand up two separate manufacturing processes.) Information technology companies—including Apple, IBM, and Microsoft—pushed back against the encroachment of government-developed decryption out of fear that it would fuel customer distrust of the devices (and therefore undermine market share). Additionally, companies feared for their market share abroad, worrying that facilitating government decryption would harm their products' export prospects.[164]

At the heart of this standoff was a concept called "key escrow." Two government agencies would hold the "keys" that could be used to decrypt what the clipper chip encoded.[165] At the beginning of an encrypted communications session, the chip would copy the encryption key and send it to the government, which would hold the key in "escrow" until the government had an authorization—such as a court-approved wiretap—to access the key that would decode the encrypted communication.[166] In 1994, the US attorney general designated the National Institute for Standards and Technology and the Department of Treasury as the key escrow agents.[167]

The US government also attempted to use export controls as tools to dissuade private industry from incorporating strong encryption. While companies could imbue their products, for domestic use, with strong encryption, the ability to export only a watered-down version in order to sell them abroad would—like the purchasing power of the US government—drive them to adopt a lower standard of encryption across their products.[168] Industry chipped away at this restriction. In 1996 the US government conditionally raised the limit on exportable encryption. Companies that planned to develop a key recovery system within two years could export 56-bit encryption; companies that did not plan to develop such a recovery system were limited to exporting 40-bit encryption; and products that included a key recovery feature could export encrypted products with no limits on bit length.[169]

Ultimately the US government backed down from its demands. In 1998, it instead stated that it wanted a "good faith dialogue" to develop "cooperative solutions." (Such words often serve as euphemisms for conceding defeat.) Consistent with this the government announced, in 1999, that US industry could export encryption items with no limit on bit length, after a technical review, to individuals, firms, and other nongovernment end-users in any country except for a state sponsor of terrorism.[170]

One interesting narrative in the prolonged standoff between the public and private sectors is how private industry may take advantage of unpopular US policies to advance its own standing. During a conference sponsored by cybersecurity company RSA, computer hardware, software, and telecommunications companies articulated their desire to adopt an industry, rather than government, coding standard.[171] RSA was, in fact, jockeying to become holder of the encryption standard, and as of 1994 this standard had been adopted by information technology companies including Apple, AT&T, Lotus, and Microsoft.[172]

Telecommunications are not the only area where the US government has collided with industry; the government has also clashed with the private sector on the related issue of internet traffic. In 2000 the FBI deployed its "Carnivore" email surveillance system. EarthLink, an internet service provider, allowed the FBI—which had a court order—to install the Carnivore hardware at an EarthLink hub in California.[173] However, it turned out that Carnivore was not compatible with EarthLink's current operating system and therefore EarthLink had to install an older version of its system in order to facilitate Carnivore. This development caused EarthLink's remote access servers to crash. One wonders whether the issue would have attracted as much attention if the FBI had avoided using such an incendiary name for this technology and stuck to Carnivore's other moniker, the far less provocative DCS 1000.[174]

The US government continued to remain concerned about how the increasing inaccessibility of communications could endanger US national security. As early as 1998—while the clipper chip controversy ebbed away from the government's preferred outcome—the FBI advised Congress about 183 cases in which advanced telecommunications equipment had impaired or prevented the execution of a court order.[175] In 2010 the issue cropped up again, indicated by the inclusion of $9 million in the FBI's budget for a "Going Dark Program."[176] In 2011 an FBI official explained to Congress that the Going Dark problem was not one of legal authorities but rather of the practical difficulty in intercepting the communications and related data once courts have authorized collection of it.[177] As then–FBI director James Comey explained in 2016, the encryption problem was twofold: the encryption used to lock devices when they were at rest as well as the encryption that encoded communications in transit.[178] In 2015 there was clipper chip déjà vu when Michael Rogers, then director of the National

Security Agency (NSA), suggested that technology companies create a digital key to open any smartphone or locked device and divide the key into pieces so that no single entity could unilaterally opt to use it.[179]

The issue blew up in early 2016 when Apple and US government officials found themselves at loggerheads over accessing information on an iPhone in the wake of a terrorist attack in San Bernadino, California. In 2014, Apple had already announced that it would make it impossible for the company to turn over data from most iPhones or iPads to authorities, even when those authorities had a search warrant.[180] This amounted to closing the handcuffs and swallowing the key. In 2016, Apple made clear that this was about market share when its senior counsel advised Congress that "hundreds of millions of law abiding citizens trust Apple's products with the most intimate details of their daily lives."[181] When a federal judge ordered Apple to help the FBI unlock an iPhone—with an operating system that preceded the impossible-to-decrypt feature—in early 2016, Apple insisted that providing the keys to its technology would compromise its users' security.[182] Like RSA, Apple used the incident to bolster its standing with consumers. In response to the judge's demand, Tim Cook, the CEO of Apple, directed a 1,100-word letter to Apple customers stating that the ruling represented a "chilling" breach of privacy, and vowed to appeal the decision.[183] Unlike Rogers's 2015 suggestion regarding an encryption key, then–FBI director James Comey indicated that he did not want a permanent compromise but rather good corporate citizenry. According to Comey, the FBI was simply "asking Apple to take the vicious guard dog away; let [the FBI] try and pick the lock."[184]

Finally, government assistance to the tech sector can arguably have political consequences. For instance, In-Q-Tel provided venture capital to help establish Palantir.[185] Therefore it is hardly surprising that Palantir developed its flagship software with the CIA in mind.[186] One of Palantir's cofounders is Peter Thiel, who has become an outspoken political figure on the right (and the alternative right as indicated by his attendance at the DeploraBall).[187] The point here is not to debate the merits of Thiel's political leanings but simply to point out that recipients of government funding for their projects can leverage government resources without being beholden to the same restrictions as government employees (e.g., the Hatch Act). While—for a host of reasons—it would not be good policy for the government to impose any sort of ideological litmus test on its investments, it does suggest that the impact of government sponsorship on the development of political figures is an unexplored avenue of inquiry.

Foreign Relations

Industry is in the business of carving out market share rather than prioritizing national security. Consequently, it may, on occasion, work at cross purposes to the interests of the United States. As of 2018, Google had started developing a search application, known as Dragonfly, that would restrict content banned by China, and had demonstrated this service to Chinese government officials.[188] It appears that some of the same work force that decried Google's work for the Pentagon remained consistent and also excoriated Google's efforts to ingratiate itself with China. Approximately 1,400 employees signed a letter stating that Google's attempt to court the Chinese authorities raised "urgent moral and ethical issues."[189] In 2019 a Google vice president advised the Senate Judiciary Committee that it had "terminated" Dragonfly.[190]

However, although internal and external scrutiny seemed to scuttle Dragonfly, Google has contributed to China's technological advancement in multiple ways, particularly in the field of artificial intelligence (AI), which has significant implications for the future of conflict. In 2017, Google announced the creation of its China AI Center, which would partner with multiple Chinese universities.[191] Among these is Tsinghua University, which, as of 2015, had worked with another US technology firm—Microsoft—for approximately twenty years.[192] Alarmingly, even as Tsinghua is benefiting from US expertise, it is also home to a high-end laboratory for military intelligence and research funded by the Central Military Commission of the People's Liberation Army.[193]

The threat to US industry—as well as to US national security, which increasingly relies on industry's development—from engaging in projects on foreign soil is not a new one. For instance, the Commission to Review Department of Defense Security Policies and Practices (the Stilwell Commission), which issued its report in 1985, assessed that co-production arrangements with a foreign country could result not only in the compromise of a specific item but also in the siphoning of "know-how" necessary to manufacture that item in large quantities.[194] Nearly forty years after the publication of the Stilwell Commission report, the siphoning of knowledge is all too alive and well. China is probably the most egregious practitioner of this approach; it premises external actors' foreign direct investment in that country upon the investment aligning with Chinese industrial policy priorities. Not surprisingly, in 2018, for instance, numerous US firms operating in China had been pressured to transfer technology.[195] This practice is even more troubling as the geopolitical rivalry between China and the United States increases and both sides clamor for the know-how that the private sector has developed.

In addition to providing China access to their work by going abroad, US companies are making themselves vulnerable to Chinese interference by partnering with them on US soil. For instance, Baidu, a Chinese information technology company, has invested significant sums in US firms pursuing artificial intelligence.[196] This contributes to the appearance that China is making a multipronged effort to access US AI expertise. The likelihood that China's commercial sector is serving as a cat's paw for the interests of China's government is indicated by the companies' politicization. A 2017 Chinese law requires that any Chinese organization or citizen must cooperate in national intelligence work.[197] According to recent guidelines issued by the Chinese regime, its businesses must build up Communist Party organizations, and entrepreneurs must "identify politically, intellectually, and emotionally" with the party.[198] Furthermore, the Chinese government requires all companies in China to have an internal Chinese Communist Party committee.[199] China has even imposed its political control on foreign companies seeking to do business in China, asking firms engaged in joint ventures to afford Communist Party cells a role in those companies' business decisions.[200]

Critical Infrastructure

In addition to developing technologies that will impact US elements of national power, the private sector directly controls much of the critical infrastructure on which the United States relies. According to the Department of Homeland Security, critical infrastructure sectors are those in which "assets, systems, and networks, whether physical or virtual, are considered so vital to the United States that their incapacitation or destruction would have a debilitating effect on security, national economic security, national public health or safety, or any combination thereof."[201] The private sector controls approximately 85 percent of US critical infrastructure.[202]

The areas covered under the definition of critical infrastructure continue to evolve as new technologies and ways of doing business create associated vulnerabilities. For instance, in January 2017 the Department of Homeland Security designated US election systems as a subsector of the government facilities critical infrastructure sector.[203] The DHS decision was prompted by a 2016 intelligence community assessment on foreign interference in the US political process.[204] Social media—especially its role in disseminating disinformation—is emerging as another set of entities that experts have suggested needs to be considered critical infrastructure.[205] Both the DHS and the FBI have taken actions that would seem to validate the need to consider this approach. The FBI's Foreign Influence Task Force has provided actionable intelligence to technology companies in order to facilitate safeguarding of their platforms.[206] Furthermore, in 2017 the acting secretary of the DHS traveled to Silicon Valley to meet with tech companies regarding the exploitation of their platforms by threat actors and advised Congress that the DHS would continue to press companies to identify and remove terrorist content.[207] The potential for disruption to the fundamental processes of governance certainly makes social media a candidate for designation as critical infrastructure.

Government Efforts to Communicate with the Telecommunications Sector

The federal government has attempted to work with the telecommunications sector to acquire the technical information that will facilitate intelligence collection. In 1995 the FBI planned to establish a Telecommunications Industry Liaison Unit, within its Information Resources Division, which would be responsible for, among other tasks, consulting with industry to clarify requirements and influencing standards development.[208] In 2008 the FBI, in discussing its "Going Dark" problem, indicated that it would "institute greater and broader industry liaison, particularly with [internet protocol–based] communications service providers and manufacturers."[209] Then, in 2011, the US government indicated that it planned to establish a Data Communications Assistance Center, which would facilitate the sharing of technology between law enforcement agencies.[210] The center, which operates under the auspices of the Department of Justice, is the government's "first ever attempt to develop a center of electronic surveillance knowledge management."[211]

Notes

1. Bureau of Industry and Security, Office of Export Enforcement, *A Brief History of United States Export Controls,* n.d., https://www.governmentattic.org/34docs/BIShistoryABHOUSEC_undated.pdf.

2. Ibid.

3. Congressional Research Service, *The International Emergency Economic Powers Act: Origins, Evolution, and Use* (Washington, DC, 2019).

4. Ibid.

5. Public Law 703, July 2, 1940, https://www.loc.gov/law/help/statutes-at-large/76th-congress/session-3/c76s3ch508.pdf.

6. Department of State, Regulations Governing the Exportation of Articles and Materials Designated in the President's Proclamation of July 2, 1940, Issued Pursuant to the Provisions of Section 6 of the Act of Congress Approved July 2, 1940, https://history.state.gov/historicaldocuments/frus1931-41v02/pg_214.

7. Bureau of Industry and Security, *A Brief History of United States Export Controls.*

8. US Congress, *Investigation and Study of the Administration, Operation, and Enforcement of the Export Control Act of 1949 and Related Acts,* report of the Select Committee on Export Control, 87th Congress (Washington, DC, 1962), https://books.google.com/books?id=iNQz97wiQbgC&pg=RA105-PA59&lpg=RA105-PA59&dq=secretary+of+commerce+revoked+soviet+license+1961&source=bl&ots=hZNjDZORfi&sig=ACfU3U1AaiQ2_7nTvwGC0s2EWNCd_Ag8Ow&hl=en&sa=X&ved=2ahUKEwiE1OPKoqPqAhV3mHIEHeGAAdEQ6AEwCXoECA0QAQ#v=onepage&q=secretary%20of%20commerce%20revoked%20soviet%20license%201961&f=false.

9. Congressional Research Service, *The Export Administration Act: Evolution, Provisions, and Debate* (Washington, DC, 2003), https://fas.org/sgp/crs/RL31832.pdf.

10. Bureau of Industry and Security, *A Brief History of United States Export Controls.*

11. Congressional Research Service, *The Export Administration Act.*

12. Congressional Research Service, *US Export Control System and the Export Control Reform Initiative* (Washington, DC, 2020).

13. Congressional Research Service, *Transfer of Defense Articles: Sale and Export of U.S.-Made Arms to Foreign Entities* (Washington, DC, 2020), https://crsreports.congress.gov/product/pdf./R/R46337.

14. Congressional Research Service, *US Export Control System.*

15. General Accounting Office, *What Would Be the Impact of Raising or Repealing the Commercial Arms Sales Ceiling?* (Washington, DC, 1980), https://www.gao.gov/assets/130/128529.pdf.

16. General Accounting Office, *US Munitions Export Controls Need Improvement* (Washington, DC, 1979), https://www.gao.gov/assets/130/126431.pdf.

17. *Second Annual Report of the National Munitions Control Board for the Year Ending November 30, 1937,* https://tinyurl.com/4s54xnjv.

18. Department of State, *The Neutrality Acts, 1930s* (Washington, DC, undated), https://history.state.gov/milestones/1921-1936/neutrality-acts.

19. Congressional Research Service, *The International Emergency Economic Powers Act: Origins, Evolution, and Use* (2019).

20. Congressional Research Service, *The International Emergency Economic Powers Act: Origins, Evolution, and Use* (Washington, DC, 2020), https://crsreports.congress.gov/product/pdf./R/R45618.

21. Congressional Research Service, *The International Emergency Economic Powers Act: Origins, Evolution, and Use* (2019).

22. Ibid.

23. Ibid.

24. Ibid.

25. Export Control Reform Act of 2018.

26. 50 USC chap. 58: "Export Control Reform," https://uscode.house.gov/view.xhtml.?path=/prelim@title50/chapter58&edition=prelim.

27. Executive Order 11858, "Foreign Investment in the United States," https://www.archives.gov/federal-register/codification/executive-order/11858.html.

28. Department of the Treasury, *Summary of the Foreign Investment Risk Review Modernization Act of 2018* (Washington, DC, undated), https://www.treasury.gov/resource-center/international/Documents/Summary-of-FIRRMA.pdf.

29. Congressional Research Service, *The Committee on Foreign Investment in the United States (CFIUS)* (Washington, DC, 2020), https://crsreports.congress.gov/product/pdf./RL/RL33388.

30. James K. Jackson, *The Committee on Foreign Investment in the United States* (Washington, DC: Congressional Research Service, 2014).

31. Congressional Research Service, *The Committee on Foreign Investment in the United States (CFIUS).*

32. Jackson, *The Committee on Foreign Investment in the United States.*

33. James K. Jackson, *The Exon-Florio National Security Test for Foreign Investment* (Washington, DC: Congressional Research Service, 2013).

34. Jackson, *The Committee on Foreign Investment in the United States.*

35. Jackson, *The Exon-Florio National Security Test for Foreign Investment.*

36. Jackson, *The Committee on Foreign Investment in the United States.*

37. Jackson, *The Exon-Florio National Security Test for Foreign Investment.*

38. Congressional Research Service, *The Committee on Foreign Investment in the United States (CFIUS).*

39. Department of Commerce, *Bureau of Export Administration: Improvements Are Needed in Programs Designed to Protect Against the Transfer of Sensitive*

Technologies to Countries of Concern (Washington, DC, 2000), https://www.oig.doc.gov/OIGPublications/IPE-12454.pdf.

40. Jackson, *The Committee on Foreign Investment in the United States.*

41. Ibid.

42. Ibid.

43. Ibid.

44. General Accounting Office, *Foreign Investment* (Washington, DC, 1995), https://www.gao.gov/assets/230/221994.pdf.

45. Jackson, *The Committee on Foreign Investment in the United States.*

46. Department of Commerce, *Bureau of Export Administration.*

47. David A. Sampson, deputy secretary, Department of Commerce, testimony before the US Senate Committee on Banking, Housing, and Urban Affairs, October 20, 2005, https://www.banking.senate.gov/imo/media/doc/sampson.pdf.

48. Bureau of Industry and Security, *Fiscal Year 2019 President's Submission,* http://www.osec.doc.gov/bmi/budget/FY19CBJ/BIS_FY19_President%27s_Budget_FINAL.pdf.

49. National Counterintelligence and Security Center, *National Counterintelligence Strategy of the United States of America, 2020–2022* (Washington, DC), https://www.dni.gov/files/NCSC/documents/features/20200205-National_CI_Strategy_2020_2022.pdf.

50. Stewart Baker, assistant secretary for policy, Department of Homeland Security, written testimony before the US Senate Committee on Banking, Housing, and Urban Affairs, October 20, 2005, https://www.banking.senate.gov/imo/media/doc/baker.pdf.

51. Jackson, *The Exon-Florio National Security Test for Foreign Investment.*

52. Department of the Treasury, *CFIUS Reform: The Foreign Investment and National Security Act of 2007* (Washington, DC, 2008), https://www.treasury.gov/resource-center/international/foreign-investment/Documents/Summary-FINSA.pdf.

53. Jackson, *The Committee on Foreign Investment in the United States.*

54. Spencer Ante and William Mauldin, "IBM, Lenovo Deal Likely to Spark Security Review," *Wall Street Journal,* January 24, 2014.

55. Jackson, *The Committee on Foreign Investment in the United States.*

56. Department of the Treasury, *CFIUS Reform.*

57. Jackson, *The Exon-Florio National Security Test for Foreign Investment.*

58. *The Economist,* "Keep Your T-Bonds, We'll Take the Bank, Sovereign Wealth Funds," July 28, 2007.

59. Congressional Research Service, *The Committee on Foreign Investment in the United States (CFIUS).*

60. Congressional Research Service, *CFIUS Reform Under FIRRMA* (Washington, DC, 2020), https://fas.org/sgp/crs/natsec/IF10952.pdf.

61. Fred Barbash and Ellen Nakashima, "Chinese Hackers May Have Breached the Federal Government's Personnel Office, US Officials Say," *Washington Post,* July 10, 2014, https://www.washingtonpost.com/news/morning-mix/wp/2014/07/09/report-chinese-hacked-into-the-federal-governments-personnel-office; Ellen Nakashima, "Chinese Breach Data of 4 Million Federal Workers," *Washington Post,* June 4, 2015, https://www.washingtonpost.com/world/national-security/chinese-hackers-breach-federal-governments-personnel-office/2015/06/04/889c0e52-0af7-11e5-95fd-d580f1c5d44e_story.html.

62. Nicole Perlroth, "Two from China Are Charged in 2014 Anthem Data Breach," *New York Times,* May 9, 2019, https://www.nytimes.com/2019/05/09/technology/anthem-hack-indicted-breach.html.

63. US-China Economic and Security Review Commission, *2019 Report to Congress* (Washington, DC).

64. David E. Sanger, "Grindr Is Owned by a Chinese Firm, and the US Is Trying to Force It to Sell," *New York Times*, March 28, 2019, https://www.nytimes.com/2019/03/28/us/politics/grindr-china-national-security.html.

65. "China's Kunlun Tech Agrees to U.S. Demand to Sell Grindr Gay Dating App," Reuters, May 13, 2019, https://www.reuters.com/article/us-grindr-m-a-beijingkunlun/chinas-kunlun-tech-agrees-to-u-s-demand-to-sell-grindr-gay-dating-app-idUSKCN1SJ28N.

66. Department of Treasury, *Summary of the Foreign Investment Risk Review Modernization Act of 2018* (Washington, DC, undated), https://www.treasury.gov/resource-center/international/Documents/Summary-of-FIRRMA.pdf.

67. Opinion from DC Circuit in *Ralls v. CFIUS*.

68. Damian Paletta, Keith Johnson, and Sudeep Reddy, "Obama Blocks Chinese Firm from Wind-Farm Projects," *Wall Street Journal*, September 29, 2012.

69. *United States v. Hui Sheng Shen and Huan Ling Chang*, https://www.justice.gov/archive/usao/nj/Press/files/pdf.files/2012/Shen,%20Hui%20Sheng%20and%20Chang,%20Ling%20Huan%20amended%20Complaint.pdf.

70. Stephen Dockey, "Chinese Company Will Sell Wind Farm Assets in CFIUS Settlement," *Wall Street Journal*, November 4, 2015.

71. Department of Justice, "California Resident Convicted of Conspiring to Illegally Export Fighter Jet Engines and an Unmanned Aerial Vehicle to China," June 9, 2016, https://www.justice.gov/usao-sdfl/pr/california-resident-convicted-conspiring-illegally-export-fighter-jet-engines-and.

72. Michael Brown and Pawneet Singh, *China's Technology Transfer Strategy* (Washington, DC: Defense Innovation Unit, 2018), https://admin.govexec.com/media/diux_chinatechnologytransferstudy_jan_2018_(1).pdf.

73. Congressional Research Service, *CFIUS Reform Under FIRRMA*.

74. Department of Commerce, *Bureau of Industry and Security: Deemed Export Controls May Not Stop the Transfer of Sensitive Technology to Foreign Nations in the US* (Washington, DC, 2004).

75. US Congress, *Departments of State, Justice, and Commerce, the Judiciary, and Related Agencies Appropriations for 1966*, before the Committee on Appropriations, House of Representatives, 89th Congress (Washington, DC, 1965).

76. Federal Bureau of Investigation, *Exposé of Soviet Espionage* (Washington, DC, 1960), https://www.cia.gov/library/readingroom/docs/CIA-RDP65B00383R000200040033-2.pdf.

77. Department of Justice, "Singaporean National Pleads Guilty to Acting in the United States as an Illegal Agent of Chinese Intelligence," July 24, 2020, https://www.justice.gov/opa/pr/singaporean-national-pleads-guilty-acting-united-states-illegal-agent-chinese-intelligence.

78. US Congress, *Departments of State, Justice, and Commerce, the Judiciary, and Related Agencies Appropriations for 1963*, before the Committee on Appropriations, House of Representatives, 87th Congress (Washington, DC, 1962).

79. US Congress, *Departments of State, Justice, and Commerce, the Judiciary, and Related Agencies Appropriations for 1966*.

80. Federal Bureau of Investigation, Bureau Bulletin no. 17: (D) Plant Surveys, May 15, 1941, https://ia802702.us.archive.org/2/items/foia_FBI_Confidential_Informants-HQ-1a/FBI_Confidential_Informants-HQ-1a.pdf.

81. Central Intelligence Agency, *Soviet Acquisition of Western Technology and Its National Security Implications* (Langley, February 23, 1982), https://www.cia.gov/library/readingroom/docs/CIA-RDP83M00914R002000070021-4.pdf.

82. US Congress, *Investigation and Study of the Administration, Operation, and Enforcement of the Export Control Act of 1949 and Related Acts*.

83. Central Intelligence Agency, *Summary Report on Technology Transfer to Communist Countries and the Intelligence Community's Role and Effectiveness* (Langley, 1981), https://www.cia.gov/library/readingroom/docs/CIA-RDP85T00176R000900020001-5.pdf. CIA-RDP85T00176R000900020001-5.

84. Central Intelligence Agency, *The Technology Acquisition Efforts of the Soviet Intelligence Services* (Langley, 1982), https://www.cia.gov/library/readingroom/docs/CIA-RDP82M00786R000104810001-5.pdf.

85. Federal Bureau of Investigation, *Soviet Intelligence Targets in the United States, 1946–1953* (Washington, DC, 1953), https://www.governmentattic.org/2docs/FBI_Monograph_Soviet-Targets-US_1953.pdf.

86. Federal Bureau of Investigation, *Exposé of Soviet Espionage.*

87. US Congress, *Departments of State, Justice, and Commerce, the Judiciary, and Related Agencies Appropriations for 1966.*

88. Office of the National Counterintelligence Executive, *Annual Report to Congress on Foreign Economic Collection and Industrial Espionage.*

89. US Congress, *PRC Acquisition of U.S. Technology* (Washington, DC, 1999), https://www.govinfo.gov/content/pkg/GPO-CRPT-105hrpt851/pdf./GPO-CRPT-105hrpt851-1-5.pdf.

90. Federal Bureau of Investigation, *Exposé of Soviet Espionage,* 2004.

91. Gus Weiss, "The Farewell Dossier: Duping the Soviets," 2007, https://www.cia.gov/library/center-for-the-study-of-intelligence/csi-publications/csi-studies/studies/96unclass/farewell.htm.

92. Federal Bureau of Investigation, *Soviet Intelligence Targets in the United States.*

93. Federal Bureau of Investigation, *Exposé of Soviet Espionage.*

94. Weiss, "The Farewell Dossier."

95. "East-West Exchange Program," 1969, https://www.cia.gov/library/readingroom/docs/DOC_0001495226.pdf.

96. US Congress, *Departments of State, Justice, and Commerce, the Judiciary, and Related Agencies Appropriations for 1966.*

97. "East-West Exchange Program."

98. Ibid.

99. Federal Bureau of Investigation, *Soviet Military, Naval, and Air Representatives in the United States* (Washington, DC, 1955), https://ia800704.us.archive.org/12/items/SovietMilitaryNavalAndAirRepresentativesInTheUnitedStates/Soviet%20Military%2C%20Naval%2C%20and%20Air%20Representatives%20in%20the%20United%20States.pdf.

100. US Senate, *Meeting the Espionage Challenge: A Review of United States Counterintelligence and Security Programs,* report of the Select Committee on Intelligence (Washington, DC, 1986), https://www.cia.gov/library/readingroom/docs/CIA-RDP90-00530R000300620021-3.pdf.

101. US Congress, *Departments of State and Justice, the Judiciary, and Related Agencies Appropriations for 1962,* before the Committee on Appropriations, House of Representatives, 86th Congress (Washington, DC, 1961).

102. US Senate, *Meeting the Espionage Challenge.*

103. Central Intelligence Agency, *The Technology Acquisition Efforts of the Soviet Intelligence Services.*

104. Federal Bureau of Investigation, *Soviet Military, Naval, and Air Representatives in the United States.*

105. US Congress, *Departments of State, Justice, and Commerce, the Judiciary, and Related Agencies Appropriations for 1964,* before the Committee on Appropriations, House of Representatives, 88th Congress (Washington, DC, 1963).

106. US Congress, *Departments of State, Justice, and Commerce, the Judiciary, and Related Agencies Appropriations for 1966.*

107. Federal Bureau of Investigation, *Soviet Military, Naval, and Air Representatives in the United States.*

108. Federal Bureau of Investigation, *Soviet Intelligence Targets in the United States.*

109. US-China Economic and Security Review Commission, *2009 Report to Congress.*

110. US Congress, *Department of Justice Appropriation Bill for 1943,* before the Committee on Appropriations, House of Representatives, 76th Congress (Washington, DC, 1942).

111. Federal Bureau of Investigation, *Soviet Intelligence Targets in the United States.*

112. Central Intelligence Agency, *Soviet Acquisition of Western Technology and Its National Security Implications.*

113. Federal Bureau of Investigation, *Chinese Communist Intelligence Activities in the United States* (Washington, DC, 1954), https://ia801908.us.archive.org/35/items/FBIPRCSpying/fbi-prc-spying.pdf.

114. Daniel Golden, *Spy Schools* (New York: Holt, 2017).

115. John Fialka, *War by Other Means* (New York: Norton, 1997).

116. US Senate, *Threats to the US Research Enterprise: China's Talent Recruitment Plans,* staff report (Washington, DC, 2019), https://www.hsgac.senate.gov/imo/media/doc/2019-11-18%20PSI%20Staff%20Report%20-%20China's%20Talent%20Recruitment%20Plans.pdf.

117. Ellen Barry and Gina Kolata, "China's Lavish Funds Lured U.S. Scientists; What Did It Get in Return?" *New York Times,* February 6, 2020, https://www.nytimes.com/2020/02/06/us/chinas-lavish-funds-lured-us-scientists-what-did-it-get-in-return.html; Federal Bureau of Investigation, *Chinese Talent Programs 2015* (Washington, DC), https://webcache.googleusercontent.com/search?q=cache:d58XN_T2cpQJ:https://compliance.fiu.edu/documents/SPIN%2520-%2520Chinese%2520Talent%2520Program.pdf.+&cd=1&hl=en&ct=clnk&gl=us.

118. US-China Economic and Security Review Commission, *How Chinese Companies Facilitate Technology Transfer from the United States* (Washington, DC, 2019), https://www.uscc.gov/sites/default/files/Research/How%20Chinese%20Companies%20Facilitate%20Tech%20Transfer%20from%20the%20US.pdf.

119. Federal Bureau of Investigation, *Chinese Talent Programs 2015.*

120. Barry and Kolata, "China's Lavish Funds Lured U.S. Scientists."

121. US Congress, *China's Pursuit of Emerging and Exponential Technologies,* before the House Committee on Armed Services, 115th Congress (Washington, DC, 2018), https://www.govinfo.gov/content/pkg/CHRG-115hhrg28966/pdf./CHRG-115hhrg28966.pdf.

122. Barry and Kolata, "China's Lavish Funds Lured U.S. Scientists."

123. Christopher Wray, director of the Federal Bureau of Investigation, "The Threat Posed by the Chinese Government and the Chinese Communist Party to the Economic and National Security of the United States," July 7, 2020, https://www.fbi.gov/news/speeches/the-threat-posed-by-the-chinese-government-and-the-chinese-communist-party-to-the-economic-and-national-security-of-the-united-states.

124. US Senate, *Threats to the U.S. Research Enterprise.*

125. US-China Economic and Security Review Commission, *2019 Report to Congress.*

126. Barry and Kolata, "China's Lavish Funds Lured U.S. Scientists."

127. US Senate, *Threats to the U.S. Research Enterprise.*

128. Department of Justice, "Harvard University Professor Indicted on False Statement Charges," June 9, 2020, https://www.justice.gov/opa/pr/harvard-university-professor-indicted-false-statement-charges.

129. Department of Justice, "Former Emory University Professor and Chinese Thousand Talents Participant Convicted and Sentenced for Filing a False Tax Return," May 11, 2020, https://www.justice.gov/usao-ndga/pr/former-emory-university-professor-and-chinese-thousand-talents-participant-convicted.

130. 104th Congress Rec. S737-S742 (1996).

131. US Congress, *Corporate and Industrial Espionage and Their Effects on American Competitiveness,* before the Committee on International Relations, House of Representatives, 106th Congress (Washington, DC, 2000), https://www.govinfo.gov/content/pkg/CHRG-106hhrg68684/pdf./CHRG-106hhrg68684.pdf.

132. Economic Espionage Act of 1996, 18 USC 1831–1837, "Prospective Policy," February 27, 2013.

133. Statement of Adam S. Hickey, deputy assistant attorney general, National Security Division, Department of Justice, before the Senate Judiciary Committee at a hearing titled "Dangerous Partners: Big Tech and Beijing," March 4, 2020, https://www.judiciary.senate.gov/imo/media/doc/Hickey%20Testimony.pdf.

134. H.R. Rep. no 788, 104th Congress, 2nd session (1996).

135. Ibid.

136. Susan W. Brenner and Anthony C. Crescenze, "State-Sponsored Crime: The Futility of the Economic Espionage Act," *Houston Journal of International Law* no. 28 (2006): 389.

137. Commission on the Theft of American Intellectual Property, *The IP Commission Report* (Washington, DC: National Bureau of Asian Research, 2013).

138. Michael J. de la Merced and Peter Lattiman, "Appeals Court Limits Federal Law Used in Goldman Programmer Case," *New York Times,* April 12, 2012.

139. Congressional Research Service, *The Global Research and Development Landscape and Implications for the Department of Defense* (Washington, DC, 2018).

140. Congressional Research Service, *Federal Research and Development (R&D) Funding, FY 2020* (Washington, DC, 2020), https://crsreports.congress.gov/product/pdf./R/R45715.

141. Congressional Research Service, *Federal Research and Development (R&D) Funding.*

142. Congressional Research Service, *The Global Research and Development Landscape.*

143. US Congress, *China's Pursuit of Emerging and Exponential Technologies.*

144. Dan Steinbrock, *The Challenges for America's Defense Innovation* (Washington, DC: Information Technology and Innovation Foundation, 2014).

145. Central Intelligence Agency, *Semiannual Report to the Director of Central Intelligence, July–December 2000,* https://www.cia.gov/library/readingroom/docs/DOC_0001311476.pdf.

146. Department of Defense, *Report of the Defense Science Board Task Force on Basic Research* (Washington, DC: Office of the Undersecretary of Defense, for Acquisition, Technology and Logistics, 2012).

147. Central Intelligence Agency, *Semiannual Report to the Director of Central Intelligence.*

148. Ibid.

149. Department of Defense, *Report of the Defense Science Board Task Force on Basic Research.*

150. Honorable Tara O'Toole, MD, MPH, undersecretary for science and technology, Department of Homeland Security, testimony before the US Senate, Homeland Security and Governmental Affairs Committee, July 17, 2013.

151. Jay Solomon, "Investing in Intelligence: Spy Agencies Seek Innovation Through Venture-Capital Firm," *Wall Street Journal,* September 12, 2005.

152. Jack Corrigan, "The Pentagon's Startup Outreach Office Is No Longer an Experiment," August 9, 2018, https://www.nextgov.com/cio-briefing/2018/08/pentagons-startup-outreach-office-no-longer-experiment/150408; Patrick Tucker, "Pentagon Shakes Up Silicon Valley Outreach," May 11, 2016, https://www.defenseone.com/technology/2016/05/pentagon-shakes-silicon-valley-outreach/128198.

153. Tucker, "Pentagon Shakes Up Silicon Valley Outreach."

154. Corrigan, "The Pentagon's Startup Outreach Office."

155. Josh Hicks, "Homeland Security Is Laying Roots in Silicon Valley, and You Might Not Like Its Reasons," *Washington Post,* April 22, 2015.

156. Department of Homeland Security, "Silicon Valley Innovation Program," https://www.dhs.gov/science-and-technology/svip.

157. Jack Corrigan, "DHS Startup Accelerator Awards Its First Final-Phase Contract," April 30, 2018, https://www.nextgov.com/emerging-tech/2018/04/dhs-startup-accelerator-awards-its-first-final-phase-contract/147864; Jack Corrigan, "DHS Is Exploring How Blockchain Can Stop Counterfeits and Forgeries," December 4, 2018, https://www.nextgov.com/emerging-tech/2018/12/dhs-exploring-how-blockchain-can-stop-counterfeits-and-forgeries/153273; Jack Corrigan, "DHS Contract Will Help Drones Automatically Spot Border Threats," May 10, 2018, https://www.nextgov.com/emerging-tech/2018/05/dhs-contract-will-help-drones-automatically-spot-border-threats/148088.

158. Hannah Denham, "No Tech for ICE: Protesters Demand Amazon Cut Ties with Federal Immigration Enforcement," *Washington Post,* July 12, 2019.

159. Tajha Chappellet-Lanier, "After Protest, Open Source Software Company Chef Will Let ICE Contract Expire," *Fedscoop,* September 23, 2019.

160. Daisuke Wakabayashi and Scott Shane, "Google Will Not Renew Pentagon Contract That Upset Employees," *New York Times,* June 1, 2018, https://www.nytimes.com/2018/06/01/technology/google-pentagon-project-maven.html.

161. Shirin Ghaffary, "Google Employees Are Demanding an End to the Company's Work with Agencies Like CBP and ICE," August 14, 2019, https://www.vox.com/2019/8/14/20805562/human-rights-concerns-google-employees-petition-cbp-ice.

162. John Markoff, "Electronics Plan Aims to Balance Government Access with Privacy," *New York Times,* April 16, 1993.

163. Markoff, "Electronics Plan"; Edmund L. Andrews, "U.S. Plans to Push Giving F.B.I. Access in Computer Codes," *New York Times,* February 5, 1994.

164. Andrews, "U.S. Plans to Push Giving F.B.I. Access."

165. General Accounting Office, *Information Superhighway: An Overview of Technology Challenges* (Washington, DC, 1995), https://www.govinfo.gov/content/pkg/GAOREPORTS-AIMD-95-23/pdf./GAOREPORTS-AIMD-95-23.pdf.

166. Kristin Finklea, *Renewed Crypto Wars* (Washington, DC: Congressional Research Service, 2016), https://fas.org/sgp/crs/misc/IN10440.pdf.

167. General Accounting Office, *Information Superhighway.*

168. Steven Levy, "Battle of the Clipper Chip," *New York Times,* June 12, 1994.

169. Congressional Research Service, *Encryption Technology: Congressional Issues* (Washington, DC, 1998).

170. Ibid.

171. John Markoff, "Industry Defies U.S. on Data Encryption," *New York Times,* January 14, 1994.

172. Levy, "Battle of the Clipper Chip."

173. Nick Wingfield, Ted Bridis, and Neil King Jr., "Earthlink Says It Won't Install Device for FBI," *Wall Street Journal,* July 14, 2000.

174. Federal Bureau of Investigation, "Carnivore/DCS 1000 Report to Congress," February 24, 2003, https://www.epic.org/privacy/carnivore/2002_report.pdf.

175. US Congress, *Departments of Commerce, Justice, and State, the Judiciary, and Related Agencies Appropriations for 1998,* before a subcommittee of the Committee on Appropriations, House of Representatives, 105th Congress (Washington, DC, 1997).

176. Charlie Savage, "U.S. Tries to Make It Easier to Wiretap the Internet," *New York Times,* September 27, 2010.

177. US Congress, *Going Dark: Lawful Electronic Surveillance in the Face of New Technologies,* before the Committee on the Judiciary, House of Representatives, 112th Congress (Washington, DC, 2011), https://www.govinfo.gov/content/pkg/CHRG-112hhrg64581/pdf./CHRG-112hhrg64581.pdf.

178. US Senate, *Current and Projected National Security Threats to the United States,* before the Select Committee on Intelligence, 114th Congress (Washington, DC, 2016), https://www.govinfo.gov/content/pkg/CHRG-114shrg20544/pdf./CHRG-114shrg20544.pdf.

179. Ellen Nakashima and Barton Gellman, "As Encryption Spreads, U.S. Grapples with Clash Between Privacy, Security," *Washington Post,* April 10, 2015.

180. Craig Timberg, "Apple Will No Longer Unlock Most iPhones, iPads for Police, Even with Search Warrants," *Washington Post,* September 18, 2014, https://www.washingtonpost.com/business/technology/2014/09/17/2612af58-3ed2-11e4-b03f-de718edeb92f_story.html.

181. US Congress, *The Encryption Tightrope: America's Security and Privacy,* before the Committee on the Judiciary, House of Representatives, 114th Congress (Washington, DC, 2016), https://www.govinfo.gov/content/pkg/CHRG-114hhrg98899/pdf./CHRG-114hhrg98899.pdf.

182. Eric Lichtblau, "Judge Tells Apple to Help Unlock iPhone Used by San Bernadino Gunman," *New York Times,* February 16, 2016, https://www.nytimes.com/2016/02/17/us/judge-tells-apple-to-help-unlock-san-bernardino-gunmans-iphone.html.?searchResultPosition=5.

183. Eric Lichtblau and Katie Benner, "Apple Fights Order to Unlock San Bernadino Gunman's iPhone," *New York Times,* February 17, 2016, https://www.nytimes.com/2016/02/18/technology/apple-timothy-cook-fbi-san-bernardino.html.?searchResultPosition=6.

184. US Congress, *The Encryption Tightrope.*

185. Murad Ahmed, "Palantir Goes from CIA-Funded Start Up to Big Business," *Financial Times,* June 24, 2015, https://www.ft.com/content/926af768-1a4c-11e5-a130-2e7db721f996.

186. Cade Metz, Erin Griffith, and Kate Conger, "What's a Palantir? The Tech Industry's Next Big IPO," *New York Times,* August 26, 2020, https://www.nytimes.com/2020/08/26/technology/palantir-ipo.html.?searchResultPosition=4.

187. Metz, Griffith, and Conger, "What's a Palantir?"; Andrew Marantz, *Antisocial* (New York: Viking, 2019).

188. Li Yuan and Daisuke Wakabayashi, "Google Seeking a Return to China, Is Said to Be Building a Censored Search Engine," *New York Times,* August 1, 2018, https://www.nytimes.com/2018/08/01/technology/china-google-censored-search-engine.html.

189. Kate Conger and Daisuke Wakabayashi, "Google Employees Protest Secret Work on Censored Search Engine for China," *New York Times,* August 16, 2018, https://www.nytimes.com/2018/08/16/technology/google-employees-protest-search-censored-china.html.?searchResultPosition=1.

190. Davey Alba, "A Google VP Told the US Senate the Company Has 'Terminated' the Chinese Search App, Dragonfly," *BuzzFeed,* July 16, 2019.
191. Elsa B. Kania, *China's Threat to American Government and Private Sector Research and Innovation Leadership,* testimony before the House Permanent Select Committee on Intelligence, July 19, 2018, https://docs.house.gov/meetings/IG/IG00/20180719/108561/HHRG-115-IG00-Wstate-KaniaE-20180719.pdf.
192. Melissa Korn, "Microsoft Brings U.S. and China Universities Together," *Wall Street Journal,* June 19, 2015.
193. Kania, *China's Threat.*
194. US Senate, *Meeting the Espionage Challenge: A Review of United States Counterintelligence and Security Programs,* report of the Select Committee on Intelligence (Washington, DC, 1986), https://www.cia.gov/library/readingroom/docs/CIA-RDP90-00530R000300620021-3.pdf.
195. US-China Economic and Security Review Commission, *2019 Report to Congress.*
196. Brown and Singh, *China's Technology Transfer Strategy.*
197. Richard McGregor, "How the State Runs Business in China," *The Guardian,* July 25, 2019, https://www.theguardian.com/world/2019/jul/25/china-business-xi-jinping-communist-party-state-private-enterprise-huawei.
198. Chris Buckley and Keith Bradsner, "China's Communists to Private Business: You Heed Us, We'll Help You," *New York Times,* September 17, 2020, https://www.nytimes.com/2020/09/17/business/china-communist-private-business.html.
199. US Congress, *Investigative Report on the U.S. National Security Issues Posed by Chinese Telecommunications Companies Huawei and ZTE,* report by Chairman Mike Rogers and Ranking Member C. A. Dutch Ruppersberger of the Permanent Select Committee on Intelligence, House of Representatives, 112th Congress (Washington, DC, 2012).
200. Ashley Feng, "We Can't Tell If Chinese Firms Work for the Party," *Foreign Policy,* February 7, 2019, https://foreignpolicy.com/2019/02/07/we-cant-tell-if-chinese-firms-work-for-the-party.
201. "Critical Infrastructure Sectors," https://www.cisa.gov/critical-infrastructure-sectors.
202. Advisory Panel to Assess Domestic Response Capabilities for Terrorism Involving Weapons of Mass Destruction, *Fourth Annual Report to the President and the Congress* (Santa Monica, CA: Rand, 2002), https://www.rand.org/content/dam/rand/www/external/nsrd/terrpanel/terror4.pdf.
203. Congressional Research Service, *Critical Infrastructure: Emerging Trends and Policy Considerations for Congress* (Washington, DC, 2020).
204. Congressional Research Service, *The U.S. Election Assistance Commission: Overview and Selected Issues for Congress* (Washington, DC, 2019), https://crsreports.congress.gov/product/pdf./R/R45770.
205. Charles Clancy and Emily Frye, "Is It Time to Designate Social Media as 'Critical Infrastructure'?" *The Hill,* July 27, 2020, https://thehill.com/opinion/cybersecurity/509154-is-it-time-to-designate-social-media-as-critical-infrastructure.
206. US Congress, *Global Terrorism: Threats to the Homeland,* pt. 2, before the Committee on Homeland Security, House of Representatives, 116th Congress (Washington, DC, 2020), https://www.govinfo.gov/content/pkg/CHRG-116hhrg40463/pdf./CHRG-116hhrg40463.pdf.
207. US Senate, *Threats to the Homeland,* before the Committee on Homeland Security and Governmental Affairs, 115th Congress (Washington, DC, 2017), https://www.govinfo.gov/content/pkg/CHRG-115shrg29657/pdf./CHRG-115shrg29657.pdf.

208. US Congress, *Departments of Commerce, Justice, and State, the Judiciary, and Related Agencies Appropriations for 1996,* pt. 2, before a subcommittee of the Committee on Appropriations, House of Representatives, 104th Congress (Washington, DC, 1995).

209. Federal Bureau of Investigation, *"Going Dark": Law Enforcement's Need to Preserve Lawful Intercept Capabilities* (2008), https://ia800402.us.archive.org/10/items/FBIGoingDark/20110322_fbi_going_dark_finalrelease.pdf.

210. US Congress, *Going Dark.*

211. "About NDCAC," https://ndcac.fbi.gov/about/about-the-ndcac; US Senate, *Commerce, Justice, Science, and Related Agencies Appropriations, FY 2016,* US Senate Appropriations Committee, 114th Congress (Washington, DC, 2016), https://www.govinfo.gov/content/pkg/CHRG-114shrg93106/pdf./CHRG-114shrg93106.pdf.

3

Disrupting the Theft of Assets

STATUTORY AND DE FACTO RULES OF THE ROAD GOVERN THE RELATIONSHIP between the US government and industry. Foreign powers have consistently attempted to circumvent and subvert these in order to access knowledge and technology that will enhance their interests. The private sector has been a long-standing target because the capabilities it can provide have the potential to enhance the elements of national power that state and nonstate actors can wield against US interests. Developing awareness of—and disrupting—these threats is largely the job of the Federal Bureau of Investigation (FBI). Disruption of the threat has taken two primary forms: investigation/prosecution as threat actors, and proactive outreach to vulnerabilities.

Private Sector Knowledge: A Long-Acknowledged Vulnerability

The US government's concerns about the theft of private sector information with implications for US national security is a long-standing and valid one. In 1950, for instance, the FBI advised Congress that certain foreign espionage agents had attempted to obtain information regarding US industrial resources.[1] Nearly a decade later, an FBI official noted the "intense efforts [the Soviets were] making to obtain industrial secrets and processes in [the United States]."[2] The official assessed, alarmingly, that "the Soviets [had] the most developed and best industrial-spying system in the world."[3] In 1961, then–FBI director J. Edgar Hoover advised Congress that Soviet and Eastern-bloc agents were "assiduously" collecting information about strategic industrial potential. As an example, Hoover specifically noted the targeting of aircraft production.[4] In 1965, Hoover again noted the Soviets' collection against US industry.[5] As the Cold War progressed, the threat to US industry continued. In 1979, then–FBI director William Webster indicated that there was a continuing threat from foreign intelligence services that were pursuing scientific and technological information.[6]

Threats seemed to multiply after the Cold War ended. In 1996, the Central Intelligence Agency (CIA) advised the Senate Select Committee on Intelligence that foreign government–orchestrated theft of US corporate science and technology data was the greatest espionage threat to US economic competitiveness. The governments most aggressive in this type of espionage included China, Iran, Cuba, Russia, Israel, and France.[7] Then–FBI director Louis Freeh also indicated the extent of the threat to US industry when he advised Congress, in 1997, that there were twenty-three countries that maintained active and clandestine economic espionage pursuits in the United States. Freeh explained that the objectives of these operations were to "turn over those trade secrets to state industries, to manufacturers, and out-compete the US companies, particularly the companies whose intellectual property or trade secret has been stolen."[8]

Foreign governments did not just steal information in order to emulate it; they increasingly used it to develop applications to compete against the United States. In 1981, the CIA assessed that although the Soviet practice during the 1950s and 1960s had been simply to copy Western technology, the Soviets had become "far more discriminating" and chose only the design elements and engineering approaches that best fit their capabilities.[9] (This has a corollary in China's more recent practice of allowing foreign direct investment only from companies that align with its own industrial priorities.) With this head start, foreign perpetrators of industrial espionage could offer new and improved versions of US products against their US originators. The FBI pointed out in 1998 that foreign collectors were "aggressively" targeting bid, contract, customer, and strategy information.[10] Such information would certainly help a foreign competitor to undercut its US competition.

After 9/11, while the US intelligence community retooled for the war against terrorism, traditional state-sponsored collection against the US private sector continued. In its 2003 report, the US National Counterintelligence Executive (NCIX) assessed that representatives of more than ninety countries targeted US technologies and corporate trade secrets in both 2002 and 2003.[11] Approximately a decade later, then–director of national intelligence James Clapper advised Congress that Russia and China were aggressive and successful purveyors of economic espionage against the United States.[12]

A Variety of Threat Actors

The traditional image of a spy—including one engaged in economic espionage—is that of a foreign diplomat. As Hoover noted in 1961, many of the Soviet and satellite agents pursuing information about US technology operated with diplomatic immunity from the United Nations in New York and from their respective embassies in Washington, DC.[13] Then–FBI director Clarence Kelley confirmed this image when, in 1974, he noted that Soviet intelligence officers operated under diplomatic cover from Soviet platforms in Washington, San Francisco, and New York.[14] Kelley, in 1977, noted that "the extent of hostile intelligence operations is directly proportionate to the number of foreign officials permanently assigned to the United States."[15] Nearly a decade later, a congressional study assessed that the "foundation for domestic counterintelligence is systematic collection on a foreign country's official representatives in the United States."[16] "Diplomat" definitely equaled "duplicity."

The role of diplomats in the conduct of intelligence activities should not be surprising. The job of a diplomat is to advance their government's informational advantage either through acquisition of data or by shaping perceptions. Additionally, foreign governments have augmented their collection of proprietary, private sector information through state-controlled businesses. What better way for a foreign government to reach targeted industries than to give the impression, through commercial proxies, that there is business to be done (and a profit to be made). This also changes the dynamic for spying. An individual who might never consider handing over US secrets to a foreign government might be more willing to provide a foreign competitor with an unethical advantage for a fee, even if this amounts to selling out a US capability to a US competitor or even an adversary.

The Soviet Union's Amtorg Trading Corporation provides a clear historical example of how a foreign government can use a state-run commercial entity to engage in collection against private industry. In 1924—nearly a decade before Washington and Moscow established formal diplomatic relations—the Soviets established Amtorg, in New York, to act as an importer and exporter for the Soviet Union.[17] However, only in 1949 did Amtorg have to register with the US Department of Justice as a foreign agent.[18]

Since access to industry—and technologies of interest—was a core function of Amtorg, it is logical that Amtorg provided a platform for both licit and illicit business. The Soviets, since the establishment of Amtorg, used it as a cover for industrial espionage activities directed against US interests.[19] In 1965, the FBI assessed that more than half of the Soviet nationals whom Amtorg employed were either known or suspected intelligence officers.[20] Access included not only contacts with US industrial concerns but also the ability to purchase US goods for delivery to the Soviet Union. During the late 1940s, Amtorg placed orders for equipment, useful in atomic energy work, from the US private sector. Such above-board attempts at acquisition complemented the clandestine collection that Soviet agents had conducted against the US Manhattan Project. More prosaically, Amtorg placed orders for samples of nearly every household item manufactured in the United States, which the Soviet Union—according to an FBI informant—duplicated.[21]

Amtorg did not just rely on Soviet intelligence officers in furtherance of collection; it also utilized witting Americans in its efforts. For instance, Cyril Lamkin, a charter member of the Communist Party of the United States, was employed by Amtorg and did "secret work" while employed by the corporation.[22] The FBI, in 1962, assessed that Amtorg employees also sought out Americans in the private sector "who sometimes place their desire for a quick profit above their regard for the internal security of the United States."[23]

Additionally, the Soviet Union made effective use of its satellite countries' commercial presence to acquire technology. For instance, the Soviets were impressed by Poland's success in acquiring Western technology and subsidized this activity through direct contributions to the Polish budget.[24] According to a CIA assessment, the KGB constantly pressured the Polish Ministry of Internal Affairs to obtain Western technological expertise in the fields of microelectronics and computers, aircraft mainframes and engines, avionics, and military-related technologies.[25]

The collection of this technological knowledge was at the heart of a Polish spy case that used the country's ostensibly commercial representation in the United States. Marian Zacharski, the president of the Polish American Machinery Corporation (POLAMCO), developed a relationship with William Holden Bell, a project manager of the Radar Systems Group at Hughes Aircraft in California.[26] POLAMCO, a machine exporter based in Elk Grove Village, Illinois, was largely owned by the Polish government, and Zacharski was an officer of the Polish Intelligence Service (PIS).[27] Zacharski had arrived in Los Angeles, California, in 1976, as the West Coast branch manager for POLAMCO, and the PIS had assigned him to spot and recruit agents within California's aerospace industry.[28] In June 1981, the FBI confronted Bell, but not before he had compromised Hughes's "quiet radar" and other systems.[29] Poland also used the guise of commerce in its recruitment of James Durward Harper, who provided the PIS with information regarding the survivability of the Minuteman missile system and about US defenses against attack by ballistic missiles. Harper's PIS contact, Zdzislaw Przychodzien, was overtly an official in Poland's Ministry of Machine Industry, but was in reality a lieutenant colonel in the PIS. The PIS got the technology and Harper got life in prison.[30]

Foreign governments have continued to direct corporations, engaged in sensitive fields, on US soil. In 2010, for instance, the Russian firm TENEX established the TENAM Corporation in the United States. TENEX was a wholly owned subsidiary of JSC Atomenergoprom, which, in turn, was a wholly owned subsidiary of the State Atomic Energy Corporation (nesting dolls are, after all, uniquely Russian) and functioned as the sole supplier and exporter of Russian Federation uranium and uranium-enrichment services to nuclear power companies worldwide.[31] In 2015, the United States sentenced Vadim Mikerin, the president of TENAM, to four years in prison for conspiracy to commit money laundering in connection with his role in arranging corrupt payments to secure improper advantages for US companies that did business with TENEX.[32] While TENAM was not engaged in collection, its activities are fundamentally similar to the examples cited earlier, in that its executive employed illicit activities, against US interests, in the operation of a foreign state-run company.

Foreign diplomats and officials affiliated with other state-run organizations might as well wear name tags saying "Hi, I'm a spy." Their purpose—after all—is to provide their governments with an informational advantage either through collecting information or by peddling information in furtherance of a desired US policy outcome. However, in 2007, the FBI stated that "the most notable trend" in foreign government-sponsored economic espionage was the movement away from the direct involvement of intelligence services.[33] Previously, in 2005, the FBI had made a similar observation about its "growing concern" regarding certain intelligence services that supplemented their collection capabilities in the United States by using nontraditional actors.[34] This should not have come as a surprise. In 1998, the Bureau had advised Congress that there had been numerous cases illustrating that individuals outside of the formal intelligence service apparatuses had engaged in clandestine activity that was "inimical to the security and economic well-being of the United States."[35] These statements, over the course of a decade, seemed to continually identify the same problem as a new concern.

These wide-eyed announcements in fact are not new at all. Hoover, in discussing the East-West Exchange Program in 1965, noted that Soviet scientists who had visited the United States under the program's auspices were required by the KGB to submit comprehensive reports on the technical aspects of their trips, including research being conducted and the status of particular projects.[36] Nearly two decades later, the CIA assessed that exchange programs that gave Soviet scientists and engineers entry to the United States were among the most important sources of technology loss, because of the "hands on" experience and collegial working relations with US counterparts that Soviet participants enjoyed.[37] The US intelligence community—and the FBI specifically—also encountered this problem with China. In 1999, the Bureau advised Congress that most of Beijing's intelligence collection was conducted by a wide variety of individuals who were not professional intelligence officers.[38]

The FBI's Approach to Countering the Threat Against Private Industry

Counterintelligence is a long-standing FBI mission. Its formal origin is in a September 6, 1939, directive, from then-president Franklin D. Roosevelt, that invested the FBI with responsibility for investigative work in matters regarding espionage, sabotage, and violations of neutrality regulations.[39] The FBI almost immediately entered into countering the intelligence threat to industry through the Plant Survey Program, which it initiated in October 1939 to make commercial entities aware of the potential for espionage against their information.[40] Plant surveys were a relatively short-lived function of the FBI, which turned the project over to the military in 1942.[41] However, these surveys were early indicators that the Bureau was attuned to the threat that foreign actors posed to US business. It continued to demonstrate its commitment to the private sector. For instance, in 1953, the FBI indicated that it had been involved with reviewing the loss of documents from General Electric's Syracuse and Schenectady plants.[42] Hoover's observations regarding Soviet industrial targets further confirmed the FBI's awareness that intelligence collection inimical to US national security went well beyond government information.

The next milestone for the FBI's counterintelligence program was in the mid-1970s. Investigations of US intelligence activities prompted the US attorney general to promulgate the first set of foreign counterintelligence guidelines, which took effect in May 1976.[43] Then–FBI director Clarence Kelley suggested that the Bureau's counterintelligence operations were more than a US-Soviet game of spy-versus-spy when, in 1977, he noted that the extent of hostile intelligence operations was "not necessarily related to prevailing diplomatic postures or to the international political climate."[44] This somewhat cryptic remark suggested that foreign governments might be in pursuit of objectives (e.g., acquisition of private sector information) beyond geopolitics. However, hard power, rather than economic competition, continued to be the prevailing framework for counterintelligence. In 1977, FBI counterintelligence investigations were primarily a defensive reaction to the initiatives of twenty-three hostile intelligence services operating against the United States on behalf of the thirteen communist countries with representatives in the United States.[45] Kelley's successor, William Webster, suggested the importance of a focus on economic espionage when he noted, in 1980, that counterintelligence efforts provide insights about a foreign government's interests, particularly when it came to the technology field.[46] (Webster echoed a much earlier Hoover analysis that identifying the objectives—among other things—was more important than identifying the malefactor.)[47] The sluggishness of the US counterintelligence enterprise to combat threats to US industry was apparent nearly a decade later when a report by the House Permanent Select Committee on Intelligence suggested that an area of "possible fruitful investigations" was the development of opportunities for countering espionage directed at US firms possessing critical technology.[48] This suggested that Webster's philosophy had not taken hold.

Changes that occurred as the Cold War waned led to a rethinking of US counterintelligence. In mid-1989, the administration of then-president

George H. W. Bush directed the intelligence community to provide an assessment of what the intelligence threat would look like in the 1990s and to define a counterintelligence/security countermeasures response.[49] Bush endorsed the assessment's findings via a national security directive he issued on October 5, 1990. The directive also had implications for the government's role vis-à-vis industry in its statement that "U.S. technologies, both classified and proprietary, will remain a high priority for those seeking a competitive edge in international markets."[50]

Consistent with the tenor of the times, the FBI reassessed its own approach to counterintelligence. As the Bureau advised Congress in 1989, its investigative responsibilities had expanded due to the increased access to the United States by foreign nationals from a broadening array of countries.[51] Then–FBI director William Sessions stated, in 1992, that in 1989 he "undertook to examine the foreign counterintelligence effort in [the FBI's] Intelligence Division and shake it up completely."[52] Sessions's objective was to ensure that the Bureau "had a mechanism that allowed [it] to identify and to meet any foreign counterintelligence threat wherever it came from."[53] As he noted in 1991, the threats were myriad and the FBI needed to prevent technology and other critical information from being "drained away" from the United States.[54]

Sessions's review prompted the creation of a new paradigm for counterintelligence: the National Security Threat List (NSTL). The NSTL—which the attorney general approved, and the FBI implemented, in 1992—consisted of two components: country threats and issue threats.[55] Country threats were "foreign governments and entities whose intelligence activities are so hostile or of such concern to the national security of the United States that counterintelligence or monitoring activities directed against such countries are warranted."[56] Issue threats were "categories of activity that pose a significant threat, or are of such concern to the national security of the United States when engaged in by any foreign power or entity that counterintelligence or monitoring actions directed against such activities are warranted."[57] The FBI assessed that the most significant change associated with the implementation of the NSTL concept was the creation of "a mechanism for neutralizing intelligence activity conducted by any foreign power targeting key U.S. national security issues."[58] According to Sessions, the NSTL issue threats allowed the FBI to be "much more focused" because it would be dealing with the specifics of what a foreign actor was attempting to obtain, rather than simply the actor's origin.[59] This line of thinking was consistent with Webster's remark, more than a decade before, about the explanatory value of an adversary's collection against technology.[60]

Creation of the NSTL represented a departure from the Cold War approach to counterintelligence. Sessions noted, in 1992, that the NSTL included both former hostile countries as well as those that would be considered to be among the friends of the United States.[61] The NSTL—in the field of counterintelligence (as it also provided a rubric for assessing terrorism concerns)—determined the level of threat that a country or entity posed based on four criteria: the observed level of intelligence activity; the nature of intelligence collection's target; the capability of the country or entity for intelligence collection; and the political or military alignment of the country or entity.[62]

The NSTL also represented a further step toward incorporating the private sector into an understanding of national security. Initially, the NSTL's threat issues included the targeting of core US technologies.[63] However, the General Accounting Office was unimpressed by the US government's efforts to protect proprietary information, stating, in 1992, that the CIA and the FBI were not sufficiently coordinated to adequately protect US industry against economic espionage.[64] The FBI refined this, following the passage of the Economic Espionage Act in 1996. According to a 1998 Bureau publication, the FBI had revised the NSTL in concert with the US intelligence community and other elements of the US government. The updated version of the NSTL included three issue threats pertinent to industry: economic espionage, proliferation, and targeting of the national information infrastructure. Economic espionage concerned foreign power–sponsored or foreign power–coordinated intelligence activity directed at targets including US corporations or persons involving the unlawful or clandestine targeting or acquisition of sensitive financial, trade, or economic policy information, proprietary economic information, or critical technologies. Additionally, the economic espionage facet of the NSTL included an influence aspect: "the unlawful or clandestine targeting or influencing of sensitive economic policy decisions." The proliferation NSTL threat issue acknowledged that US corporations could be the target of activity directed at the illicit movement of weapons of mass destruction or conventional weapons. Finally, targeting the national information infrastructure had significant implications for the private sector—since the private sector was responsible for much of the infrastructure in question. From a counterintelligence perspective, the most pertinent aspect of this issue threat was the "unauthorized monitoring of computer, cable, satellite or telecommunications systems" and the "unauthorized disclosure of proprietary . . . information stored within or communicated through computer, cable, satellite or telecommunications systems."[65]

While the NSTL provided a rubric for assessing threats, the FBI still had to find a way to engage in collection against the country and issue threats. In 1994, the Bureau initiated its Economic Counterintelligence Program.[66] The Bureau's intent was to detect and neutralize threats, either sponsored or coordinated by foreign powers, against US economic interests. There was good news, according to the FBI: "this focused effort resulted in a dramatic increase in FBI investigations"; as well as bad news:

"a realization that existing legal remedies at the federal level were insufficient to address the scope and nature of the economic espionage activities."[67] In the absence of a federal statute to prosecute economic espionage, the FBI's program lacked sufficient teeth.[68] Passage of the Economic Espionage Act in 1996 gave the Bureau primary jurisdiction over the issue.[69]

Once the Economic Espionage Act took effect, it seemed as though the Bureau was primed to go after threats to the US private sector. By its own account, despite the deficiencies in prosecutorial options prior to 1996, it had nevertheless handled hundreds of matters pertaining to hostile economic intelligence activities.[70] The FBI's 1998 strategic plan seemed to reflect its new posture by including economic security as a top priority.[71] By the late 1990s, the cyber environment was creating new opportunities for malefactors. According to the National Counterintelligence Executive's 1998 annual report to Congress, "the growing use of computer networks and telecommunications for commerce and the storage and transmittal of sensitive information provides increased opportunities for technical collection."[72] In 1998, the Bureau confirmed NCIX's assessment, telling Congress that many of the efforts, by foreign entities, to target US economic interests involved the use of computer technology, either to obtain information or to facilitate more traditional forms of espionage.[73]

As the FBI approached the twenty-first century, it began to refine its approach to counterintelligence. In 1999, it created a dedicated Counterintelligence Division.[74] Two years later, in 2001, it proposed a counterintelligence initiative that would enhance personnel and financial resources. Under this plan, the FBI would enhance its investigative activities in the field and would also improve national-level program management and coordination of field investigative activities.[75]

Starting in 2002—consistent with the approach favored by then-director Robert Mueller III—the Bureau placed increased emphasis on the role of headquarters in counterintelligence. During that year, David Szady, the assistant director of the Counterintelligence Division, announced the creation of a more centralized and nationally directed counterintelligence effort. This, according to Szady, diverged from the Bureau's historical approach in which field divisions established local priorities assigned resources accordingly. Szady believed that this emphasis on headquarters would ensure that the FBI would be more proactive and predictive in protecting US critical national assets. Furthermore, Szady advised Congress that the FBI would completely integrate its national counterintelligence strategy with NCIX.[76]

It is worth noting that the reorganization that Szady touted in 2002 would establish a top-heavy, even arrogant culture, which blew up spectacularly in 2019. In 2002, Szady had assured Congress that "centralization cements accountability regarding counterintelligence program direction, control and leadership."[77] However, the Crossfire/Hurricane investigation—an inquiry into alleged malfeasance by the presidential campaign of Donald Trump—showed that this mentality could take the organization in the exact opposite direction. Headquarters opted to conduct the investigation from FBI headquarters rather than from one or more of the FBI's field offices. An inquiry by the Department of Justice's inspector general suggested that the running of the case from headquarters was hardly transparent, as the resulting report recommended that "the FBI should develop protocols and guidelines for staffing and administrating any future sensitive investigative matters from FBI Headquarters."[78] The FBI had already drawn scrutiny from the inspector general for its handling of the "Midyear Exam" investigation of then–secretary of state Hillary Clinton's usage of unclassified private servers for work-related email correspondence. The investigation—again rather than being run from a field office—was handled as a "special" investigation by FBI headquarters.[79]

Combating the Threat

Counterintelligence is primarily the responsibility of the FBI, although the US Department of Defense (DoD) has historically played—and continues to play—a role when targets have a nexus to DoD entities. The FBI was historically able to leverage other government agencies as well as industrial concerns that were beholden to government contracts to disrupt intelligence activities. Its plant informant program, in the run-up to World War II, looked for signs of trouble in key defense industries. During the Cold War it worked with other government agencies to maintain awareness of what threat actors might be attempting to exfiltrate from the United States via diplomatic privilege or through the mail.

Tripwires: Getting in Front of the Threat Through Advance Warning

Since the beginning of its history as a counterintelligence service, the FBI has implemented several initiatives meant to provide it with early warning of intelligence threats to the private sector. In late 1939 it became responsible for the security of industrial sites that the Departments of War and Navy had certified to be of vital importance to national defense.[80] The establishment of plant informants was one way through which the Bureau implemented its responsibility. Confidential plant informants were "individuals who because of previous or present affiliations or associations and their location in industrial facilities are in a position to furnish reliable information to the Bureau concerning any persons or organizations engaging in activities inimical to the national defense."[81] Per FBI instructions, each field office was supposed to develop plant informant coverage in every industrial facility that the army and navy identified as significant to their missions.[82]

As of late 1940, the FBI's Executive Conference believed that "it was preferable to develop confidential informants within each department of the plant."[83] The informants should consist of laborers who were working in the assembly lines, machine shops, and other productive areas of the plants.[84] In 1941, FBI headquarters instructed the field to begin developing confidential informants in "shadow plants." These facilities, also known as "defense plants," were constructed at government expense, owned by the government, and operated by private industry.[85]

Additionally, the Bureau emphasized the need to develop informants in industries where it suspected communist infiltration had occurred. The FBI cautioned its personnel that informants were only to be used once a reasonable suspicion that subversive activities were occurring had been established. However, Hoover was emphatic that the FBI should "not, under any circumstances, contemplate the investigation of any labor union or any labor organization."[86] Driving this point home even further, the FBI notified its personnel that agents should specifically advise their plant informants that the Bureau was not interested in employer-employee relationships and that the FBI's interest was limited to obtaining information regarding possible espionage, sabotage, any violation with the Bureau's investigative jurisdiction, or information that would be of interest to the FBI in its safeguarding of US internal security.[87]

Once the government determined plants to be military reservations, responsibility for the investigation of sabotage and espionage transferred to the War Department.[88] In early 1942, the FBI notified its personnel that the War Department had assumed responsibility for coverage, including the operation of informants, at these facilities. The Bureau instructed its field offices to provide the War Department with information about the plant informants whom field offices had developed, although hedged that if a plant informant might be of value on matters not occurring within the plant, or if the field office desired to retain the informant's services for any reason, the field office should continue to handle the informant and not reveal the informant's existence to the War Department.[89] (One Bureau official bluntly advised Hoover that a certain aspect of the turnover had been delayed so that field offices could review their informants' production, since "it is not desired that any really valuable informants be lost to the Bureau.")[90] It is worth noting that, even though plant surveys ended in 1942, the Bureau maintained official contact with plant managers on issues of intelligence collection.

Management of the plant informant program was a sprawling process. In 1941, field offices submitted semimonthly reports on the 5th and 20th of each month that listed the number of plant informants developed in each location that the War and Navy Departments had identified as a priority. A unit at headquarters tabulated this information, which ultimately made its way to Hoover himself.[91] A few months later, this process was scaled back to once per month with a report that provided the total number of plant informants in each location; the location and total number of employees in the plant; and a total of all confidential plant informants developed in all prioritized industrial facilities in the field office's area of responsibility.[92] By 1944, the FBI had reduced this requirement, mandating only a quarterly report from field offices on its plant informants. These reports had to contain the name and address of each industrial facility of importance to the national defense, the number of employees, and the number of plant informants whom the respective field office had developed at each facility.[93] As of January 1941, the FBI had accumulated 7,000 plant informants and anticipated that the number would increase to 20,000.[94] The scope of informant coverage broadened in 1942 when, in January of that year, the Bureau provided instructions to all field office special-agents-in-charge regarding establishing informant coverage "in all facilities wherein materials for use in National Defense may be stored or temporarily located."[95] By November 1942, the FBI had developed 15,247 informants in 2,445 industrial plants.[96] Two years later, the FBI had coverage in 13,300 facilities.[97]

Recruitment of informants was a tricky business. According to instructions from mid-1941, agents were not to use contacts with plant management to identify potential informants—the officials were not even supposed to have knowledge of who the informants were—and agents should instead make direct contact with prospective informants.[98] However, the Bureau changed its approach to this relatively quickly. This is indicated by instructions to the field, in February 1942, stating that management of manufacturing plants and other facilities of importance to the national defense could be asked for the names of prospective confidential plant informants.[99]

In addition to plant informants, the FBI developed "sources of information." These were essentially passive collectors—consisting of officials and other persons serving in supervisory or administrative capacities."[100] In September 1940, FBI field offices received instructions to develop confidential sources in defense plants of interest to the military.[101] According to Bureau instructions promulgated in early 1941, increasing the number of informants, rather than sources, was an "imperative" task at the present time.[102] However, several months later, the FBI "deemed urgent" the development of sources of information in supervisory and administrative roles who could provide the Bureau with any indicators that national defense statutes were being violated. Interestingly, although the FBI set up well-defined procedures (and thus demanded additional resources) for administering the plant informant program, sources of information (e.g., presidents, secretaries, plant treasurers) who did not care whether their identity was known would only be considered as sources of information, and the Bureau did not require information about their existence except when the sources of information provided reporting.[103]

Although the FBI's Plant Survey Program ended in early 1942, the plant informant program did not follow suit. In 1942, Hoover directed field offices to draw up lists of their confidential plant informants and provide them to the War Department, as part of the handoff of plant survey work. However, Hoover advised office that "in any instance where the informant may be of value . . . relative to matters not occurring within the plant, or if you desire to retain his services for any reason, he should be continued as an informant of your office."[104] The FBI even delayed the final turnover from January to March, since it would "be necessary for the Field to go over the production of their various informants. . . . This [would] take considerable time in view of the fact that it is not desired that any really valuable informants be lost to the Bureau."[105]

The not-quite-wholesale turnover of plant informants has created some confusion about when the plant informant program actually ended. A late 1945 memo notes the termination of the plant informant program, a development seemingly corroborated by a February 4, 1946, communication indicating that the discontinuation of the plant informant program necessitated a change in the *National Defense Manual.*[106] However, according to a 1951 communication, the Internal Security Unit of the FBI's Domestic Intelligence Division continued to handle copies of current plant informant quarterly reports.[107] Field offices continued to operate the plant informant program. An inspection report on the Bureau's Birmingham office made reference, in its administrative data, to plant informants. The office's special-agent-in-charge selected the plants of interest based on their relevance to "any possible future national emergency."[108] Furthermore, a 1967 inspection of the Domestic Intelligence Division recommended streamlining the plant informant program.[109] According to the Church Committee's staff reports, the plant informant program continued to officially exist until 1969.

In addition to informants and sources of information, the FBI also sent its own personnel, undercover, into industrial plants. In September 1939, the FBI's Executive Conference advised the FBI director that

> as a basic and general principle in selecting men for specialized work in munitions factories, airplane factories and similar commercial occupations due consideration must be given to a man's background and qualifications and his industrial or commercial experiences prior to his entering the service of the Bureau. Most certainly the services of Accountants who had prior industrial experience would be considered seriously by the Bureau in those instances where it is necessary to put a man in a commercial plant.[110]

Furthermore, in November 1939, the FBI's Executive Conference considered, but rejected, a proposal that would have trained a number of stenographers in espionage and sabotage matters so that they could obtain employment in munitions and war material plants "in order to serve in an undercover capacity as informants and observers in such plants."[111] The conference did not object in principle but expressed its belief that getting agents trained in this area was the greater priority.[112] However, the FBI would revisit the idea of placing agents undercover in industrial facilities during the postwar era.

After World War II, the FBI took on new responsibilities to protect the US atomic energy program. One aspect of this was an expansion of the plant informant program.[113] In 1949, an FBI communication indicated that its field offices had instituted an ongoing effort to develop informants in the vital atomic energy installations. Contractors—as in the case of World War II–era shadow plants—ran a number of these sites. FBI headquarters directed the field officers in El Paso, Texas; Seattle, Washington; Knoxville, Tennessee; New York, New York; Albany, New York; Boston, Massachusetts; Chicago, Illinois; and Los Angeles and San Francisco, California, to make every effort to promote an informant reporting on communism into a position with the Atomic Energy Commission (AEC). According to the Bureau, the objective of this project was "detecting and investigating attempts to obtain restricted data."[114] As described by a 1951 Bureau communication, the "program of developing informants among employees at AEC installations is kept constantly vitalized and is closely supervised in order to insure complete coverage."[115] The scope of the program is apparent in a communication from that year stating that the FBI had developed 3,846 plant informants at 517 atomic energy plants.[116]

Then, in 1949, the FBI revisited the concept of sending FBI agents on undercover assignments to pose as employees of the facilities in which the Bureau had an interest. In its consideration of the concept, the Bureau pondered two possibilities: either having an agent apply for a job directly with an AEC installation without disclosing their identity, or arranging with a trusted, high-level AEC official to place an agent in a position with a facility. Officials at FBI headquarters considered the former option to be preferable, but did note slight trepidation at the thought that "public disclosure of the program would result in outcries of Gestapo tactics."[117] While no one wants to be called the Gestapo, the FBI was perhaps more sensitive to this in 1949, since a few years prior its bid to become a global intelligence service had been scuttled, in part, by concerns about the emergence of a too-powerful agency that would resemble a US Gestapo.[118]

Once it had decided upon its course of action, the FBI had to surmount another hurdle: finding employees who could credibly apply for AEC jobs. The Bureau's Administrative Division reviewed the vocation records of suitable special agents—72 chemists, 310 physicists, 115 electricians, 105 engineers, 12 biologists, 19 mathematicians, and 18 metallurgists—and compiled a list of 150 agents who possessed either the applicable training or experience in a technical field.[119] (This was probably a difficult task. After all, according to Hoover's testimony to Congress in 1950, the FBI's personnel were either lawyers or certified public accountants.)[120] A further complication to finding appropriate candidates was stated by a Bureau official who explained that "obviously many of the Agents allocated to the various technical fields would neither be adaptable nor available for an undercover job."[121] From this list, the Bureau selected fifteen agents it believed would make passable candidates for employment at AEC facilities in Los Alamos, New Mexico; Oak Ridge, Tennessee; and Hanford, Washington.[122] The agents would remain on the Bureau's rolls but would be paid by the civilian contractors responsible for running the AEC facilities. Of course, government being government, no one was going to make any additional money off of this assignment. Agents would submit any salary payments, in excess of actual expenses to the FBI, via cashier checks, which would then be deposited with the US Department of the Treasury.[123]

Ultimately, several Bureau agents submitted their applications for ostensible career changes. Armand Cammarota sought employment at Oak Ridge Atomic Energy Plant—through Carbide and Carbon Chemicals Corporation, the plant's primary contractor—as a laboratory technician, with the understanding that the items that hostile services would target included new developments in the field of atomic weapons and data and information concerning the amount and extent of the stockpile.[124] Richard Frye pursued a job at Los Alamos as a chemist.[125] The Bureau planned to maintain contacts with these undercover agents via the local field offices, which would make arrangements similar to those that the Bureau used for handling informants.[126]

There was just one problem: the agents could not get the needed jobs. Carbide and Carbon Chemicals notified Cammarota that they could afford him no encouragement with respect to employment. Frye encountered similar rejection. The University of California—which was the prime contractor for the AEC's Los Alamos Laboratory—notified Frye that there were no present openings available and there would not be any in the foreseeable future. Even if Bureau agents had obtained undercover jobs, the AEC's security setup would—paradoxically—make those agents' national security mission difficult if not impractical. As an FBI official observed, the AEC's security setup was "compartmentalized and any activity of an undercover agent outside of a small section of a plant would arouse suspicion."[127]

In addition to its informant program and its aborted undercover enterprise, the FBI developed a third type of tripwire: overt liaison. Hoover, in 1951, advised Congress—with the aid of a map illustrating all AEC facilities and security offices throughout the United States—that the FBI had to maintain liaison with the AEC locations.[128] Furthermore, the Bureau made semiannual contacts with law enforcement agencies operating in the vicinity of AEC facilities to ensure that the Bureau promptly received information regarding the installations' security.[129]

As part of its efforts to provide better security for the AEC, the FBI engaged in multiple educational initiatives to ensure that it was operating efficiently in this area. For instance, in 1951 the FBI noted that selected agents—located in offices that covered major AEC installations—had attended special training courses provided by the AEC about the accountability system used at the AEC facilities.[130] Additionally, FBI headquarters expressed interest, in 1947, about the possibility of having an official assigned to the Bureau's Atomic Energy Section assist with an inspection of a field office. This, according to an FBI official, would benefit the section by providing firsthand knowledge of issues that the field encountered with atomic energy–related cases.[131]

The overt liaison has a present-day corollary. In 2003, the FBI established its Agents in Laboratories Initiative. The defining feature of this program is the assignment of FBI agents to Department of Energy (DoE) nuclear weapons and science laboratories.[132] The agents assigned possess academic credentials in mechanical and nuclear engineering.[133] (The Bureau has diversified the competencies for which it hires since Hoover's insistence on lawyers and accountants.) The purposes of this program are both to raise counterintelligence awareness among DoE personnel and to broaden the FBI's access to intelligence within the labs.[134] (The Department of Energy is one of the sixteen intelligence agencies under the Office of the Director of National Intelligence.)

Additionally, the FBI has also worked with the DoE in other liaison capacities. For instance, the Bureau, in conjunction with the DoE and the Nuclear Regulatory Commission, established a Nuclear Site Survey.[135] The FBI provided little additional information about this program when it mentioned it to Congress in 1998, but in 2002, illustrating its role in preempting another terrorist attack, it noted that it had instituted a Nuclear Site Security Program to improve liaison between FBI field offices and critical nuclear facilities in those offices' areas of responsibility.[136] In 2007, the FBI further elaborated that, as part of the Nuclear Site Security Program, it was coordinating with the DoE and the Department of Defense in a proactive effort to prevent criminal/terrorist activities that may be directed against those sites. This program included routine liaison functions—including intelligence-sharing and threat briefings—as well as joint training and exercises that usually focused on emergency response coordination to disrupt an incident.[137]

Co-opting Customs

The FBI had to contend with the thorny problem of foreign governments acquiring restricted items via their diplomatic personnel in the United States. As Hoover noted to the attorney general in 1952, "many of the items subject to export controls are sold openly in the United States markets to any purchaser having the means to buy."[138] Hoover pointed out that there was "little likelihood that Soviet bloc officials would admit to Customs officers that their baggage under diplomatic seal contain[ed] contraband."[139] By 1952, the FBI had already identified efforts on the part of Soviet, Czechoslovakian, and Hungarian diplomats to evade export controls by shipping items out of the country under diplomatic cover.[140]

By the time Hoover made his assessment, elements of the Bureau had already established mechanisms to ensure that foreign governments were not exploiting their diplomatic privileges to ship restricted items out of the United States. Since 1948, the FBI's Baltimore office and the Customs House in that city had reached an "informal working arrangement" by which, if during the course of normal operations Customs examined suspicious shipments of Soviet-bloc cargo, Customs advised the Baltimore office so that the FBI could make a visual inspection of the potential contraband. In 1967, the Baltimore office's special-agent-in-charge characterized these inspections as having been very limited and confined to visual examination of electronic equipment, radio equipment, and firearms.[141] The FBI's Washington, DC, field office had a similar arrangement with Customs. Notification by Customs made it possible for FBI agents to observe items and make notes or photograph them.[142]

However, the FBI maintained plausible deniability about its involvement with the scrutiny of diplomatic shipments. Customs personnel—not FBI agents—were responsible for conducting inspections and searches.[143] When, for instance, the Washington, DC, field office indicated an interest in viewing a shipment, this was done with "the explicit understanding that Customs in the course of its normal business opens the shipment" and "the FBI's interest is maintained in confidence."[144] According to the Washington, DC, field office's special-agent-in-charge, "the opening of the shipments is always left to the judgement of Customs which exercises its prerogative to examine shipments."[145]

Use of International Mail

The FBI historically identified the use of international mail as a means for the transmission of sensitive information to hostile foreign countries. In a 1954 assessment, the Bureau noted the existence of an organized communications system between a San Francisco business and Hong Kong–based firms with communist connections. According to the Bureau, this system relied on indirect mailing methods, through neutral countries.[146] Consistent with this concern, the FBI initiated several mail-opening projects to thwart the outbound flow of knowledge. Between 1964 and 1966, the FBI operated a project known as Survey no. 6 in San Francisco, California, which focused on mail leaving the United States for a Far East Asian country. The project's objectives included identifying efforts to persuade scientists and other persons of Asian descent residing in the United States to return to that country.[147] Additionally, the FBI engaged in the opening of mail to and from a specific Asian country if the letter had a nexus to a university, scientific, or technical facility.[148]

Furthermore, the FBI recognized that incoming Chinese directions also relied on communications via mail. In a 1954 assessment, the Bureau noted that "although interest has been indicated in radio communications between the Chinese Communists and points in the United States, the actual unofficial or nondiplomatic communications have centered in the use of the mails and personal couriers."[149] It is possible that the FBI had this in mind when it established a program to screen incoming mail from Asia. Between 1956 and 1966, the FBI's San Francisco presence engaged in the opening and photographing of first-class letter mail from a Far East Asian country, with the intent of identifying individuals in the United States who—based on their foreign contacts—constituted a threat to US internal security.[150]

The CIA also conducted mail openings, in San Francisco, of items from Asia. According to the Rockefeller Commission, the objective of the San Francisco operations was to "obtain technical intelligence concerning foreign censorship, secret writing and the like."[151] A 1969 operation known as KM/Sourdough within the Plans (later Operations) Directorate, and West-Pointer by the Office of Security, covered mail coming into San Francisco from communist China and Vietnam.[152] The CIA's Far East Division and Technical Services Division ran the project jointly. Rather than maintain a permanent presence in San Francisco, the CIA ran Sourdough over the course of four three-week trips between September 1969 and October 1971.[153] This operation, in addition to furnishing information for the CIA, also provided leads for the FBI.[154] However, according to the Church Committee, the Bureau was not actually aware of the project, never levied any requests on it, and received only sanitized domestic intelligence leads from Sourdough.[155]

Additionally, the FBI, from an outpost in Hawaii, participated in the screening of mail during the mid-1950s. In early 1955, an FBI agent in Honolulu who was focusing on counterintelligence matters learned that a CIA officer was working with Customs to review intercepted mail from Asia in an attempt to identify incoming political propaganda. The Bureau's agent notified headquarters of this project and—with the CIA officer's concurrence—the FBI assigned an agent trained in mail-opening techniques to the effort for several months.[156]

The FBI, not surprisingly, given what it knew about the Soviet Union's efforts to exfiltrate technological knowledge from the United States, aimed similar activities against mail from that country. In 1952, the Agency initiated Pointer, a program that surveyed mail between the United States and the Soviet Union.[157] The CIA's Office of Security started this project in response to a request from the Soviet Russia division. The office copied names and addresses of mail to and from the Soviet Union.[158] The Rockefeller Commission characterized this as a counterintelligence operation, designed to identify US persons who were cooperating with Moscow.[159] By 1955, it was the CIA's counterintelligence staff, led by the legendary James Jesus Angleton, who saw the greatest value in the project (renamed HT/Lingual) as a counterintelligence tool.[160] Under this latter program, the CIA not only copied the exterior of mail but also opened and copied the contents of select pieces of international mail.[161]

Initially, the FBI was unaware of the CIA's HT/Lingual operation and discovered it only in the process of establishing its own mail-opening project. In 1958, the New York field office's FBI special-agent-in-charge made preliminary inquiries about establishing this project, only to learn that the US Post Office could not cooperate because "something had happened in Washington on a similar matter."[162] It was in this roundabout way that the Bureau became aware that the Agency already had a mail-opening system in place and—although such an activity seemed to impinge on the FBI's jurisdiction—the FBI acquiesced and agreed to a procedure, code-named Hunter, by which it could levy requirements on the CIA's operation. Between 1958 and 1973, the FBI provided the CIA with a variety of requirements including any traffic that involved an addressee or addressor who was "employed in a sensitive industry, i.e., missile field"; traffic concerning current or former Soviet researchers; and traffic regarding individuals known to be of interest to the Soviets because of their specialized knowledge.[163]

Enlisting the American Public

The US government's counterintelligence efforts are not limited to operations where it controls the playing field, such as overt liaison to industrial facilities, which must cooperate with their government patrons. Foreign actors are interested in material outside of the government's control and therefore US government agencies must find ways to work with the entities that have legitimate control over, or access to, that data. FBI intelligence collection in this area has historically used variations on methodologies used vis-à-vis government or government-affiliated enterprises. For instance, the FBI has used—or at least attempted to use—tripwires, individuals who are in a position, as a result of their responsibilities, to identify activities that might provide insights about foreign threat activities. It has also used time-honored spy-craft such as double-agentry.

One of the challenges in protecting the private sector is that, increasingly, the information at stake is not classified. Until 1996, economic espionage against companies' proprietary information was difficult to prosecute because it was not a theft from the government. However, foreign intelligence services have a long history of collecting open-source, nonclassified information, as this data helps them to establish a baseline knowledge and further refine their targeting of specific information. Disrupting this early collection of seemingly innocuous data can help the US government outflank foreign collectors' zeroing in on more sensitive knowledge. Unfortunately, the keepers of that knowledge do not always recognize its significance and may perceive US counterintelligence operations as unwarranted snooping.

Tripwires (or Tripping-Up) Against Open-Source Collection

Knowledge about foreign operatives' collection activities directed at open-source material can yield indicators of areas in which foreign adversaries and competitors might subsequently seek sensitive data. By developing knowledge about publicly available information, foreign actors can conserve resources—and reduce risk of detection—by focusing on filling specific gaps that an overview of the literature has defined (i.e., steps *x* and *z* of a proprietary chemical compound are known, but step *y* is not; an intelligence collector seeking to put the pieces together would focus its clandestine collection against solving for *y*). The collection of open-source material is a counterintelligence problem, since, by virtue of being public it is unrestricted. Two problems emerge from this reality. First, the material can often be obtained from any number of sources, meaning it is an unrealistic expectation that the US government can effectively harden all potential vulnerabilities against foreign collection. The most effective approach then is to enlist the assistance of personnel associated with those vulnerabilities. This, however, leads to the second problem: the unwillingness, whether prompted by distrust, ideology, or a combination thereof, of certain elements outside of government to assist US intelligence agencies.

The Soviet Union provided multiple instances that illustrate the importance that foreign intelligence services place on the value of open-source intelligence (OSINT). One defector noted that the availability of information in the United States obviated much of the need for hazardous and time-consuming clandestine operations. A different defector estimated that the office of the Soviet military attaché in the United States was able to legally obtain 95 percent of the material that furthered its intelligence objectives.[164] For instance, in 1950, the office of the Soviet naval attaché in Washington, DC, indicated an interest in catalogs, from General Electric, Raytheon, RCA, Bendix, and Westinghouse, containing information about radar equipment.[165]

A significant amount of OSINT collection appears to have been an effort to remain ahead of US technological development. For instance, in 1944, the Soviet Government Purchasing Commission—a wartime body—in Washington, DC, ordered copies of 5,810 patents, and the New York office of this commission ordered two copies of 18,000 patents.[166] (The collection of patents would remain a foreign intelligence objective for decades. In 2016, then–director of national intelligence [DNI] James Clapper advised the US Congress about foreign governments' collection of patent data.)[167] After World War II, Amtorg explicitly advised the American Society for Testing Materials that it desired to replenish Soviet industrial libraries with current publications.[168] Several decades later, the Soviets were still using OSINT collection to target specific technologies. According to a leading expert on Soviet computers, the Soviets annually acquired thousands of Western research papers, new product announcements, and product descriptions of new applications.[169]

Soviet officials' collection of publicly available materials indicated Moscow's emphasis on OSINT. Officials subscribed to a wide variety of US newspapers, magazines, technical journals, and scientific publications.[170] For instance, in 1959, personnel at the Soviet Military, Naval, and Air Attaché Offices subscribed to forty-four newspapers and fifty-eight magazines of a technical, scientific, military, and general news nature.[171] According to a 1982 CIA assessment, Soviet intelligence officers subscribed heavily to science and technology periodicals, which they relayed to their headquarters for translation and analysis or from which they pulled information over time to write reports.[172] As technology changed how information was presented, the Soviets modified their collection accordingly. A 1982 CIA assessment noted that the Soviets subscribed to at least one privately owned US microfilm information management system that contained more than 7,000 unclassified documents published by various US government agencies. (Although the documents were governmental in nature, the private sector played a definitive role in making these available to a customer who happened to be the primary US adversary.) The same assessment stated that the Soviets had also gained access to Western commercial computer databases.[173] In 1986, a congressional report noted the role of Soviet access to computerized Western reference systems as conduits for collection.[174] By 2016, little had changed. Then-DNI James Clapper advised Congress that foreign countries were openly purchasing access to US research through aggregated indexes.[175]

Although this collection might seem benign, the Soviets' efforts to hide their acquisitions suggest that they understood how OSINT collection could provide US intelligence with a trail of breadcrumbs leading to more nefarious activities. Officials often subscribed under false names, without noting their affiliation, and sometimes through the use of intermediaries.[176] For instance, Soviets instructed collaborators to use the cover of a school or commercial firm in order to make requests for specific publications.[177]

The Soviets' long history of attendance at industry events indicates that Moscow believed OSINT collection warranted resources. According to a 1953 FBI assessment, Soviet representatives had been attending electronics conferences, conventions, and exhibitions.[178] Two Soviets, in attendance at a 1959 convention of the Western Electric convention in Los Angeles, California, collected such a massive amount of literature that one of them dropped the material with a check stand before resuming his collection. By the end of this convention, the Soviets had acquired approximately 250 pounds of material.[179] Soviets, according to 1965 congressional testimony, augmented their collection of data by-the-pound with photography of material on display.[180] Exploitation of trade events continued for decades. For instance, a 1982 CIA assessment stated that Soviets—ostensibly as part of trade promotion efforts—regularly attended high-technology trade shows, with the actual intent of acquiring emerging technological know-how before it was deemed sensitive for military reasons.[181]

Although the People's Republic of China did not establish diplomatic relations with the United States until 1979, China had been exploiting US OSINT data almost as far back as the establishment of that country's authoritarian regime of despotic ideologues (e.g., Mao) in 1949. For instance, in late 1950, a Chinese government delegation to the United Nations attempted to gather a significant amount of publications regarding scientific, political, and national defense matters. Several years later, Chinese in California who were associated with communist activities attempted to convey several hundred pounds of technical papers back to China.[182] More than half a century later, little has changed. According to a 2017 report by the US government's United States–China Economic and Security Review Commission, China used a vast open-source collection apparatus to acquire data concerning foreign technologies that China had not yet been able to develop.[183]

US intelligence saw the availability of information to foreign actors as a definite problem. As early as 1949, the Soviets indicated interest in developing contacts in US libraries and records sections. During that same year, a Soviet operative acknowledged the value of these repositories when the operative tasked a contact to attend an exhibit at the New York Public Library and, as the FBI later put it, "look for anything of interest and get it."[184] Hoover, in 1965, warned that Soviet-bloc personnel made extensive use of technical information that they could access through US libraries.[185]

Having identified that the Soviets saw OSINT as a target for collection, the FBI looked for ways to combat this activity and—in the process—derive intelligence information. As of July 1960, the FBI had identified a library-card application that had a previously unknown address associated with the name of a subject. As the result of further investigation, the FBI determined that the subject possessed implements of espionage (e.g., materials for secret writing and a shortwave radio receiver protected by multiple locks). The Bureau obtained an admission from the subject that he was, indeed, operating in the United States at the behest of East German and Soviet superiors.[186]

Soviet activities during the late 1970s demonstrated a persistent effort to collect open-source information from US libraries. In October 1979, two

Soviet officials assigned to Washington, DC, visited a library in Ely, Nevada. The two travelers requested books on industry in Las Vegas, Nevada, an area restricted to the Soviets, not because they would drink the gambling establishments out of vodka but because of the city's proximity to the Nevada nuclear site and to Nellis Air Force Base. During their visit to Ely, the Soviets located and copied a 300-page environmental impact statement for the Nevada nuclear site. The Russians then proceeded to make several additional stops in the area and in the course of these made inquiries about industry in the region, which was a potential site for experimental missiles.[187]

These indicators, across multiple decades, provided US intelligence with a reason to believe that foreign powers were indeed exploiting the ease of access to information in the United States. Based on this reality, the FBI attempted to establish a new set of tripwires that might help identify the presence—and targets—of foreign intelligence officers.

Noting the role that libraries played—especially in the pre-internet days—the FBI viewed them as potential tripwires that could provide an early warning about foreign intelligence operatives' targets. An antecedent for these efforts can be found in a 1941 interaction between an FBI agent and a reference librarian.[188] The librarian, at the Business Branch of the Newark Public Library in New Jersey, suggested to the agent that this location—as well as other business libraries—contained information that might be valuable to individuals engaged in subversive activities.[189] This information included army and navy contracts as well as details concerning industrial facilities in the United States such as their locations and manufacturing abilities. According to the librarian, some of the individuals who demonstrated interest in this material were not gathering the information to advance their education, since they did not appear to be of the "intellectual type." The helpful librarian even provided the FBI with a list of the business libraries, in which the Bureau might be interested, across the country.[190]

"Marian the Librarian" (the pseudonym being a nod to *The Music Man*) prompted the Newark FBI field office's special-agent-in-charge to provide J. Edgar Hoover with a proposal. According to the agent: "It is believed that the Bureau might desire to direct the field offices covering the cities in which these libraries are located to make some arrangements with the libraries whereby they would maintain a register of persons desiring information which might be considered highly technical of a vital nature."[191] The agent assessed that such registers "may prove valuable to the Bureau in its investigations of espionage and sabotage" but did concede that "some of the requests of these libraries would be legitimate."[192]

Several decades later, the FBI launched the first of several initiatives consistent with this early proposal. According to a Bureau official, the FBI had started a program of outreach to libraries, between 1973 and 1976.[193] The official noted that "very, very helpful things came out of it."[194] During the 1980s, unfortunately, librarians began to turn against the FBI's efforts, with library directors at universities in Maryland, Texas, Michigan, Wisconsin, and Utah claiming that the FBI had attempted to recruit library staff and collect reading lists.[195]

It was an FBI program in New York that drew the most attention. The Library Awareness Program, which the New York field office started in 1985, was an initiative directed at contacting twenty-one scientific and technical libraries within the New York City area.[196] Through this program, the FBI hoped to counter Soviet intelligence officers who were attempting to develop sources and contacts among librarians in the greater New York area.[197] The FBI contacted the librarians at these specialized libraries to educate them about how they and their libraries were, or had historically been, significant targets for collection by Soviet intelligence. Additionally, the Bureau hoped to identify intelligence officers, their assets, their methodology, and their tradecraft.[198] A Bureau official put the Library Awareness Program into context, explaining to Congress that although the information available to Soviet intelligence in specialized and technical libraries was not classified, restricted, or unlawful to collect and maintain, the Soviets employed tactics and methodology to collect the information that showed "a blatant disregard for American laws and the personal rights of American citizens."[199]

Unfortunately, a lot had changed since 1941 and librarians were adamantly opposed to serving as sentries for the FBI. In response to the FBI's efforts, the acting vice president for information services at Columbia University claimed that Bureau agents had suggested that she keep an ear out for individuals with accents or "foreign sounding" names.[200] Furthermore, librarians claimed that the Bureau had been unnecessarily aggressive in its attempts to sensitize libraries to the threat from foreign intelligence services. According to the executive director of the Association of Research Libraries, FBI agents tended to approach student assistants or clerical workers rather than a member of the library's professional staff or the library administration.[201]

Whether warranted or not, the criticism of the program changed the FBI's tone about the initiative. In 1988, then–FBI director William Sessions agreed to curtail some of the program's more controversial aspects.[202] In 1992, Sessions advised Congress that there was no longer any suggestion that the FBI had improperly employed its outreach to libraries.[203] Years later, the Bureau still seemed leery about creating an impression that it was snooping on legitimate activity. In 2000, an FBI deputy assistant director assured Congress that it was not the intent of the law to prevent foreign diplomats from collecting open-source information, whether it was in libraries or on the internet.[204]

Double-Agentry

Operation of double-agents is a subset of human intelligence (HUMINT) collection. While informants provide reporting, they do not always make the other side believe they are working for that side. The information on how the US government has conducted these operations to protect private industry is limited. However, one significant case demonstrated how the FBI was able to foil a Soviet plot to penetrate a US company and obtain sensitive information.

The Soviets had long used their positions at the United Nations in New York as cover for their intelligence officers, and that is where this episode began. Gennadiy Fedorovich Zakharov was ostensibly a scientific affairs officer assigned to the Center for Science and Technology for Development at the UN Secretariat; he had been in that position since 1982. Claiming that his job was to collect and appraise information about scientific developments for developing countries, Zakharov visited a number of institutions, including libraries. It was while visiting Queens College that Zakharov noticed a student's solicitation for work doing retrieval of scientific and technological literature.[205] Zakharov approached the student—Leakh Bhoge, a permanent resident alien from Guyana—who was in the third year of majoring in computer science.[206]

Bhoge's area of study made him an ideal candidate for Zakharov to maneuver into a company working on cutting-edge technology. What Zakharov did not know is that Bhoge almost immediately contacted the FBI to report this incident. Although Zakharov did not initially identify himself as a Soviet, he did advise Bhoge that he was employed at the United Nations and requested assistance with obtaining material regarding robotics and computer technology. Zakharov provided Bhoge with a list of specific microfiche that Bhoge was to obtain for Zakharov. Between May 1983 and March 1985, Zakharov and Bhoge met on numerous occasions and Bhoge, at Zakharov's direction, stole unclassified microfiche from various libraries and information centers.[207]

As Bhoge approached graduation, Zakharov directed Bhoge's career toward access to information for which the Soviets were looking. Zakharov encouraged Bhoge to apply for a job with a high-tech business. In late 1985, Bhoge acquired a position with a company that manufactured unclassified precision components for use in aircraft engines and in radar that defense contractors such as Bendix and General Electric built. It was hardly surprising that once Bhoge got the job, Zakharov shifted the focus of his collection from microfiche on technical subjects to documents relating to the company's manufacturing activities. In mid-1986 Zakharov upped the stakes again. He had already obtained an agreement from his putative source to enter into a clandestine intelligence relationship and in May 1986 dictated an agreement to Bhoge that would have Bhoge work on behalf of the Soviets for ten years, at which point it could be reconsidered and renegotiated. There was a catch. Bhoge had to acquire classified material that was not available to the Soviet Union.[208]

The operation concluded when the FBI, which had been an unseen participant, finally moved in on Zakharov. In August 1986, Zakharov met Bhoge on a subway platform and paid him $1,000 for three documents, including information about an air force jet.[209] Zakharov probably did not expect what happened next. Two FBI agents who had been posing as lovers made their way to Zakharov and arrested him.[210]

There is an unfortunate epilogue to this story. Zakharov had originally noted what he thought was an ideological motivation for Bhoge's spying, as Zakharov was aware that Bhoge was critical of US policies and racial prejudices in the country.[211] According to Zakharov, Bhoge would be motivated—in part—by his ability to hurt the United States.[212] Unfortunately, despite Bhoge's assistance to the FBI, he walked away from the operation feeling little better about the United States than when he began assisting the Bureau. Bhoge, when interviewed by the media, said that although the FBI provided him $20,000 and assistance with obtaining citizenship, it failed to come through on promises of $100,000, assistance with finding a good job, and a medal. When asked whether he would do it again, Bhoge told the media, "not under the circumstances, not under the conditions [in which he] worked. Not with the treatment [he] was given by the FBI."[213]

Counterintelligence Awareness

It is unrealistic to think that the government's investigative techniques can eliminate the threat to the private sector. The scope of vulnerabilities is simply too broad and too amorphous. Potential targets are numerous and widespread. Furthermore, the work they do may or may not draw foreign intelligence interest, depending on a country's collection requirements. Such requirements may change as governments opt to pursue certain technologies over others and as new technologies become available that may introduce heretofore unidentified potential. Furthermore, industry's ways of doing business—for example, use of international supply chains—may create new vulnerabilities. All of these factors suggest that the private sector needs to acknowledge its role as a front-line combatant against compromise of knowledge and technologies that have implications for elements of US national power. The government has accepted this paradigm, as indicated by its steady introduction of initiatives meant to make the private sector a partner in thwarting foreign threats.

During World War I, US intelligence developed a system for protecting industry against foreign subversion and sabotage. According to historian Joan Jensen, the initial impetus for government involvement with the protection of industry came from the private sector, which sought the federal government's assistance with safeguarding industrial facilities. Between 1917 and 1918 the War Department created a Plant Protection Section as an organization within the Military Intelligence Division and subsequently integrated this section into its own structure.[214]

Following World War I, the responsibility for plant protection shifted from the military to a civilian agency. In 1931 the military agreed that, in the event of another emergency, the FBI would take responsibility for plant protection activities.[215] As European tensions became open hostilities, the US government began to consider the threat to strategic industrial facilities. Consistent with the 1931 agreement, the Intelligence Divisions of the War and Navy Departments sought assistance from the FBI with developing a program under which the Bureau would inspect the facilities of industries responsible for providing products essential to the operations of the departments. The Intelligence Divisions of the two departments and the Interdepartmental Industrial Mobilization Committee developed a list of these facilities for the FBI.[216]

The Bureau quickly established a program to ensure that these industrial facilities were hardened—and capable of maintaining a defensive stance—against espionage. According to instructions that the FBI issued in August 1939, each field office was to develop appropriate liaison relations with the army inspectors covering each of the plants located within the offices' territories in order to ensure that the FBI would expeditiously receive notifications of all espionage and sabotage matters.[217] (This was consistent with President Franklin D. Roosevelt's 1939 direction that put the Bureau in charge of counterintelligence matters.) Then in September 1939, the FBI initiated a program to survey the key industrial facilities of US basic industries, which the military had identified as being of vital importance, throughout the United States.[218] Bureau agents made a physical inspection of the plants, their facilities, and sources of supply in order to ensure the "safe and continuous output of the unit for such periods as War and Navy Departments would be interested in its products."[219] (The "sources of supply" aspect was a concern, as the FBI was aware that the Germans had conducted industrial espionage against the United Kingdom by using service agencies including those providing mundanities such as towels and mimeograph services.)[220]

The survey process was an early example of using a "red team" (i.e., thinking like the adversary) analysis. At an immediate, operational level, special agents inspected sites, identified vulnerabilities, and provided recommendations on how to mitigate threats.[221] Prior to conducting a survey, the FBI sent a specially selected group of Bureau officials to various parts of the country in order to make a complete study of the protective methods and facilities of representative industrial institutions. Furthermore, the FBI consulted specialists in various technical fields for information and advice.[222] Once agents had completed their assessments, they provided the results to the FBI, which in turn provided the plant with recommendations for improvements.[223]

Agents selected for this work came from both FBI headquarters and various field offices. These individuals received training, during a several-day in-service, on topics including Bureau policy on plant protection; specific examples of methods and types of espionage reported to the Bureau to date; laboratory aspects of plant protection; approved protection devices and policy; and the submission of appropriate recommendations to plant management.[224] As Hoover noted to Congress, "it was necessary to train all of the investigative personnel of the F.B.I. in this work, due to the technical nature of the facilities involved and the necessity of painstaking, uniform, and careful procedure in the conduct of these surveys."[225]

This program was outlined in the *Plant Protection Manual,* which the Bureau prepared to guide this program. The manual included basic information about the FBI's role in plant protection and the cooperative functions of the Bureau.[226] Furthermore, the manual contained guidance specific to implementing effective counterintelligence. Topics included the safeguarding of documents and materials in plants in order to prevent thefts.[227] Additional guidance regarding promptness in reporting irregularities; searches of persons; searches of places; patrol work; and supervising visitors to the plant was also clearly relevant to protecting information.[228]

The Bureau provided guidance, in 1939, for agent interaction with plant officials. It noted that it was "highly desirable for Agents during the survey to discuss with the plant protection officials the apparent weaknesses of the present plant protection set-up. This procedure affords the Agent an opportunity of ascertaining what steps have been taken in the past to attempt to improve the protection organization and the results of these attempts. This procedure affords the Agent an opportunity of receiving suggestions from the plant protection officials."[229] Hoover made very clear that the FBI's role was not one of furnishing protection but rather to serve as an adviser on technical matters related to plant protection.[230]

In addition to the vulnerabilities of facilities, agents provided guidance regarding operational security. According to a set of instructions that the FBI furnished to its field offices in 1941, it was important for plant officials with access to national defense information to guard this information from individuals who did not have a need to know.[231] Relatedly, the FBI acknowledged that the individual without a need to know might be lurking in the wires. Agents were to caution plant officials about furnishing information via telephone.[232]

Longer-term efforts to engage plant management in maintaining awareness of potential threats complemented the survey work. In mid-1940, the FBI's Executive Conference (comprising the Bureau's assistant directors) determined that the booklet "Suggestions for Protection of Industrial Facilities" should be given out when requested by officials. These books, provided after the plant inspection, would provide the special-agent-in-charge with "the basis for continuing the contact with the plants." However, the Bureau did indicate concern that if the books circulated too widely and a plant obtained one prior to a plant survey, the plant might put into effect recommendations "which an Agent would be in a better position to make to more effectively protect the plant's facilities."[233] (Hoover emphatically noted that the *Plant Protection Manual* was not to be given to plants.)[234]

Plant survey work was a massive undertaking for the Bureau. According to Hoover, approximately 12,000 plants might eventually become part of the war effort.[235] By 1940, the Bureau had already surveyed 540 plants that were under contract to the army and navy.[236] The military established a list of 1,602 facilities and by 1941 the FBI had surveyed and submitted recommendations covering 1,150 plants.[237] By 1942, Hoover advised Congress that the Bureau had conducted surveys of 2,279 companies; he noted that the companies had a number of subsidiary plants—some widely separated—and so the actual number of plants, according to Hoover, far outnumbered the 2,279 commercial concerns.[238] In addition to work on behalf of the US military, the Plant Survey Program also protected the interest of an ally. In 1941, the Bureau advised its field offices that a "List of Important Plants Having Contracts with the British Purchasing Commission" should be added to the list of priority facilities.[239]

The FBI not only provided specific surveys and liaisons but also conducted and participated in broader trainings for plant officials. The early instructions in this field indicated that the Bureau would help organize schools for plant protection officers and would assist in teaching certain selected subjects for a period not exceeding one day. However, by 1940, the FBI's duties prevented personnel from being able to conduct these schools.[240] Although the Bureau was unable to sponsor specific courses, the Executive Conference, in May 1940, determined that the Bureau would accept invitations to address plant executives and provide general advice on how to "focus their attention in the plants on problems which would reduce the likelihood of espionage or sabotage."[241] The *Plant Protection Manual* highlighted the value of external expertise—after all, the Bureau could not be an expert in everything—by encouraging the training of plant managers by non-FBI instructors on plant organization and records; fire protection; plant rules; physical setup of the plants; and first aid.[242]

Instructional opportunities allowed the Bureau to remain engaged with industry. The FBI's Executive Conference attempted to finesse these speaking engagements and noted: "It is recognized that some degree of stimulation of these invitations should be undertaken. It should be done slowly, however, and the first two approved suggestions were that an oral contact be made with the secretary of the Manufacturers' Association and, second, that [a Bureau official] orally contact [an official of the War and Navy Joint Industrial Board] so that these two individuals through their contacts can let it be known that such services are available."[243] (However, the Bureau also made it clear that these trainings were not linked to specific plant surveys. In 1941, the Executive Conference gave consideration to a suggestion that the Bureau follow up its plant survey recommendations with group conferences. The conference opposed this specifically because "if such conferences were held, we would be vulnerable to attack by unscrupulous labor groups having subversive tendencies, by their making statements that we are entering into the labor field problem.")[244]

According to the FBI's own account, the Plant Survey Program produced successes. A Bureau memorandum noted that through the surveys industrial concerns had become aware of their vulnerabilities to espionage and sabotage. The private sector was able to act on this information and take action to counter attempts that subversive individuals might take in furtherance of hindering the US war effort.[245] In one instance, the Bureau determined that West Coast aircraft factory draftsmen would obtain plans from the foreman without any inquiry by the foreman as to the need to know or any record kept of who obtained the plans. Hoover cited this instance as an example of the activities that the FBI had corrected.[246] (It is interesting to note that in 2001, Ronald Dick, the head of the FBI's National Infrastructure Protection Center—which focused on government engagement of commercial entities—noted "the need for a military public private sector partnership similar to that in the days of World War II.")[247]

FBI plant survey work assisted and laid the groundwork for other agencies' efforts in this field. According to the Bureau, it made the confidential *Plant Protection Manual* available to army and navy intelligence officers.[248] Furthermore, despite concerns about it circulating too broadly, the FBI made "Suggestions for Protection of Industrial Facilities" available at the request of the Department of Commerce's Foreign and Domestic Commerce Division for distribution to the department's field district offices.[249] Additionally, the FBI observed that it made "the vast amount of information and knowledge concerning plant protection, a comparatively new field," available to multiple US government agencies in order to facilitate those agencies' implementation of their own survey and protective programs.[250]

The FBI's role in plant survey work was a busy, although relatively short-lived, one. On June 1, 1941, the Navy Department instituted its own program to conduct surveys of the industrial facilities responsible for the production of material in which the navy had a specific interest.[251] The War Department followed suit not long thereafter, announcing that it would assume responsibility on January 5, 1942, for surveys of the facilities that were of importance for War Department procurement.[252] In November

1941, the Bureau instructed its field offices to discontinue the assignment of plant survey cases and to expedite the completion of any surveys currently in progress.[253] (The War Department did not take responsibility of "shadow plants"—facilities operated by private management that were financed and constructed by the federal government—until March 1942.)[254] As the Bureau's Plant Survey Program concluded, Hoover observed that it had been "a remarkable accomplishment & the Field & the National Defense Divisions are to be commended."[255]

Although the plant surveyed ended, the FBI continued to play a role in securing US infrastructure. On multiple occasions, the Bureau undertook surveys at the specific request of other government agencies. For instance, the FBI conducted two surveys of facilities operated by Pan American Air Ferries—on behalf of the War Department—in connection with the movement of bombers from the United States to locations in Africa and Iraq. Additionally, at the direction of Roosevelt and in conjunction with the Office of Naval Intelligence, the FBI conducted a survey of waterfront facilities at the Port of New York.[256]

However, the Bureau was definitely trying to absent itself from this work. Hoover, in 1942, in response to a request for a survey, observed that "I doubt if we should start this. I thought we were getting out of the plant survey work instead we are going back into it. I don't like this at all."[257] By the time of the Korean conflict, the FBI was even more adamant about its departure from plant survey work. The Bureau noted that the responsibility for securing defense plants remained with the Munitions Board and had not been transferred back to the FBI from the military services. In the memo explaining this position, the FBI's bitterness about its removal from the plant survey field was obvious. It noted that the Bureau had "developed and pioneered" the plant survey work, whereas the Munitions Board had, up to this point, done nothing more than to work on a plant protection manual.[258]

Despite its withdrawal from plant survey work, the FBI, for the first several decades of the Cold War, engaged in a variety of activities meant to harden US targets against foreign intelligence activities. These activities fell into three categories: developing counterintelligence awareness among key constituencies that might be targeted for recruitment; disrupting attempts by subversive entities to infiltrate US entities; and using the media to create broad awareness of how foreign intelligence services were targeting the US private sector.

The FBI was aware that the Soviet Union was targeting US expertise. Hoover, in 1950, advised Congress that certain foreign espionage agents were collecting information on atomic research and the identities of atomic scientists in the United States.[259] In a 1953 assessment, the Bureau assessed that Soviet intelligence had asked a contact for a list of scientists in the United States who had advocated for outlawing atomic weapons.[260]

The FBI instituted a program to counter this targeting. In 1949, it initiated a program to contact scientists on a selected basis to advise them about the possibility that they might be contacted by the Soviets.[261] According to a 1953 memorandum, the Bureau had instituted a program to contact scientists on a selected basis.[262] Approximately a decade later, the Bureau made reference to similar activity, noting a program specifically directed at alerting key scientists in the San Francisco area to the possibility that they might be targeted for recruitment to conduct espionage.[263] (In the course of such contacts, the FBI looked for assistance from these scientists as informants and also sought scientists who might work as double-agents on the Bureau's behalf.)[264]

Preventing the infiltration of US entities—some of which would now be called critical infrastructure—was another objective toward which the FBI worked. Under the Responsibilities Program, which the Bureau initiated as the result of a 1951 meeting between Hoover and a committee of governors, the FBI provided information to state and local officials when a subject on the Security Index was employed in a public utility or in a public or semipublic organization. Prior to provision of information, a field office had to furnish a statement to FBI headquarters regarding the reliability and discretion of the individual to whom the field would provide the information. Once headquarters approved the field office's proposal, the office would disseminate the information orally. The FBI justified this program under the Bureau's overall responsibility for the country's internal security: it therefore clearly had a responsibility for the safeguarding of public utilities, public organizations, and semipublic organizations that served large segments of the country's population.[265]

Similar to the Responsibilities Program, the FBI conducted inquiries to thwart foreign access to material related to weapons of mass destruction. In the course of applicant investigations on individuals seeking positions with the Atomic Energy Commission, the FBI made a detailed review of the investigation in cases where it discovered subversive, derogatory information. The Bureau provided the AEC with the derogatory information about the applicant. If the AEC nevertheless cleared the individual to work, the FBI made arrangements to pay additional attention to these individuals.[266] (The Department of Energy—successor to the AEC—was notorious for lax counterintelligence, which, based on the AEC's apparent willingness to employ questionable individuals, may have been an inherited pathology.)[267]

The FBI also provided awareness to industry through participation in relevant industry associations. For instance, FBI agent Fern Stukenbroeker delivered a talk in 1965 to a luncheon at the eleventh annual seminar of the American Society for Industrial Security (ASIS). The Society provided very positive feedback on this address, complimenting Stukenbroeker's talk as being "particularly effective in relating the espionage threat to government and industrial leaders present responsible for the protection of classi-

The FBI's Continued Role in Preventing Infiltration of Sensitive Facilities

Threat actors will always attempt to infiltrate facilities where they can do harm, as indicated by the FBI's concerns about individuals with questionable backgrounds taking positions with the AEC. While foreign intelligence services' illicit forays certainly continue to be a concern, terrorism is an equally pressing problem. The FBI's Criminal Justice Information Services Division, through the work of its Bioterrorism Risk Assessment Group (BRAG) plays a significant role in keeping malign actors from accessing dangerous implements.[268] The Public Health Security and Bioterrorism Act of 2002 drives BRAG's mission, which BRAG commenced in April 2003.[269] Like the FBI's Responsibilities Program, BRAG's role is to inform other government agencies about the threat of infiltration. Specifically, BRAG is responsible for conducting security risk assessments on individuals identified by the US Department of Agriculture's Animal and Plant Health Inspection Service as well as the Department of Health and Human Services' Centers for Disease Control and Prevention as having a need to possess, use, or transfer biological select agents and toxins.[270] A security risk assessment is a database check ensuring that an individual is not disqualified under any of the Patriot Act's prohibitions, including being an agent of a foreign power.[271] Once BRAG completes its assessment, it provides the results to the sponsoring agency, which closes the loop, in writing, by providing its decision regarding denial or approval of the candidate.[272]

fied and proprietary information."[273] Stukenbroeker met with similar success during a 1971 ASIS conference where he delivered both the opening and principal address and participated in a panel discussion.[274]

In addition to engagement with industry organizations, the FBI conducted outreach to specific companies of interest. In 1965, Stukenbroeker spoke to approximately 500 employees of the Western Electric Company about the communist threat and the continuing need to safeguard all classified and proprietary information.[275] Stukenbroeker in 1967 delivered a presentation about Soviet espionage to approximately 200 employees of the American Telephone & Telegraph Company.[276]

Print media also played a significant role in the FBI's efforts to curb foreign intelligence services. The Crime Records Division played an important role in this area, as it had the responsibility for disseminating information to cultivate a favorable public image of the FBI.[277] In 1959, the FBI established this division, which would consolidate the Crime Records Section, FBI tours, and certain other aspects of work for which the Records and Communications Division had been responsible.[278] As of 1963, the organization of the Crime Records Division was as follows: front office; Crime Research Section; Correspondence and Tours Section; and Uniform Crime Reporting Section.[279]

It was the Crime Record Division's public relations program that helped to combat foreign intelligence activities. According to congressional testimony, one section of the division had the responsibility of assembling material necessary for a public relations program, including information authors and newspapermen.[280] During the 1960s, the division was responsible for multiple articles warning of the foreign intelligence threat to business, which appeared in high-profile publications.

A piece under Hoover's byline was an early example of this approach. The *Nation's Business,* in its May 1962 issue, carried the cautionary story from the FBI director titled "Why Reds Make Friends with Businessmen."[281] Hoover commended Stukenbroeker (who was apparently quite good at navigating the relationship between the FBI and private industry) for a "splendid job" in the article.[282]

Approximately two years later, the FBI ran another significant warning about the foreign intelligence threat to private industry. The February 1964 issue of the *Harvard Business Review* featured a sixteen-page article again under Hoover's byline (and developed by Stukenbroeker) titled "The U.S. Businessman Faces the Soviet Spy."[283] According to an FBI official, the article discussed how Soviet agents would "hoodwink" US businessmen and provided suggestions about ways in which US businessmen could help the Bureau and the United States writ large.[284] Hoover advised Congress that the intent of the article was "to set forth in language that could be understood by a businessman of the country, the things that can be done, what the picture is, and what they ought to know in dealing with Soviet bloc representatives."[285] The FBI was quite impressed with its accomplishment, assessing it as being published by the "foremost magazine of American business and industry" and representing a "new concept in the format of the [magazine] which beforehand steadfastly published only articles of business policies or procedures."[286]

The article was a hit, both within the Bureau and externally. Hoover commended Stukenbroeker in a personal note stating that "the effectiveness of this article can be attributed to a large degree to your personal direction, your exemplary performance in the basic research, and your liaison with representatives of the magazine."[287] Externally, the article not only reached the *Harvard Business Review's* 100,000 readers but also was reprinted in dozens of publications.[288]

Warnings to the private sector were not limited to the topic of Soviet spies. In 1966, under Hoover's byline, the *Nation's Business* published "How Red China Spies on the U.S." The article cautioned readers that China, through requests to US entities, was "gathering valuable information" on multiple topics, including those in the industrial and economic fields. According to Hoover, readers should handle requests from foreign entities with caution and, before fulfilling those requests, ask themselves, "What do I know about the source of this order? Would I be harming my country if I placed this information in communist hands? Should this order be brought to the attention of official agencies, particularly the special unit of the Department of Commerce which has been designated to handle such matters?"

In at least one instance, the Bureau's development of stories had a distinctly operational aspect. An astute FBI agent noted that the Association of the United States Army (AUSA) would hold its annual meeting in August 1960 at the Sheraton Park Hotel in Washington, DC. From his previous experience, the agent surmised that representatives of Soviet and Soviet-bloc country embassies would attend the event. The agent suggested that activities of Soviet-bloc representatives gathering quantities of material during the event might make a good feature story and that the Bureau could make arrangements with a press photographer to unobtrusively get pictures of the Soviets and their ilk in action.[289]

The FBI decided it liked the idea. Two agents from the Washington, DC, field office received assignments to act as spotters during the AUSA meeting and, using prearranged signals, cued the press photographer to catch the Soviet and Soviet-bloc country officials in the act. The *Washington Evening Star* ran the story under the headline "No Need for Spying—We Give Reds What They Want."[290] The article chronicled how Lieutenant Colonel Viktor Lobanov and Lieutenant Colonel Vsevolod Tovma, both of whom were assistant military attachés at the Soviet embassy; Vsevolod Generalov and Avgust Yashin, assistant air attachés at the embassy; and Edward Gordon and Roman Misztal, officials from the Polish embassy, acquired "briefcase loads of documents about every phase of America's defense industry."[291] At some point during the day, Lobanov and Tovma disposed of what was in their briefcases and came back for a second helping of material.[292]

Apart from publicly identifying that hostile foreign governments targeted industry events, the article highlighted several specific counterintelligence concerns regarding tradecraft and targets. Neither Generalov nor Yashin wore identification of any sort. Had they identified themselves as representatives of a hostile government, it is possible that exhibitors might have been more circumspect about interacting with these officials. Additionally, the convention—which featured 144 exhibitors—provided a firsthand look at military equipment that the public (including the roving foreign officials) had never seen before including a T-114 tracked reconnaissance vehicle that could replace an armored car or light tank; a new configuration of the Nike-Zeus missile; an SD-5 drone; a Red Eye hand-carried antiplane weapon; and a working model of a Fadac computer.[293]

Formalization of Counterintelligence Initiatives

The FBI's efforts during the 1950s and 1960s, when viewed retrospectively, amounted to a three-pronged information-sharing program. In the 1970s, the Bureau initiated the first of several successive, formalized programs, which resembled elements of earlier activities including the Plant Survey Program, directed at enhancing the private sector's awareness of intelligence threats. These programs adapted to reflect the changing nature of the relationship between industry and national security. Establishment of these programs also highlighted government redundancy, since they existed alongside similar Department of Defense activities. The United States has attempted to coordinate these functions—starting in the 1990s—with a succession of coordinating bodies, the most recent of which is the National Counterintelligence and Security Center.

Development of Counterintelligence Awareness Program

It was not until the late 1970s that the FBI developed a formal initiative similar in expansiveness or formalization to the Plant Survey Program. In 1976, an FBI field office established a program to systemically contact all of the defense contractors within the territory of that office. The office's intent was to alert the contractors to the threat from foreign intelligence services and provide a point of contact at the Bureau to which a company could report any out-of-the-ordinary activity.[294] Then, in 1978, the FBI launched its Development of Counterintelligence Awareness (DECA) program at an organization-wide level.[295] FBI headquarters provided field offices with information, materials, and speaker support to facilitate a specific request or need.[296]

Although FBI headquarters had assumed ownership of the program, the field remained responsible for implementing it. Each field office had a DECA coordinator who maintained regular liaison with companies located in the offices' territories. Coordinators provided briefings, videotapes, pamphlets, and other material that was supposed to help the private sector become more cognizant of and alert to foreign economic espionage.[297] According to an FBI headquarters official, company employees enjoyed hearing Bureau "war stories" and usually responded "very favorably."[298] Through DECA, the Bureau identified and forged relationships with companies that held US government contracts.[299] Given the extensive number of contractors within the United States, the FBI prioritized those businesses that dealt in products presumed to have been targeted by foreign intelligence services.[300]

DECA had aspects of previous FBI counterintelligence awareness activities. These included conversations with company security officers, hearkening back to the Plant Survey Program; talks to large groups of employees, calling to mind the publications and presentations of earlier in the Cold War; and even a foreign intelligence threat information journal—at the classified and unclassified levels—called *DECA Notes,* which, as an educational aid for companies, recalled the Plant Survey Program's "Suggestion for the Protection of Industrial Facilities."[301] The FBI, as it had done through presentations during the 1960s and early 1970s, maintained its relationship with American Society for Industrial Security as part of DECA. The Society strongly endorsed the program, and individual ASIS members participated in the exchange of information through the DECA program.[302]

The focus of DECA shifted during its existence, in parallel with changes in the relationship between industry and national security. Initially, the FBI emphasized DECA's outreach to facilities cleared to secret and top secret levels.[303] However, as early as 1982, the FBI was working to expand the types of companies, especially firms dealing with emerging technologies, with which DECA representatives were in contact. As a member of the FBI headquarters' Intelligence Division explained, a significant amount of sensitive research and development information was publicly available.[304] (It is interesting that the official seemed to be discovering a new vulnerability—after all, Soviet and Polish officials had viewed newly unveiled military equipment at the AUSA event in 1960.)

DECA provided an opportunity for the government to collect information that increased its understanding of threat actors. Contacts that the FBI developed through DECA, according to a declassified US government document, produced a significant number of reports about hostile intelligence service activity. Based on this information, the Bureau was in turn able to pursue investigations.[305] During a 1986 congressional hearing, Philip Parker, a deputy assistant director in the FBI's Intelligence Division, acknowledged that DECA had produced "some important information that may not have otherwise come to [the FBI's] attention."[306] A decade later, the National Counterintelligence Center made a similar assessment, stating in its annual report to Congress that DECA furnished investigative leads from corporations regarding illicit collection efforts by foreign government and corporate entities against US economic and technological information.[307]

The DECA program did not just serve the parochial interests of the FBI. A 1978 Central Intelligence Agency memo noted that a "strong DECA

Speaking to the Private Sector

Counterintelligence outreach by government to industry—especially as noncleared corporate entities become increasingly integral to US national security—is a challenge. As the Plant Survey Program's instructions specifically noted, there were topics that were not to be addressed with the individuals whom the FBI was supposed to be briefing. DECA coordinators' briefings were supposedly tailored to the specific needs and concerns of each company.[308] Although the FBI might have understood vulnerabilities in the private sector and communicated awareness of these to audiences, there are indicators that it was less well postured to explain the nature of threat for which the private sector should be looking. For instance, a 1986 Senate Select Committee on Intelligence report noted that the FBI needed to develop threat awareness briefings that were tailored to convey the specific characteristics of the intelligence threat from China. These briefings, according to the report, "should alert American citizens to the risks of giving assistance to [Chinese] nationals who may have espionage assignments."[309]

program throughout U.S. industry will serve the security interests of the entire Intelligence Community."[310] The memo's author, in his observations to the director of central intelligence (DCI), noted that the Agency was relying on the Bureau to provide CIA contractors with information about industrial espionage and counterintelligence. Therefore, according to the memo, the DCI should voice his support of the goals of the DECA program in the course of his discussion with the director of the FBI.[311] Several years later, the CIA was continuing to steer its contacts to the DECA program. In a speech to ASIS, John McMahon, the deputy director of central intelligence, advised attendees to make use of DECA, which McMahon characterized as "an excellent program to help contractors protect themselves."[312] The CIA also provided active assistance to DECA. According to the National Counterintelligence Center's 1995 report to Congress, the CIA provided information to the FBI for use in the DECA program.[313]

Awareness of National Security Incidents and Response Program

A new program succeeded DECA in the mid-1990s. In 1996—the same year that Congress passed the Economic Espionage Act—the FBI established its Awareness of National Security Incidents and Response (ANSIR) program.[314] The ANSIR program was directed at raising US corporations' understanding of vulnerability to economic espionage.[315] Its emphasis was on the "techniques of espionage," including "dumpster diving" and planting wireless microphones in corporate boardrooms.[316] Similar to DECA, each field office had an ANSIR coordinator who worked directly with corporate security directors in their respective offices' regions.[317] Criteria to serve as an ANSIR coordinator included experience in national security investigations, advanced counterintelligence and counterterrorism training, and computer literacy.[318] Like their predecessors, each ANSIR coordinator was a member of ASIS, which helped to bridge the public-private gap.[319] It was mandatory that a special agent fill the role, since, according to the deputy assistant director in charge of the program, "decades of experience with the ANSIR audience has shown that the private sector prefers discussing national security issues with an individual who has operational experience." This seemed to be a holdover of the attitude, indicated by the "war stories" remark, that characterized DECA.[320]

Despite ANSIR's sweeping nature, the FBI afforded it extremely limited resources. A single supervisory special agent, in the FBI's National Security Division, was responsible for the program at the institutional level. Furthermore, the agents who served as ANSIR coordinators in the FBI's field offices were supposed to spend no more than 10 percent of their time on this program. Another limitation, one that differentiated ANSIR's functions from those of the Plant Survey Program, was the limited services it provided to the private sector. Whereas the Plant Survey Program assessed facilities' vulnerabilities, the FBI, as of 2001, lacked the resources to conduct physical, personnel, or informational evaluations for the private sector.[321]

ANSIR kept some of the practices established by predecessor programs and established new ones. ANSIR-FAX—a facsimile transmission system—joined the other forms of media that the FBI used to disseminate information to the private sector. As of 1998, ANSIR-FAX disseminated unclassified counterintelligence and terrorism threat warning information to approximately 25,000 individuals.[322] In 1997, ANSIR leveraged the internet to distribute information via ANSIR-Email—which the FBI viewed as the principal method of disseminating awareness information and anticipated would provide unclassified threat and warning information to more than 100,000 recipients.[323] This number may reflect a 1998 report that the FBI had purchased a list of 100,000 system administrators and others from CorpTech, a Massachusetts research firm, which it used to reach—or according to some individuals, spam—these individuals with messages about terrorism. Despite ANSIR's focus on unclassified communications, an individual who was annoyed by being included on the distribution list received a reply from an FBI spokesperson that it would be unusual to send this type of information via insecure means.[324] The FBI suspended ANSIR-Email in 2003.[325]

Spy-Catcher and Movie Producer

The FBI has, for decades, made adept use of multiple types of media—including print and film—to educate the American public about, and enlist its assistance in, furthering national security. Within this broader endeavor, the Bureau has directed attention to reaching the private sector.

Historically, the FBI had a great deal of success enlisting A-list names. One need only look at *The FBI Story,* released in 1959, featuring Jimmy Stewart. The movie was based on Don Whitehead's 1956 book of the same name. Whitehead was a two-time Pulitzer Prize winner and although the FBI provided him with material and reviewed the manuscript, it did not edit Whitehead's writing.[326] According to the Bureau's files, Fern Stukenbroeker, in the Crime Records Section, did a significant amount of original research for the book and played "devil's advocate" when reviewing the manuscript.[327]

In adapting the book to film, the FBI exerted more control. The Bureau's agreement with Warner Brothers stipulated that the FBI would have full control over the approval of people—including not only the actors but also the production personnel—who would be "intimately associated" with the movie, in order to avoid "potential embarrassment" from the employment of a security or criminal subject with the film.[328]

It appears that the FBI's experience with Warner Brothers was a positive one. Approximately five years after *The FBI Story* premiered, the Bureau partnered with Warner Brothers—after waiting out 600 offers from other producers—to develop the television series titled simply *The FBI,* starring Efrem Zimbalist Jr., which first aired in 1965. The series was definitely a partnership. Although it was the project of a major studio, the FBI's agreement with the show's producer stipulated that the Bureau would have complete approval over the series' scripts, personnel, and sponsorship.[329]

The show became sufficiently iconic that the FBI, nearly two decades later, was able to use Zimbalist in one of its early efforts to connect with industry. To raise awareness among private sector entities, the FBI launched a number of television and radio spots, featuring Zimbalist warning about hostile foreign intelligence services, in the Silicon Valley region.[330]

More recent FBI efforts to reach the increasingly important private sector have, by comparison, been professionally underpowered. Its most significant undertakings have been several short, publicly released dramatizations. The Bureau has directed several of these movies at private sector audiences.

The most relevant film in this series is titled *The Company Man: Protecting America's Secrets.* According to the FBI, the purpose of the film—which the Bureau, in conjunction with the National Counterintelligence and Security Center, released in 2015—was to raise audiences' awareness about the threat from economic espionage and the theft of trade secrets.[331] Its basis is an actual FBI trade secrets case that uncovered Chinese executives' attempts to suborn an American into providing information about engineering plans for top-of-the-line insulation technology.[332] FBI personnel initiated a campaign—implemented by the Strategic Partnership Coordinators (the successors to ANSIR coordinators)—in which the FBI personnel showed the film and then held short discussions with audience members afterward.[333] The FBI's hope was that this campaign would spur the private sector to generate referrals of incidents to the Bureau.[334] *Foreign Policy* called the script—written by Sean Paul Murphy—"something of a clunker."[335]

Murphy was also responsible for the FBI-produced *Game of Pawns,* which was part of the Bureau's "ramped up efforts to educate American university students preparing to study abroad about the dangers of knowingly or unknowingly getting caught up in espionage activities."[336] Certainly this has implications for private industry, which may unwittingly hire such newly minted students into positions of trust. The film, when the Bureau released it for public consumption in 2013, fell far short of *The FBI Story.*[337] Instead, *Time* referred it as "a bit of a stinker" featuring "clunky dialogue."[338] (Select lines include: "You are lucky Shanghai is a big city, or I'm afraid you'd run out of girlfriends" and "Well, you may not believe this, but your brother is back and he brought home the bacon.") The *Washington Post* referred it as "strikingly cheesy" and "obviously low budget."[339]

In 2020, the Bureau and National Counterintelligence and Security Center again returned to Murphy's barely damp well for *The Nevernight Connection.* This new production featured the story of a former US Navy officer, active in the consulting world, who was targeted via a professional social media platform by China, and arrested for helping that country to access sensitive US military information. The Bureau planned to make *Nevernight* publicly available in order to warn various audiences, including executives at top US technology companies, of China's intelligence collection techniques.[340] *Wired* scratched its editorial head over this project, stating that "the FBI has made a short film of its very own, for some reason" and described the film's "modest production values."[341]

Recent FBI efforts to engage the public—particularly the private sector—through entertainment do not come close to the Hollywood partnerships that the Bureau proved itself capable of forging to develop both *The FBI Story* and *The FBI*. Certainly there is enough talent remaining that would be eager to work with the Bureau on an A-list project. The FBI deserves more than claptrap, direct-to-DVD-level hackery.

Counterintelligence Strategic Partnerships

After 9/11, the FBI revised its approach to private industry. The Bureau, in 2002, advised Congress that the Bureau's Counterintelligence Division was reorienting itself in order to work more closely with the private sector.[342] According to the National Counterintelligence Executive, the Bureau continued to operate its ANSIR program into 2004.[343] As of 2005, FBI field offices were establishing relationships with the private sector, specifically "high-tech, cutting edge technology companies" at the executive level that it referred to as "business alliances."[344] The business alliance concept made the private sector the "first line" of counterintelligence against foreign intelligence threats.[345] This approach made sense. Since private industry was increasingly in the crosshairs of foreign adversaries seeking proprietary information, it was sometimes the first to encounter threats and thus was well-positioned to provide the US government with an informational advantage about threat actors' objectives as well as the tactics and methodologies that threat actors used in pursuit of those objectives.

The business alliance concept had evolved by the early 2010s. Then–FBI director Robert Mueller III advised Congress, in 2013, that the FBI needed to establish "structured partnerships" with the private sector.[346] In 2014, the FBI described its Counterintelligence Strategic Partnerships Program (CISPP) as an initiative to combat illicit collection by foreign actors targeting private industry to acquire, among other things, sensitive technologies, advanced scientific research, and trade secrets. The CISPP became the mechanism for handling the FBI's business alliances.[347] Field offices were responsible for much of the CISPP work. As of 2013, the FBI had CISPP coordinators in each field office.[348] Unfortunately, according to the 9/11 Review Commission, the FBI tended to undervalue liaison positions such as the CISPP coordinators.[349] As of 2014, there were approximately eighty special agents serving as strategic partnership coordinators. These agents were responsible for working with approximately 15,000 contacts nationwide.[350]

CISPP activities did not occur in a vacuum, and some degree of overlap with other FBI programs was inevitable. In 2014, the Bureau attempted to mitigate this by creating a director of private sector engagement to develop a more integrated strategy for interfacing with the private sector.[351]

The associated Office of Private Sector described itself as an "entity within the FBI that coordinates—and has a 360-degree understanding of—the Bureau's engagement with the American business community."[352] In 2018, according to the Federal Register, the office planned to conduct a survey of entities, including businesses, to measure the effectiveness of its own engagement efforts.[353] It would be interesting to know what the results of this looked like in comparison to the office's predecessors (e.g., DECA, ANSIR, business alliances, CISPP; see Table 3.1).

In the field, the Office of Private Sector seems to have replaced the CISPP with a new creation. According to a fact sheet from the office, it intended to "redesign legacy partnerships," thereby reinventing the formative concept of the Plant Survey Program, DECA, ANSIR, the CISPP yet again.[354] Each field office, under Office of Private Sector purview, had a single private sector coordinator who was responsible for maintaining an understanding of the FBI's engagement with private industry and connecting commercial entities with the Bureau entities responsible for the issue of concern to the private sector.[355]

According to the FBI, counterintelligence-specific outreach seems to be under Office of Private Sector purview. The office is supposed to facilitate the emergence of an "FBI voice."[356] FBI testimony by the assistant director for the Bureau's Counterintelligence Division to Congress in 2019 seemed to confirm this relationship. According to the assistance director, the Counterintelligence Division's Engagement Office worked with the Office of Private Sector to strengthen engagement and promote messaging on key threats.[357]

The Office of Private Sector had a theoretically hazy idea of what it was supposed to accomplish. In its fact sheet, the office stated that it determined "how threats to the private sector align with the FBI's threats, priorities, and strategy."[358] At best, this is an infelicitously written statement. At worst, it reads like cherry-picking intelligence by looking for those developments that happen to align with the Bureau's understanding of the threat,

Table 3.1 FBI Efforts to Engage the Private Sector

1939	FBI Plant Survey Program
1978	Development of the Counterintelligence Awareness (DECA) Program
1985	Library Awareness Program
1996	Awareness of National Security Incidents and Response (ANSIR) Program
1998	Infraguard
2003	Agents in Laboratories Initiative
ca. 2005	Business Alliances
ca. 2014	Counterintelligence Strategic Partnership Program (CISPP)
ca. 2014	Creation of the Office of Private Sector

rather than using developments to assess how the Bureau might need to adjust its awareness of how threats (and opportunities for intelligence collection) have evolved.

Department of Defense Counterintelligence and the Private Sector

The FBI, through its DECA program, partnered with the Department of Defense. In 1972, the DoD created the Defense Investigative Service (DIS) to carry out certain functions that the individual military services and DoD agencies had performed, with the expectation that this consolidation would result in financial savings.[359] (Worth noting is that one of the functions for which the DIS had responsibility was background investigations and, in the case of contractors, providing the results of these inquiries to the Defense Industrial Security Clearance Office.)[360]

One of the DIS's functions was counterintelligence. In 1985, the Commission to Review Department of Defense Security Policies and Practices recommended that the DIS, in conjunction with the FBI and the military departments, increase the size, effectiveness, and coordination of the security awareness program in the private sector.[361] In 1993, the DIS established a counterintelligence office in order to sensitize cleared companies to the foreign intelligence threat. As part of its counterintelligence function, the DIS participated in DECA briefings throughout the United States.[362]

According to a 1998 report, the DIS's counterintelligence work included providing information to industry as well as gathering counterintelligence information from industry.[363] Both of these functions sound duplicative of—rather than complementary to—those of DECA. Furthermore, there is some confusion about which agency was supposed to take the lead vis-à-vis industry. According to a 1996 congressional briefing, the DIS was a participant in DECA briefings. However, a previous description of the relationship, by the director of central intelligence, portrayed DECA as supplementary to the DIS's work.[364] Finally, there seemed to be a profusion of potentially duplicative resources. DECA published a foreign intelligence threat information journal titled *DECA Notes*.[365] Contractors also received DIS security awareness bulletins.[366] These overlapping functions hinted at a long-standing problem and one that persists to the present day: How can the US government coordinate effective outreach to industry?

Defense Security Service

In the late 1990s, the DoD evolved its role in protecting its private sector partners. According to a 1997 DoD memorandum, the Defense Investigative Service was renamed the Defense Security Service (DSS) in acknowledgment that the DIS had taken on "broader missions and functions."[367] The DSS's work vis-à-vis the private sector is multipronged. It administers the National Industrial Security Program, initiated by the DoD in 1993, which provides assistance to US contractor facilities that are cleared for access to classified information.[368] Furthermore, the DSS has—similar to several other government agencies—taken on the role of educator to industry. As of 2005, it had provided threat information and counterintelligence briefings to nearly 12,000 cleared defense contractors. Based on contacts with contractors, the DSS tracks and analyzes how the threat to US technologies has changed.[369]

The DSS has also been engaged in broader, interagency initiatives to combat the illicit acquisition of sensitive technology and knowledge by foreign adversaries. In May 2016, the National Counterintelligence Security Center assigned a senior officer to serve as a liaison to the DSS.[370] Additionally, the DSS is an important player in preventing technology transfer. Its Counterintelligence Directorate provides a liaison officer to the Department of Homeland Security (DHS)–led Export Enforcement Coordination Center.[371]

National Counterintelligence and Security Center

Multiple entities within the US national security community conduct outreach to the private sector in an effort to counteract (and sometimes exploit) foreign threats that target technological progress, critical infrastructure, and so forth. The Office of the Director of National Intelligence (ODNI)—which leads the other sixteen members of the US intelligence community—engages in its own dialogue with industry through the ODNI's various components.

Prior to the Intelligence Reform and Terrorism Prevention Act of 2004, the director of central intelligence led both the Central Intelligence Agency and the intelligence community. However, passage of the act wrested this overarching authority away from the CIA and made it just like any other federal bureaucracy. The legislation vested the newly created director of national intelligence with the interagency responsibilities that were previously the domain of the DCI. (The staff of the Office of the Deputy Director of Central Intelligence for Community Management was transferred to the ODNI. That said, according to intelligence scholar Gregory Treverton, the staff was "nowhere up to the task of helping DNIs actually manage the community.")[372] The DNI concept was not an entirely new one. It dates to at least 1980, when Congress included a similar position in the proposed National Intelligence Act.[373]

Organization of the DNI was also rooted in post-9/11 reform efforts. According to Treverton, the "most sweeping change" introduced by the 9/11 Commission—the work of which informed the Intelligence Reform and Terrorism Prevention Act—was the creation of national intelligence centers, organized around specific issues, under the authority of the DNI.[374] As of

2019, these were the National Counterterrorism Center, the National Counterproliferation Center, the National Counterintelligence and Security Center, and the Cyber Threat Intelligence Integration Center.[375] All of these entities are responsible for addressing issues with direct significance to private industry.

The National Counterintelligence and Security Center (NCSC) is of particular importance. At least as far back as 1990, the idea for an interagency counterintelligence center had been in circulation. Speaking to Congress during that year, Kenneth deGraffenreid, a former National Security Council staffer, proposed the creation of a National Counterintelligence and Security Access Center that would provide macro-level counterintelligence security vulnerability and threat analysis as well as coordinate strategic security and counterintelligence operations and activity.[376] In 1994, the US government established an organization—the National Counterintelligence Center (NACIC)—similar to the one that deGraffenreid had described.[377] Creation of NACIC came in the wake of a presidential review of US counterintelligence prompted by the espionage investigation of Aldrich Ames, a CIA turncoat.[378] NACIC became responsible, among other tasks, for developing all-source assessments of foreign intelligence and other related threats to US national and economic security.[379] Leadership of NACIC rotated between agencies. Initially, a senior FBI official helmed the center, with a military counterintelligence component executive as deputy. The chair rotated among the FBI, CIA, and Department of Defense, with the FBI serving as deputy when it was not acting as chair.[380]

In 2000, the US government created a new position and office that superseded NACIC. With a presidential decision directive titled "Counterintelligence for the 21st Century," the White House created the National Counterintelligence Executive (NCIX). The newly created NCIX functioned as the "substantive leader of national-level counterintelligence."[381] Selection was by the National Counterintelligence Board of Directors—which the presidential directive also created—with the concurrence of the attorney general, the DCI, and the secretary of defense. The Counterintelligence Executive headed the newly created Office of the National Counterintelligence Executive (ONCIX), which superseded NACIC.[382] It was not until 2002, however, that the NCIX was established, in law, by the Counterintelligence Enhancement Act.[383]

ONCIX brought together multiple aspects of counterintelligence under one office. These included strategic planning (including production of "The National Threat Identification and Prioritization Assessment" on an annual basis); strategic analysis (including counterintelligence damage assessments and lessons-learned papers); collection and targeting coordination (including the development of counterintelligence investigative, operational, and collection objectives and priorities that implement the National Counterintelligence Strategy); and program budget and evaluation.[384]

How the US government thought about—and developed an infrastructure to address—counterintelligence continued to evolve as the decade progressed. In 2010, then-DNI James Clapper announced the merger of ONCIX with the DNI's Special Security Center and the Center for Security Evaluation. The ONCIX functions became part of the National Counterintelligence and Security Center, which the DNI created in September 2014.[385]

From early on, NACIC and its successors have pursued initiatives to engage the private sector and harden targets within it against exploitation by foreign intelligence actors. In October 1994, NACIC held its first meeting of representatives from government agencies having awareness programs that focused on educating the private sector. This interagency interaction was the beginning of what would become NACIC's Awareness Working Group, which, according to a Department of Defense study, successfully reduced the duplication of effort among agencies. The working group also participated in developing analytical products—on specific collection techniques and the foreign entities' methods of operations—for private sector consumption. In addition to reducing overlap among agencies, the working group emphasized the integral role of the private sector as a significant consumer of—and contributor to—government intelligence.[386] According to its 1995 report to Congress, NACIC had completed a survey of US industry's counterintelligence needs and had implemented initiatives to provide more timely and relevant threat information to the private sector.[387]

ONCIX implemented similar activities. According to information presented during a 2003 congressional hearing, NCIX led regional, unclassified conferences for corporations on a routine basis and consciously made information publicly available for interested companies.[388] (This function built upon work that NACIC had initiated during 1995, which ultimately led to its sponsoring of twelve regional seminars addressing issues of concern to the private sector.)[389] Additionally, NCIX made specific efforts to ensure awareness among critical infrastructure entities, the vast majority of which were private sector–controlled. This was required by an executive order, titled "Critical Infrastructure Protection in the Information Age," mandating that NCIX coordinate with the Critical Infrastructure Protection Board to address threats from hostile foreign intelligence services to programs under the board's auspices.[390]

The NCSC has continued these efforts to protect US industry from foreign compromise, in ways that reflect evolving threats. According to the NCSC, it is statutorily responsible for providing counterintelligence outreach to US private sector entities that are vulnerable to foreign intelligence penetration.[391] Specifically, it is responsible for "the dissemination to the public of warnings on intelligence threats to the United States."[392] Picking up on a function that the government has filled through varying efforts since the FBI's Plant Survey Program, the NCSC has conducted vulnerability assessments of private sector entities to enable timely, effective countermeasures.[393] Furthermore, the NCSC has continued to conduct traditional outreach including engagements through trade associations (also historically a target of foreign intelligence collection) to raise awareness and share best practices. However, the NCSC has also acknowledged that conducting "effective and sustained" outreach to the private sector is a challenge.[394]

The NCSC has had to grapple with new threats created and exacerbated by the globalization of technology and capital. In 2013, the NCSC established the first DNI policy to address supply-chain risk management.[395] Additionally, according to the National Counterintelligence Strategy, which the NCSC produces, one of the counterintelligence tasks is to "identify and counter foreign investments in the United States that pose a national security threat." In order to counter such threats, the 2020–2022 strategy notes that the US government would work with private industry to better track foreign investment and to "understand, share, and potentially mitigate counterintelligence issues arising from these investments."[396]

Notes

1. US Senate, *Departments of State, Justice, Commerce, and the Judiciary Appropriations for 1951,* pt. 1, before the Committee on Appropriations, 81st Congress (Washington, DC, 1950).

2. Federal Bureau of Investigation, A. H. Belmont, memorandum to the director, director's brief for President Eisenhower on Khrushchev, September 5, 1959, https://ia801308.us.archive.org/4/items/KHRUVIS19571959/KHRUVIS_1957-1959.pdf.

3. Federal Bureau of Investigation, "A. H. Belmont to the Director."

4. US Senate, *Departments of State, Justice, the Judiciary, and Related Agencies Appropriations, 1962,* before a subcommittee of the Committee on Appropriations, 87th Congress (Washington, DC, 1961).

5. US Congress, *Departments of State, Justice, and Commerce, the Judiciary, and Related Agencies Appropriations for 1966,* before a subcommittee of the Committee on Appropriations, House of Representatives, 89th Congress (Washington, DC, 1965).

6. US Senate, *Departments of State, Justice, and Commerce, the Judiciary, and Related Agencies Appropriations for Fiscal Year 1980,* before a subcommittee of the Committee on Appropriations, 96th Congress (Washington, DC, 1979).

7. US Senate, *Special Report of the Select Committee on Intelligence, January 4, 1995–October 3, 1996,* 105th Congress (Washington, DC, 1996), https://www.govinfo.gov/content/pkg/CRPT-105srpt1/pdf./CRPT-105srpt1.pdf.

8. US Senate, *Oversight of the Federal Bureau of Investigation,* before the Committee on the Judiciary, 105th Congress (Washington, DC, 1997).

9. Central Intelligence Agency, *Summary Report on Technology Transfer to Communist Countries and the Intelligence Community's Role and Effectiveness* (Langley, 1981), https://www.cia.gov/library/readingroom/docs/CIA-RDP85T00176R000900020001-5.pdf.

10. US Senate, *Current and Projected National Security Threats to the United States,* before the Select Committee on Intelligence, 105th Congress (Washington, DC, 1998), https://www.govinfo.gov/content/pkg/CHRG-105shrg51954/pdf./CHRG-105shrg51954.pdf.

11. National Counterintelligence Executive, *Annual Report to Congress on Foreign Economic Collection and Industrial Espionage, 2003* (Washington, DC), https://www.hsdl.org/?view&did=464996.

12. US Senate, *Current and Future Worldwide Threats to the National Security of the United States,* before the Committee on Armed Services, 112th Congress (Washington, DC, 2012), https://www.govinfo.gov/content/pkg/CHRG-112shrg79855/pdf./CHRG-112shrg79855.pdf.

13. US Senate, *Departments of State, Justice, the Judiciary, and Related Agencies Appropriations, 1962.*

14. US Senate, *Departments of State, Justice, and Commerce, the Judiciary, and Related Agencies, Appropriations for Fiscal Year 1975,* pt. 1, before a subcommittee of the Committee on Appropriations, 93rd Congress (Washington, DC, 1974).

15. US Senate, *Departments of State, Justice, and Commerce, the Judiciary, and Related Agencies Appropriations for Fiscal Year 1978,* pt. 6, before a subcommittee of the Committee on Appropriations, 95th Congress (Washington, DC, 1977).

16. US Senate, *Meeting the Espionage Challenge: A Review of United States Counterintelligence and Security Programs,* report of the Select Committee on Intelligence (Washington, DC, 1986), https://www.cia.gov/library/readingroom/docs/CIA-RDP90-00530R000300620021-3.pdf.

17. US Senate, *Exposé of Soviet Espionage 1960* (Washington, DC, 1960), https://www.cia.gov/library/readingroom/docs/CIA-RDP65B00383R000200040033-2.pdf.

18. US Congress, *Departments of State and Justice, the Judiciary, and Related Agencies Appropriations for 1962,* before a subcommittee of the Committee on Appropriations, House of Representatives, 86th Congress (Washington, DC, 1961).

19. US Congress, *Departments of State, Justice, and Commerce, the Judiciary, and Related Agencies Appropriations for 1963,* before a subcommittee of the Committee on Appropriations, House of Representatives, 87th Congress (Washington, DC, 1962).

20. US Congress, *Departments of State, Justice, and Commerce, the Judiciary, and Related Agencies Appropriations for 1966,* before a subcommittee of the Committee on Appropriations, House of Representatives, 89th Congress (Washington, DC, 1965).

21. Federal Bureau of Investigation, *Soviet Intelligence Targets in the United States, 1946–1953* (Washington, DC, 1953), https://www.governmentattic.org/2docs/FBI_Monograph_Soviet-Targets-US_1953.pdf.

22. Federal Bureau of Investigation, SAC Chicago, memorandum to the director, December 22, 1961, https://vault.fbi.gov/solo/solo-part-35-of-44/view.

23. US Congress, *Departments of State, Justice, and Commerce, the Judiciary, and Related Agencies Appropriations for 1963.*

24. Central Intelligence Agency, *The Technology Acquisition Efforts of the Soviet Intelligence Services* (Langley, 1982), https://www.cia.gov/library/readingroom/docs/CIA-RDP82M00786R000104810001-5.pdf.

25. Ibid.

26. Defense Personnel Security Research Center, *Espionage and Other Compromises of National Security* (Monterey, CA: Defense Personnel Security Research Center, 2009), https://fas.org/irp/eprint/esp-summ.pdf.

27. James Coates and Rogers Worthington, "How Spy Ring Went Shopping and Almost Stole the US Store," *Chicago Tribune,* October 23, 1983, https://www.cia.gov/library/readingroom/docs/CIA-RDP90-00552R000302530042-4.pdf; Defense Personnel Security Research Center. *Espionage and Other Compromises of National Security.*

28. US Senate, *Meeting the Espionage Challenge: A Review of United States Counterintelligence and Security Programs,* report of the Select Committee on Intelligence, 99th Congress (Washington, DC, 1986), https://www.intelligence.senate.gov/sites/default/files/publications/99522.pdf.

29. Defense Personnel Security Research Center, *Espionage and Other Compromises of National Security.*

30. Department of Defense Industrial Security Review Committee, *Analysis of the Effectiveness of the Department of Defense Industrial Security Program and Recommendations for Program Improvement,* report to the deputy undersecretary of defense for policy (Washington, DC, 1984), https://apps.dtic.mil/dtic/tr/fulltext/u2/a196076.pdf.

31. *United States of America v. Vadim Mikerin,* https://www.justice.gov/criminal-fraud/file/782186/download.

32. Department of Justice, "Former Russian Nuclear Energy Official Sentenced to 48 Months in Prison for Money Laundering Conspiracy Involving Foreign Corrupt Practices Act Violations," December 15, 2015, https://www.justice.gov/opa/pr/former-russian-nuclear-energy-official-sentenced-48-months-prison-money-laundering-conspiracy.

33. US Senate, *Current and Projected National Security Threats to the United States,* before the Select Committee on Intelligence, 110th Congress (Washington, DC, 2007), https://www.govinfo.gov/content/pkg/CHRG-110shrg48098/pdf./CHRG-110shrg48098.pdf.

34. US Senate, *Current and Projected National Security Threats to the United States,* before the Select Committee on Intelligence, 109th Congress (Washington, DC, 2005), https://www.govinfo.gov/content/pkg/CHRG-109shrg22379/pdf./CHRG-109shrg22379.pdf.

35. US Senate, *Current and Projected National Security Threats to the United States,* before the Select Committee on Intelligence, 105th Congress (Washington, DC, 1998), https://www.govinfo.gov/content/pkg/CHRG-105shrg51954/pdf./CHRG-105shrg51954.pdf.

36. US Congress, *Departments of State, Justice, and Commerce, the Judiciary, and Related Agencies Appropriations for 1966.*

37. Central Intelligence Agency, *The Technology Acquisition Efforts of the Soviet Intelligence Services.*

38. US Congress, *Departments of Commerce, Justice, and State, the Judiciary, and Related Agencies Appropriations for 2000,* pt. 6, before the Committee on Appropriations, House of Representatives, 106th Congress (Washington, DC, 1999).

39. US Congress, *Domestic Intelligence Operations for Internal Security Purposes,* pt. 1, before the Committee on Internal Security, House of Representatives, 93rd Congress (Washington, DC, 1974).

40. Federal Bureau of Investigation, "W. C. Hinze to Ladd: Plant Survey," memorandum, June 2, 1942 (Washington, DC: National Archives and Records Administration).

41. Ibid.

42. Federal Bureau of Investigation, director, memorandum to Alan Belmont, November 18, 1953, https://ia801802.us.archive.org/26/items/foia_Belmont_Alan_4/Belmont_Alan_4.pdf.

43. US Congress, *Departments of Commerce, Justice, and State, the Judiciary, and Related Agencies Appropriations for 1990,* pt. 2, before a subcommittee of the Committee on Appropriations, House of Representatives, 101st Congress (Washington, DC, 1989).

44. US Senate, *Departments of State, Justice, and Commerce, the Judiciary, and Related Agencies Appropriations for Fiscal Year 1978,* pt. 6.

45. US Congress, *Departments of State, Justice, and Commerce, the Judiciary, and Related Agencies Appropriations for 1978,* pt. 5, before the Committee on Appropriations House of Representatives, 95th Congress (Washington, DC, 1977).

46. US Congress, *H.R. 6588: National Intelligence Act of 1980,* before the Permanent Select Committee on Intelligence, House of Representatives, 96th Congress (Washington, DC, 1980).

47. US Senate, *Departments of State, Justice, Commerce, and the Judiciary Appropriations for 1951,* pt. 1, before a subcommittee of the Committee on Appropriations, 81st Congress (Washington, DC, 1950).

48. US Congress, *United States Counterintelligence and Security Concerns, 1986,* report of the House Permanent Select Committee on Intelligence (Washington, DC, 1987), https://www.cia.gov/library/readingroom/docs/CIA-RDP91B00390R000200160014-6.pdf.

49. US Congress, *FBI Oversight and Authorization Request for Fiscal Year 1992,* before the Committee on the Judiciary, House of Representatives, 102nd Congress (Washington, DC, 1991).

50. National Security Directive 47, "Counterintelligence and Security Countermeasures," October 5, 1990, https://www.hsdl.org/?view&did=458829.

51. US Congress, *Departments of Commerce, Justice, and State, the Judiciary, and Related Agencies Appropriations for 1990,* pt. 2.

52. US Congress, *Departments of Commerce, Justice, and State, the Judiciary, and Related Agencies Appropriations for 1993,* pt. 2B, before the Committee on Appropriations, House of Representatives, 102nd Congress (Washington, DC, 1992).

53. Ibid.

54. US Congress, *FBI Oversight and Authorization Request for Fiscal Year 1992.*

55. US Congress, *FBI Oversight and Authorization, Fiscal Year 1993,* before the Committee on the Judiciary, House of Representatives, 102nd Congress (Washington, DC, 1992).

56. Department of Justice, *Attorney General Guidelines for FBI Foreign Intelligence Collection and Foreign Counterintelligence Investigations* (Washington, DC, 1995), https://fas.org/irp/agency/doj/fbi/terrorismintel2.pdf; US Congress, *Departments of Commerce, Justice, and State, the Judiciary, and Related Agencies Appropriations for 1994,* before the Committee on Appropriations, House of Representatives, 103rd Congress (Washington, DC, 1993).

57. Ibid.

58. US Congress, *FBI Oversight and Authorization, Fiscal Year 1993.*

59. US Congress, *Departments of Commerce, Justice, and State, the Judiciary, and Related Agencies Appropriations for 1994.*

60. Ibid.

61. US Congress, *Departments of Commerce, Justice, and State, the Judiciary, and Related Agencies Appropriations for 1993,* pt. 2B.

62. US Congress, *FBI Oversight and Authorization Request for Fiscal Year 1992.*

63. Ibid.

64. General Accounting Office, *Economic Espionage: The Threat to US Industry* (Washington, DC, 1992), https://www.gao.gov/assets/110/104477.pdf.

65. Federation of American Scientists, *Awareness of National Security Issues and Response* (Washington, DC, 1998), https://fas.org/irp/ops/ci/ansir.htm.

66. US Senate, *Oversight of the Federal Bureau of Investigation,* before the Committee on the Judiciary, 105th Congress (Washington, DC, 1997).

67. US Senate, *Current and Projected National Security Threats to the United States,* 105th Congress.

68. US Senate, *Oversight of the Federal Bureau of Investigation.*

69. US Congress, *Protecting American Interests Abroad: U.S. Citizens, Businesses, and Nongovernmental Organizations,* before the Committee on Government Reform, House of Representatives, 107th Congress (Washington, DC, 2001).

70. US Senate, *Current and Projected National Security Threats to the United States,* 105th Congress.

71. 9/11 Commission, "Law Enforcement, Counterterrorism, and Intelligence Collection in the United States Prior to 9/11," Staff Statement no. 9, http://govinfo.library.unt.edu/911/staff_statements/staff_statement_9.pdf.

72. National Counterintelligence Executive, *Annual Report to Congress on Foreign Economic Collection and Industrial Espionage, 1998* (Washington, DC), https://www.hsdl.org/?view&did=463230.

73. US Senate, *Current and Projected National Security Threats to the United States,* 105th Congress.

74. 9/11 Commission, *Law Enforcement, Counterterrorism, and Intelligence Collection;* Roberto Suro, "New FBI Spy Unit Gets Reno's Approval," *Washington Post,* June 26, 1999, https://www.washingtonpost.com/archive/politics/1999/06/26/new-fbi-spy-unit-gets-renos-approval/63f6f03b-e6fe-485c-b4aa-493120b2713c; David Johnson, "F.B.I. Is Proposing a Special Division for Hunting Spies," *New York Times,* June 26, 1999, https://www.nytimes.com/1999/06/26/world/fbi-is-proposing-a-special-division-for-hunting-spies.html.

75. US Congress, *Departments of Commerce, Justice, and State, the Judiciary, and Related Agencies Appropriations for 2002,* pt. 6, before a subcommittee of the Committee on Appropriations, House of Representatives, 107th Congress (Washington, DC, 2001).

76. US Senate, *Reforming the FBI in the 21st Century,* before the Committee on the Judiciary, 107th Congress (Washington, DC, 2002).

77. Ibid.

78. Department of Justice, *Review of Four FISA Applications and Other Aspects of the FBI's Crossfire Hurricane Investigation* (Washington, DC, 2019), https://www.justice.gov/storage/120919-examination.pdf.

79. Department of Justice, *A Review of Various Actions by the Federal Bureau of Investigation and Department of Justice in Advance of the 2016 Election* (Washington, DC, 2018), https://www.oversight.gov/sites/default/files/oig-reports/o1804.pdf.

80. Federal Bureau of Investigation, D. M. Ladd, memorandum to the director, December 29, 1941.

81. Federal Bureau of Investigation, Bulletin no. 17: (J) Confidential Informants, May 15, 1941, https://ia802702.us.archive.org/2/items/foia_FBI_Confidential_Informants-HQ-1a/FBI_Confidential_Informants-HQ-1a.pdf.

82. Ibid.

83. Federal Bureau of Investigation, memorandum to the director, October 20, 1940.

84. Federal Bureau of Investigation, Bulletin no. 1: First Series 1941, January 2, 1941, https://ia802702.us.archive.org/2/items/foia_FBI_Confidential_Informants-HQ-1a/FBI_Confidential_Informants-HQ-1a.pdf.

85. Federal Bureau of Investigation, memorandum to the director, April 10, 1941; Federal Bureau of Investigation, "Proposed Revision of the *National Defense Manual,* Section 13, 'Confidential National Defense Informants,' 'Confidential Plant Informants' (n.d., ca. June 1943); Federal Bureau of Investigation, memorandum, February 26, 1942.

86. Federal Bureau of Investigation, "J. Edgar Hoover to Assistant Director E. J. Connelley," April 25, 1941, https://ia802702.us.archive.org/2/items/foia_FBI_Confidential_Informants-HQ-1a/FBI_Confidential_Informants-HQ-1a.pdf.

87. Federal Bureau of Investigation, Bureau Bulletin no. 17.

88. Federal Bureau of Investigation, memorandum, February 26, 1942 (Washington, DC: National Archives and Records Administration).

89. Federal Bureau of Investigation, memorandum, July 16, 1942 (Washington, DC: National Archives and Records Administration).

90. Federal Bureau of Investigation, D. M. Ladd, memorandum to the director, February 9, 1942 (Washington, DC: National Archives and Records Administration).

91. Federal Bureau of Investigation, memorandum to E. A. Tamm, January 27, 1941, https://ia802702.us.archive.org/2/items/foia_FBI_Confidential_Informants-HQ-1a/FBI_Confidential_Informants-HQ-1a.pdf.

92. Federal Bureau of Investigation, Bulletin no. 17.

93. Federal Bureau of Investigation, "Proposed Changes in Section 9A of the *Manual of Rules and Regulations* and Section 13 of the *National Defense Manual* Pertaining to Confidential Informants," memorandum to D. M. Ladd, May 19, 1944, https://ia802704.us.archive.org/23/items/foia_FBI_Security_Informant_Program_HQ_66-2542-3_HQ-5/FBI_Security_Informant_Program_HQ_66-2542-3_HQ-5.pdf.

94. Federal Bureau of Investigation, memorandum to E. A. Tamm, January 10, 1941, https://ia802702.us.archive.org/2/items/foia_FBI_Confidential_Informants-HQ-1a/FBI_Confidential_Informants-HQ-1a.pdf.

95. Federal Bureau of Investigation, memorandum to D. M. Ladd from R. H. Cunningham, February 11, 1942.

96. Federal Bureau of Investigation, memorandum to the director, November 13, 1942, https://ia802703.us.archive.org/32/items/foia_FBI_Security_Informant_Program_HQ_66-2542-3_HQ-3/FBI_Security_Informant_Program_HQ_66-2542-3_HQ-3.pdf.

97. Federal Bureau of Investigation, "Proposed Changes in Section 9A."

98. Federal Bureau of Investigation, Bulletin no. 17.

99. Federal Bureau of Investigation, memorandum to D. M. Ladd from R. H. Cunningham, February 11, 1942.

100. Federal Bureau of Investigation, "Proposed Revision of the *National Defense Manual,* Section 13."

101. US Senate, *Supplementary Detailed Staff Reports on Intelligence Activities and the Rights of Americans,* final report of the Select Committee to Study Governmental Operations with Respect to Intelligence Activities, book 3, 94th Congress (Washington, DC, 1976).

102. Federal Bureau of Investigation, Bulletin no. 1.

103. Federal Bureau of Investigation, Bulletin no. 17.

104. Federal Bureau of Investigation, memorandum, February 26, 1942.

105. Federal Bureau of Investigation, D. M. Ladd, memorandum to the director, February 9, 1942.

106. Federal Bureau of Investigation, D. M. Ladd, memorandum to the director, February 4, 1946; Federal Bureau of Investigation, "Master Inspection Responsibility List no. 13 dated July 31, 1945," December 11, 1945.

107. Federal Bureau of Investigation, "Bureau War Plans; Internal Security Unit," October 31, 1951; Federal Bureau of Investigation, A. H. Belmont to Ladd, "Special Agent Personnel: Advancement," Domestic Intelligence Division, January 31, 1952 (Keay personnel file FOIA release).

108. Federal Bureau of Investigation, J. Edgar Hoover, memorandum to DeLoach, July 9, 1953 (DeLoach personnel file FOIA release).

109. Federal Bureau of Investigation, "Inspection: Domestic Intelligence Division, 2/20/67–3/10/67," March 21, 1967.

110. Federal Bureau of Investigation, memorandum to the director, FBI Executive Committee, September 7, 1939.
111. Federal Bureau of Investigation, memorandum to the director, November 1, 1939.
112. Federal Bureau of Investigation, "FBI Executive Conference," November 1, 1939.
113. Federal Bureau of Investigation, A. H. Belmont to D. M. Ladd, "Undercover Assignments, Atomic Energy Commission," June 12, 1951, https://www.governmentattic.org/4docs/FBIundercoverAECfacilities_1949-1964.pdf.
114. Federal Bureau of Investigation, V. P. Keay to D. M. Ladd, "Atomic Energy Act: Informants," March 24, 1949, https://www.governmentattic.org/4docs/FBIundercoverAECfacilities_1949-1964.pdf.
115. Federal Bureau of Investigation, A. H. Belmont to D. M. Ladd, "Undercover Assignments, Atomic Energy Commission," June 12, 1951, https://www.governmentattic.org/4docs/FBIundercoverAECfacilities_1949-1964.pdf.
116. Federal Bureau of Investigation, A. H. Belmont to D. M. Ladd, "Undercover Assignments, Atomic Energy Commission," August 25, 1953, https://www.governmentattic.org/4docs/FBIundercoverAECfacilities_1949-1964.pdf.
117. Federal Bureau of Investigation, V. P. Keay to D. M. Ladd, "Atomic Energy Act: Informants," March 24, 1949.
118. David M. Barrett, *The CIA and Congress* (Lawrence: University Press of Kansas, 2005).
119. Federal Bureau of Investigation, V. P. Keay to D. M. Ladd, "Atomic Energy Act: Informants," May 2, 1949, https://www.governmentattic.org/4docs/FBIundercoverAECfacilities_1949-1964.pdf.
120. US Senate, *Departments of State, Justice, Commerce, and the Judiciary Appropriations for 1951,* pt. 1, before the Committee on Appropriations, 81st Congress (Washington, DC, 1950).
121. Federal Bureau of Investigation, V. P. Keay to D. M. Ladd, "Atomic Energy Act: Informants," March 24, 1949.
122. Ibid.
123. Federal Bureau of Investigation, Glavin memorandum to Clyde Tolson, "Atomic Energy Act: Informants," July 20, 1949, https://www.governmentattic.org/4docs/FBIundercoverAECfacilities_1949-1964.pdf.
124. Federal Bureau of Investigation, Armand A. Cammarota to V. P. Keay, "Special Assignment at the Oak Ridge Atomic Energy Plant," September 27, 1949, https://www.governmentattic.org/4docs/FBIundercoverAECfacilities_1949-1964.pdf.
125. Federal Bureau of Investigation, Richard C. Frye to V. P. Keay, "Suggested Outline of Steps to Be Taken in Connection with the Writer's Proposed Confidential Assignment," September 27, 1949, https://www.governmentattic.org/4docs/FBIundercoverAECfacilities_1949-1964.pdf.
126. Federal Bureau of Investigation, V. P. Keay to Fletcher, Special Agent Richard C. Frye, and Special Agent Armand C. Cammarota, "Special Undercover Assignment," October 4, 1949, https://www.governmentattic.org/4docs/FBIundercoverAECfacilities_1949-1964.pdf.
127. Federal Bureau of Investigation, V. P. Keay to H. B. Fletcher, "Undercover Assignment, Atomic Energy Commission," December 7, 1949, https://www.governmentattic.org/4docs/FBIundercoverAECfacilities_1949-1964.pdf.
128. Department of Justice, *Appropriations for 1952,* before the subcommittee of the Committee on Appropriations, House of Representatives, 82nd Congress (Washington, DC, 1951).
129. Federal Bureau of Investigation, A. H. Belmont to D. M. Ladd, "Undercover Assignments, Atomic Energy Commission," August 25, 1953, https://www.governmentattic.org/4docs/FBIundercoverAECfacilities_1949-1964.pdf.

130. Federal Bureau of Investigation, C. E. Henrich and A. H. Belmont, "Undercover Assignments, Atomic Energy Commission," June 5, 1951, https://www.governmentattic.org/4docs/FBIundercoverAECfacilities_1949-1964.pdf.

131. Federal Bureau of Investigation, V. P. Keay to D. M. Ladd, "Inspection of Field Offices on Atomic Energy Act Cases," December 2, 1947.

132. US Senate, *Current and Projected National Security Threats to the United States,* before the Select Committee on Intelligence, 109th Congress (Washington, DC, 2005), https://www.govinfo.gov/content/pkg/CHRG-109shrg22379/pdf./CHRG-109shrg22379.pdf.

133. US Senate, *Federal Bureau of Investigation Oversight,* before the Committee on the Judiciary, 109th Congress (Washington, DC, 2005).

134. Robert S. Mueller III, testimony before the House Appropriations Subcommittee on Science, the Departments of State, Justice, and Commerce, and Related Agencies, September 14, 2006, https://archives.fbi.gov/archives/news/testimony/the-fbi-transformation-since-2001.

135. US Senate, *Current and Projected National Security Threats to the United States,* 105th Congress.

136. US Senate, *Current and Projected National Security Threats to the United States,* 107th Congress.

137. US Senate, *Current and Projected National Security Threats to the United States,* 110th Congress.

138. Federal Bureau of Investigation, FBI director to assistant attorney general, "Searches of Diplomatic Shipments Leaving the United States," May 27, 1952, https://www.governmentattic.org/4docs/FBI-fileHQ105-18233_DiplomatoicSearches_1952-1967.pdf.

139. Ibid.

140. Federal Bureau of Investigation, D. M. Ladd, memorandum to the director, "Searches of Incoming and Outgoing Diplomatic Baggage and Shipments," October 17, 1952, https://www.governmentattic.org/4docs/FBI-fileHQ105-18233_Diplomatoic Searches_1952-1967.pdf.

141. Federal Bureau of Investigation, SAC Baltimore, memorandum to the director, "Searches of Diplomatic Shipments," May 6, 1967, https://www.governmentattic.org/4docs/FBI-fileHQ105-18233_DiplomatoicSearches_1952-1967.pdf.

142. Federal Bureau of Investigation, W. A. Branigan to W. C. Sullivan, "Searches of Diplomatic Shipments: Internal Security—Russia," September 11, 1967, https://www.governmentattic.org/4docs/FBI-fileHQ105-18233_DiplomatoicSearches_1952-1967.pdf.

143. Ibid.

144. Federal Bureau of Investigation, SAC and WFO, memorandum to the director, "Searches of Diplomatic Shipments," September 8, 1967, https://www.governmentattic.org/4docs/FBI-fileHQ105-18233_DiplomatoicSearches_1952-1967.pdf.

145. Ibid.

146. Federal Bureau of Investigation, *Chinese Communist Intelligence Activities in the United States* (Washington, DC, 1954), https://ia801908.us.archive.org/35/items/FBIPRCSpying/fbi-prc-spying.pdf.

147. US Senate, *Final Report of the Select Committee to Study Governmental Operations with Respect to Intelligence Activities,* book 3, *Supplementary Detailed Staff Reports on Intelligence Activities and the Rights of Americans,* 94th Congress (Washington, DC, 1976), https://www.intelligence.senate.gov/sites/default/files/94755_III.pdf.

148. US Senate, *Foreign Intelligence Surveillance Act of 1978,* before the Select Committee on Intelligence, 95th Congress (Washington, DC, 1978), https://www.intelligence.senate.gov/sites/default/files/hearings/s1566.pdf.

149. Federal Bureau of Investigation, *Chinese Communist Intelligence Activities in the United States.*
150. US Senate, *Supplementary Detailed Staff Reports on Intelligence Activities.*
151. Commission on CIA Activities Within the United States, *Report to the President* (Washington, DC, 1975), https://www.fordlibrarymuseum.gov/library/document/0005/1561495.pdf.
152. Mark Riebling, *Wedge: The Secret War Between the FBI and CIA* (New York: Knopf, 1994).
153. US Senate, *Supplementary Detailed Staff Reports on Intelligence Activities.*
154. Riebling, *Wedge.*
155. US Senate, *Supplementary Detailed Staff Reports on Intelligence Activities.*
156. Ibid.
157. Commission on CIA Activities Within the United States, *Report to the President.*
158. Frank Rafalko, ed., *A Counterintelligence Reader: Post World War II to Closing the 20th Century* (Washington, DC: National Counterintelligence Executive, 2004).
159. Commission on CIA Activities Within the United States, *Report to the President.*
160. John Prados, *The Family Jewels* (Austin: University of Texas Press, 2013).
161. Rafalko, *A Counterintelligence Reader.*
162. US Senate, *Supplementary Detailed Staff Reports on Intelligence Activities.*
163. Ibid.
164. US Senate, *Departments of State, Justice, the Judiciary, and Related Agencies Appropriations, 1962,* before a subcommittee of the Committee on Appropriations, 87th Congress (Washington, DC, 1961).
165. Federal Bureau of Investigation, *Soviet Intelligence Targets in the United States.*
166. US Senate, *Exposé of Soviet Espionage 1960.*
167. James R. Clapper, "Worldwide Threat Assessment of the US Intelligence Community," statement for the record, Senate Select Committee on Intelligence, February 9, 2016.
168. Federal Bureau of Investigation, *Soviet Intelligence Targets in the United States.*
169. Phillip M. Boffey, "Assessing Technology Leaks," *New York Times,* January 2, 1985, https://www.nytimes.com/1985/01/02/business/assessing-technology-leaks.html.
170. US Congress, *Departments of State, Justice, and Commerce, the Judiciary, and Related Agencies Appropriations for 1966.*
171. US Senate, *Exposé of Soviet Espionage 1960.*
172. Central Intelligence Agency, *The Technology Acquisition Efforts of the Soviet Intelligence Services* (Langley, 1982), https://www.cia.gov/library/readingroom/docs/CIA-RDP82M00786R000104810001-5.pdf.
173. Ibid.
174. US Senate, *Meeting the Espionage Challenge.*
175. Clapper, "Worldwide Threat Assessment of the US Intelligence Community."
176. US Congress, *Departments of State, Justice, and Commerce, the Judiciary, and Related Agencies Appropriations for 1966.*
177. Federal Bureau of Investigation, *Soviet Intelligence Targets in the United States.*
178. Ibid.
179. US Senate, *Exposé of Soviet Espionage 1960.*
180. US Congress, *Departments of State, Justice, and Commerce, the Judiciary, and Related Agencies Appropriations for 1966.*

181. Central Intelligence Agency, *Soviet Acquisition of Western Technology and Its National Security Implications* (Langley, 1982), https://www.cia.gov/library/readingroom/docs/CIA-RDP83M00914R002000070021-4.pdf.

182. Federal Bureau of Investigation, *Chinese Communist Intelligence Activities in the United States.*

183. US-China Economic Security Review Commission, *China's Pursuit of Next Frontier Tech: Computing, Robotics, and Biotechnology* (Washington, DC, 2017), https://www.uscc.gov/sites/default/files/transcripts/March%20Transcript.pdf.

184. Ibid.

185. US Congress, *Departments of State, Justice, and Commerce, the Judiciary, and Related Agencies Appropriations for 1966.*

186. Federal Bureau of Investigation, SAC New York, memorandum to the director, "Espionage," November 4, 1960 (Moore personnel file FOIA release).

187. Testimony of Edward O'Mally, assistant director, Intelligence Division, Federal Bureau of Investigation, https://www.cia.gov/library/readingroom/docs/CIA-RDP85M00364R001001520004-2.pdf.

188. Federal Bureau of Investigation, SAC E. E. Conroy, memorandum to Hoover, September 25, 1941, https://ia802703.us.archive.org/21/items/foia_FBI_Confidential_Informants-HQ-2/FBI_Confidential_Informants-HQ-2.pdf.

189. Ibid.

190. Ibid.

191. Ibid.

192. Ibid.

193. Linda Greenhouse, "F.B.I. Defends Library Monitoring Program," *New York Times,* July 14, 1988, https://www.nytimes.com/1988/07/14/us/fbi-defends-library-monitoring-program.html; Robert McFadden, "F.B.I. in New York Asks Librarians' Aid in Reporting on Spies," *New York Times,* September 18, 1987.

194. Greenhouse, "F.B.I. Defends Library Monitoring Program."

195. Nat Hentoff, "The FBI in the Library," *Washington Post,* July 23, 1988, https://www.washingtonpost.com/archive/opinions/1988/07/23/the-fbi-in-the-library/f0ea90c7-4c52-46c0-a546-e567220fe0a1.

196. Greenhouse, "F.B.I. Defends Library Monitoring Program"; US Congress, *FBI Counterintelligence Visits to Libraries,* testimony of James H. Geer, assistant director, Intelligence Division, Federal Bureau of Investigation, before the Subcommittee on Civil and Constitutional Rights of the Committee on the Judiciary, House of Representatives, 100th Congress (Washington, DC, 1988).

197. Ibid.

198. Ibid.

199. Ibid.

200. Hentoff, "The FBI in the Library."

201. Ibid.

202. Bill McAllister, "FBI to Limit Probes of Library Users; Program to Detect Foreign Agents Is Altered to Guard Patron Privacy," *Washington Post,* November 15, 1988.

203. US Congress, *FBI Oversight and Authorization, Fiscal Year 1993,* before the Subcommittee on Civil and Constitutional Rights of the Committee on the Judiciary, House of Representatives, 102nd Congress (Washington, DC, 1992).

204. US Congress, *Corporate and Industrial Espionage and Their Effects on American Competitiveness,* before the Committee on International Relations, House of Representatives, 106th Congress (Washington, DC, 2000), https://www.govinfo.gov/content/pkg/CHRG-106hhrg68684/pdf./CHRG-106hhrg68684.pdf.

205. Elaine Sciolino, "Zakharov Charges a 'Setup' by F.B.I.," *New York Times,* September 17, 1986.

206. US Senate, *Meeting the Espionage Challenge;* Sciolino, "Zakharov Charges a 'Setup' by F.B.I."; "FBI Broke Promises to Informant," *Chicago Tribune,* March 31, 1987.
207. US Senate, *Meeting the Espionage Challenge.*
208. Ibid.
209. "FBI Tells How Soviet Spy Operated," *Associated Press,* August 26, 1986.
210. Ibid.; "FBI Broke Promises to Informant."
211. Sciolino, "Zakharov Charges a 'Setup' by F.B.I."
212. "FBI Tells How Soviet Spy Operated."
213. "FBI Broke Promises to Informant."
214. Joan M. Jensen, *Army Surveillance in America, 1775–1980* (New Haven: Yale University Press, 1991).
215. US Senate, *Final Report of the Select Committee to Study Governmental Operations with Respect to Intelligence Activities,* book 6, *Supplementary Reports on Intelligence Activities,* 94th Congress (Washington, DC, 1976).
216. Federal Bureau of Investigation, "Plant Protection Program," July 31, 1940.
217. Federal Bureau of Investigation, memorandum, August 11, 1939.
218. US Congress, *Department of Justice Appropriation Bill for 1943,* before a subcommittee of the Committee on Appropriations, House of Representatives, 76th Congress (Washington, DC, 1942).
219. Federal Bureau of Investigation, "Plant Protection Program."
220. Federal Bureau of Investigation, memorandum to the director, February 11, 1941, https://ia902704.us.archive.org/25/items/foia_FBI_Security_Informant_Program_HQ_66-2542-3_HQ-1/FBI_Security_Informant_Program_HQ_66-2542-3_HQ-1.pdf.
221. Raymond Batvinis, *The Origins of FBI Counterintelligence* (Lawrence: University Press of Kansas, 2007).
222. Federal Bureau of Investigation, "Plant Protection Program."
223. US Congress, *Department of Justice Appropriation Bill for 1941,* before a subcommittee of the Committee on Appropriations, House of Representatives, 76th Congress (Washington, DC, 1940).
224. Federal Bureau of Investigation, "Plant Protection Program."
225. US Congress, *Department of Justice Appropriation Bill for 1943,* before a subcommittee of the Committee on Appropriations, House of Representatives, 76th Congress (Washington, DC, 1942).
226. Federal Bureau of Investigation, memorandum to the director, November 15, 1939.
227. US Congress, *Department of Justice Appropriation Bill for 1941.*
228. Federal Bureau of Investigation, memorandum to the director, November 15, 1939.
229. Ibid.
230. US Congress, *Department of Justice Appropriation Bill for 1941.*
231. Federal Bureau of Investigation, Bulletin no. 17.
232. Ibid.
233. Federal Bureau of Investigation, memorandum to the director, August 26, 1940; US Congress, *Department of Justice Appropriation Bill for 1943.*
234. US Congress, *Department of Justice Appropriation Bill for 1941.*
235. US Congress, *Department of Justice Appropriation Bill for 1942.*
236. US Congress, *Department of Justice Appropriation Bill for 1941.*
237. US Congress, *Department of Justice Appropriation Bill for 1942.*
238. US Congress, *Department of Justice Appropriation Bill for 1943.*
239. Federal Bureau of Investigation, Bulletin no. 17.
240. Federal Bureau of Investigation, memorandum to the director, Executive Conference, April 11, 1940.

241. Federal Bureau of Investigation, memorandum to the director, May 29, 1940.

242. Federal Bureau of Investigation, memorandum to the director, November 15, 1939.

243. Federal Bureau of Investigation, memorandum to the director, May 29, 1940.

244. Federal Bureau of Investigation, memorandum to the director, June 16, 1941.

245. Federal Bureau of Investigation, W. C. Hinze, memorandum to D. M. Ladd, "Plant Survey," June 2, 1942.

246. US Congress, *Federal Bureau of Investigation: Emergency Supplemental Appropriations Bill for 1940,* before a subcommittee of the Committee on Appropriations, House of Representatives, 76th Congress (Washington, DC, 1939).

247. US Senate, *Critical Infrastructure Protection: Who's in Charge?,* before the Committee on Governmental Affairs, 107th Congress (Washington, DC, 2001).

248. US Congress, *Department of Justice Appropriation Bill for 1941.*

249. Federal Bureau of Investigation, memorandum for the director, "Re: Plant Protection Pamphlet," January 21, 1941.

250. Federal Bureau of Investigation, W. C. Hinze, memorandum to D. M. Ladd, "Plant Survey," June 2, 1942.

251. Federal Bureau of Investigation, D. M. Ladd, memorandum to the director, December 29, 1941 (Washington, DC: National Archives and Records Administration).

252. US Congress, *Department of Justice Appropriation Bill for 1943;* Federal Bureau of Investigation, R. H. Cunningham, memorandum to D. M. Ladd, December 15, 1941.

253. Federal Bureau of Investigation, D. M. Ladd, memorandum to the director, December 29, 1941.

254. Federal Bureau of Investigation, memorandum, February 26, 1942.

255. Federal Bureau of Investigation, D. M. Ladd, memorandum to the director, December 29, 1941.

256. Federal Bureau of Investigation, W. C. Hinze, memorandum to D. M. Ladd, "Plant Survey."

257. Federal Bureau of Investigation, D. M. Ladd, memorandum to the director, January 1, 1942.

258. Federal Bureau of Investigation, "Plant Protection: Responsibility of Munition Board," September 27, 1950.

259. US Senate, *Departments of State, Justice, Commerce, and the Judiciary Appropriations for 1951,* pt. 1.

260. Federal Bureau of Investigation, *Soviet Intelligence Targets in the United States.*

261. Federal Bureau of Investigation, A. H. Belmont, memorandum to D. M. Ladd, "Undercover Assignments, Atomic Energy Commission," August 25, 1953, https://www.governmentattic.org/4docs/FBIundercoverAECfacilities_1949-1964.pdf.

262. Ibid.

263. Federal Bureau of Investigation, Gale to Tolson, "Inspection: Domestic Intelligence Division, November 15–December 3, 1962," December 11, 1962 (Moore personnel file FOIA release).

264. Federal Bureau of Investigation, A. H. Belmont, memorandum to D. M. Ladd, "Undercover Assignments, Atomic Energy Commission."

265. Federal Bureau of Investigation, Executives Conference, memorandum to the director, "Dissemination of Information by the Bureau Outside the Executive Departments," October 14, 1953.

266. Federal Bureau of Investigation, V. P. Keay, memorandum to D. M. Ladd, "Atomic Energy Act: Informants," March 24, 1949, https://www.governmentattic.org/4docs/FBIundercoverAECfacilities_1949-1964.pdf.

267. President's Foreign Intelligence Advisory Board, *Science at Its Best; Security at Its Worst: A Report on Security Problems at the Department of Energy* (Washington, DC, 1999), https://www.energy.gov/sites/prod/files/cioprod/documents/pfiab-doe.pdf.
268. Federal Bureau of Investigation, *The FBI's Counterterrorism Program Since September 2001* (Washington, DC, 2004).
269. Ibid.; Daniel D. Roberts, assistant director, Criminal Justice Information Services Division, Federal Bureau of Investigation, testimony before the Senate Judiciary Committee, September 22, 2009, https://archives.fbi.gov/archives/news/testimony/mission-of-fbis-bioterrorism-risk-assessment-group.
270. Federal Bureau of Investigation, *National Instant Criminal Background Check System 2013,* https://archives.fbi.gov/archives/about-us/cjis/nics/reports/2013-operations-report.
271. Vahid Majidi, assistant director, Weapons of Mass Destruction Directorate, Federal Bureau of Investigation, statement before the Senate Committee on Homeland Security and Governmental Affairs, October 18, 2011, https://archives.fbi.gov/archives/news/testimony/ten-years-after-9-11-and-the-anthrax-attacks-protecting-against-biological-threats.
272. Roberts, testimony before the Senate Judiciary Committee.
273. Letter from ASIS to J. Edgar Hoover, October 22, 1965, https://ia801803.us.archive.org/29/items/foia_Stukenbroeker_Fern_C.-4/Stukenbroeker_Fern_C.-4.pdf.
274. Federal Bureau of Investigation, SAC Memphis, memorandum to the director, February 18, 1971, https://ia801706.us.archive.org/20/items/foia_Stukenbroeker_Fern_C.-5/Stukenbroeker_Fern_C.-5.pdf.
275. A. P. Clow, Western Electric, memorandum to J. Edgar Hoover, December 20, 1965, https://ia801803.us.archive.org/29/items/foia_Stukenbroeker_Fern_C.-4/Stukenbroeker_Fern_C.-4.pdf.
276. Robert K. Matthews, AT&T, memorandum to J. Edgar Hoover, December 18, 1967, https://ia801706.us.archive.org/20/items/foia_Stukenbroeker_Fern_C.-5/Stukenbroeker_Fern_C.-5.pdf.
277. US Senate, *Foreign Intelligence Surveillance Act of 1978.*
278. Federal Bureau of Investigation, Clyde Tolson, memorandum to the director, January 27, 1959 (DeLoach personnel file FOIA release).
279. Federal Bureau of Investigation, "Inspection: Crime Records Division, April 8–19, 1963 (Wick personnel file FOIA release).
280. US Senate, *Foreign Intelligence Surveillance Act of 1978.*
281. US Congress, *Investigation and Study of the Administration, Operation, and Enforcement of the Export Control Act of 1949 and Related Acts,* report of the Select Committee on Export Control, House of Representatives, 87th Congress (Washington, DC, 1962), https://books.google.com/books?id=iNQz97wiQbgC&pg=RA105-PA59&lpg=RA105-PA59&dq=secretary+of+commerce+revoked+soviet+license+1961&source=bl&ots=hZNjDZORfi&sig=ACfU3U1AaiQ2_7nTvwGC0s2EWNCd_Ag8Ow&hl=en&sa=X&ved=2ahUKEwiE1OPKoqPqAhV3mHIEHeGAAdEQ6AEwCXoECA0QAQ#v=onepage&q=secretary%20of%20commerce%20revoked%20soviet%20license%201961&f=false.
282. Federal Bureau of Investigation, J. Edgar Hoover, memorandum to Fern C. Stukenbroeker, May 3, 1962, https://ia801708.us.archive.org/7/items/foia_Stukenbroeker_Fern_C.-3/Stukenbroeker_Fern_C.-3.pdf.
283. Federal Bureau of Investigation, M. A. Jones, memorandum to DeLoach titled "Harvard Business Review; January–February, 1964, Issue; Director's Article," January 7, 1964, https://ia801803.us.archive.org/29/items/foia_Stukenbroeker_Fern_C.-4/Stukenbroeker_Fern_C.-4.pdf.
284. Ibid.

285. US Congress, *Departments of State, Justice, and Commerce, the Judiciary, and Related Agencies Appropriations for 1965,* before a subcommittee of the Committee on Appropriations, House of Representatives, 88th Congress (Washington, DC, 1964).

286. Federal Bureau of Investigation, Jones, memorandum to DeLoach, "Harvard Business Review."

287. Federal Bureau of Investigation, J. Edgar Hoover, memorandum to Fern C. Stukenbroeker, January 10, 1964, https://ia803205.us.archive.org/11/items/foia_Stukenbroeker_Fern_C.-4/Stukenbroeker_Fern_C.-4.pdf.

288. Federal Bureau of Investigation, "Performance Rating: Fern C. Stukenbroeker—Rating Period from 4-1-63 to 3-31-64," https://ia803205.us.archive.org/11/items/foia_Stukenbroeker_Fern_C.-4/Stukenbroeker_Fern_C.-4.pdf.

289. Federal Bureau of Investigation, C. R. Davidson, memorandum to Callahan, August 11, 1960 (DeLoach personnel file FOIA release).

290. Ibid.

291. *Congressional Record,* 86th Congress, vol. 106, pt. 12 (1960), https://www.govinfo.gov/app/details/GPO-CRECB-1960-pt12/context.

292. Ibid.

293. Ibid.

294. US Congress, *Counterintelligence and National Security Information,* before a subcommittee of the Committee on Government Operations, House of Representatives, 99th Congress (Washington, DC, 1985).

295. FBI deputy assistant director Thomas E. Duhadway, testimony before the Senate Select Committee on Intelligence, October 30, 1985, https://www.cia.gov/library/readingroom/docs/CIA-RDP87M01007R000100350001-8.pdf. CIA-RDP87M01007R000100350001-8.

296. National Counterintelligence Center, *Annual Report to Congress on Foreign Economic Collection and Industrial Espionage, 1995* (Washington, DC).

297. Ibid.

298. Gregory M. Lamb, "Leaks Flow East—and West; US Industry and High-Tech Spies," *Christian Science Monitor,* December 28, 1982.

299. Duhadway, testimony before the Senate Select Committee on Intelligence.

300. Central Intelligence Agency, *Increasing International Cooperation for Multilateral Enforcement Between COCOM Member Countries* (undated), https://www.cia.gov/library/readingroom/docs/CIA-RDP84B00049R001800040009-5.pdf.

301. Duhadway, testimony before the Senate Select Committee on Intelligence; National Counterintelligence Center, *Annual Report to Congress on Foreign Economic Collection and Industrial Espionage, 1995.*

302. US Congress, *Economic Espionage,* before the Subcommittee on Crime of the Committee on the Judiciary, House of Representatives, 104th Congress (Washington, DC, 1996).

303. US Congress, *FBI Oversight and Authorization Request for Fiscal Year 1989,* before the Subcommittee on Civil and Constitutional Rights of the Committee on the Judiciary, House of Representatives, 100th Congress (Washington, DC, 1988).

304. Lamb, "Leaks Flow East—and West."

305. Central Intelligence Agency, *Increasing International Cooperation.*

306. US Congress, *Counterintelligence and National Security Information.*

307. National Counterintelligence Center, *Annual Report to Congress on Foreign Economic Collection and Industrial Espionage, 1995.*

308. Ibid.

309. US Senate, *Meeting the Espionage Challenge.*

310. Central Intelligence Agency, Robert M. Gambino, director of security, memorandum to the director of central intelligence, "Subject: Newsletter on

Industrial Security; Re: Your Memo on This Same Subject Dated 11 December 1978," December 21, 1978, https://www.cia.gov/library/readingroom/docs/CIA-RDP81-00142R000000090010-9.pdf.

311. Ibid.

312. Central Intelligence Agency, "Speech of John N. McMahon, Deputy Director of Central Intelligence Before American Society of Industrial Security," September 15, 1982, https://www.cia.gov/library/readingroom/docs/CIA-RDP91-00901R000600180005-1.pdf.

313. National Counterintelligence Center, *Annual Report to Congress on Foreign Economic Collection and Espionage, 1995.*

314. General Accounting Office, *Combating Terrorism: FBI's Use of Federal Funds for Counterterrorism-Related Activities (FYs 1995–1998)* (Washington, DC, 1998).

315. US Senate, *Current and Projected National Security Threats to the United States,* 105th Congress.

316. National Counterintelligence Center, *Annual Report to Congress 2000* (Washington, DC), https://fas.org/irp/ops/ci/docs/fy00.htm.

317. US Senate, *Current and Projected National Security Threats to the United States,* 105th Congress.

318. Defense Personnel Security Research Center, *Foreign Intelligence Threat Awareness Programs.*

319. Federation of American Scientists, *Awareness of National Security Issues and Response.*

320. Darren Tromblay and Robert Spelbrink, *Securing U.S. Innovation: The Challenge of Preserving a Competitive Advantage in the Creation of Knowledge* (Lanham: Rowman and Littlefield, 2016).

321. Michael J. Waguespack, Federal Bureau of Investigation, testimony before the House Committee on Government Reform, April 3, 2001, https://archives.fbi.gov/archives/news/testimony/fbis-ansir-program.

322. US Senate, *Current and Projected National Security Threats to the United States,* 105th Congress.

323. Waguespack, testimony before the House Committee on Government Reform; Defense Personnel Security Research Center, *Foreign Intelligence Threat Awareness Programs.*

324. "FBI Spam: Look Out for Terrorists," *Wired,* March 20, 1998, https://www.google.com/amp/s/www.wired.com/1998/03/fbi-spam-look-out-for-terrorists/amp.

325. Bureau of Justice Assistance, *Engaging the Private Sector to Promote Homeland Security: Partnerships* (Washington, DC, 2005), https://www.ncjrs.gov/pdf.files1/bja/210678.pdf.

326. Richard Gid Powers, *G-Men* (Carbondale: Southern Illinois University Press, 1983).

327. Federal Bureau of Investigation, L. B. Nichols, memorandum to Tolson, "The FBI Story," November 6, 1956, https://ia801805.us.archive.org/6/items/foia_Stukenbroeker_Fern_C.-2/Stukenbroeker_Fern_C.-2.pdf.

328. Federal Bureau of Investigation, M. A. Jones, memorandum to DeLoach, "'The FBI Story': Motion Picture Commendation Matter," June 12, 1959 (Jones personnel file FOIA release), https://ia801708.us.archive.org/2/items/foia_Jones_Milton_A._4/Jones_Milton_A._4.pdf.

329. Powers, *G-Men.*

330. Central Intelligence Agency, director of security, memorandum to deputy director of central intelligence, December 23, 1981, https://www.cia.gov/library/readingroom/docs/CIA-RDP87S00869R000200250003-5.pdf.

331. US Senate, *Economic Espionage and Trade Secret Theft: Are Our Laws Adequate for Today's Threats?,* before the Committee on the Judiciary, 113th Congress

(Washington, DC, 2014), https://www.govinfo.gov/content/pkg/CHRG-113shrg96009/pdf./CHRG-113shrg96009.pdf.

332. Ibid.; Elias Groll, "FBI Rolls Out Red Scare Film to Highlight Threat of Economic Espionage," *Foreign Policy,* July 23, 2015, https://foreignpolicy.com/2015/07/23/fbi-rolls-out-red-scare-film-to-highlight-threat-of-economic-espionage.

333. US Senate, *Economic Espionage and Trade Secret Theft.*

334. US Senate, *Threats to the Homeland,* before the Homeland Security and Governmental Affairs Committee, 115th Congress (Washington, DC, 2017), https://www.govinfo.gov/content/pkg/CHRG-115shrg29657/pdf./CHRG-115shrg29657.pdf.

335. Groll, "FBI Rolls Out Red Scare Film."

336. Federal Bureau of Investigation, "Advice for U.S. College Students Abroad: Be Aware of Foreign Intelligence Threat," April 14, 2014, http://www.fbi.gov/news/stories/2014/april/students-abroad-warned-of%20foreign-intelligence-threat.

337. US Senate, *Threats to the Homeland.*

338. "FBI Movie Warns U.S. Students Not to Spy for China," *Time,* April 16, 2014, http://time.com/64530/fbi-movie-game-of-pawns-china.

339. "A Cheesy FBI Video Hopes to Stop U.S. Students from Becoming Chinese Spies," *Washington Post,* April 15, 2014, http://www.washingtonpost.com/blogs/worldviews/wp/2014/04/15/a-cheesy-fbi-video-hopes-to-stop-u-s-students-from-becoming-chinese-spies.

340. "New FBI Film Warns About China's Recruitment of US Officials," October 1, 2020, https://www.rollcall.com/2020/10/01/new-fbi-film-warns-about-chinas-recruitment-of-us-officials.

341. "Security News This Week," *Wired,* October 3, 2020, https://www.wired.com/story/ransomware-fine-grindr-bug-joker-malware-security-news.

342. US Senate, *Counterterrorism,* before the Committee on the Judiciary, 107th Congress (Washington, DC, 2002).

343. National Counterintelligence Executive, *Annual Report to Congress on Foreign Economic Collection and Industrial Espionage, 2004.*

344. US Senate, *Current and Projected National Security Threats to the United States,* before the Select Committee on Intelligence, 109th Congress (Washington, DC, 2005); US Congress, *Science, the Departments of State, Justice, and Commerce, and Related Agencies Appropriations for 2007,* pt. 10, before the Committee on Appropriations, House of Representatives, 109th Congress (Washington, DC, 2006).

345. US Congress, *Science, the Departments of State, Justice, and Commerce, and Related Agencies Appropriations for 2007,* pt. 10.

346. US Senate, *Oversight of the Federal Bureau of Investigation,* before the Committee on the Judiciary, 113th Congress (Washington, DC, 2013); US Senate, *Economic Espionage and Trade Secret Theft.*

347. US Senate, *Economic Espionage and Trade Secret Theft.*

348. US Congress, *Commerce, Justice, Science, and Related Agencies Appropriations for 2014,* pt. 2B, before a subcommittee of the Committee on Appropriations, House of Representatives, 113th Congress (Washington, DC, 2013).

349. 9/11 Review Commission, *The FBI: Protecting the Homeland in the 21stt Century* (Washington, DC, 2015), https://www.fbi.gov/file-repository/final-9-11-review-commission-report-unclassified.pdf./view.

350. US Senate, *Economic Espionage and Trade Secret Theft.*

351. 9/11 Review Commission, *The FBI.*

352. Federal Bureau of Investigation, *Office of Private Sector: Enhancing Engagement Efforts to Stay Ahead of the Threat* (Washington, DC, 2017), https://www.fbi.gov/news/stories/office-of-private-sector.

353. *Federal Register,* August 14, 2018, https://www.govinfo.gov/content/pkg/FR-2018-08-14/pdf./2018-17406.pdf.

354. Federal Bureau of Investigation, "Office of Private Sector: Executive Fact Sheet," n.d.

355. Federal Bureau of Investigation, "Office of Private Sector," February 24, 2020.

356. Ibid.

357. John Brown, assistant director, Counterintelligence Division, Federal Bureau of Investigation, "Securing the U.S. Research Enterprise from China's Talent Recruitment Plans," statement for the record, November 19, 2019.

358. Federal Bureau of Investigation, "Office of Private Sector: Executive Fact Sheet."

359. US Congress, *United States Counterintelligence and Security Concerns, 1986,* report of the House Permanent Select Committee on Intelligence (Washington, DC, 1987), https://www.cia.gov/library/readingroom/docs/CIA-RDP91B00390R000200160014-6.pdf.

360. Commission to Review Department of Defense Security Policies and Practices, *Keeping the Nation's Secrets: A Report to the Secretary of Defense* (Washington, DC, 1985), https://www.cia.gov/library/readingroom/docs/CIA-RDP96B01172R000100090004-9.pdf.

361. Ibid.

362. US Senate, *Economic Espionage,* before the Select Committee on Intelligence and the Committee on the Judiciary, 104th Congress (Washington, DC, 1996).

363. Defense Personnel Security Research Center, *Foreign Intelligence Threat Awareness Programs: A Review,* prepared for the National Counterintelligence Policy Board (1998).

364. William J. Casey, opening remarks before Senate Select Committee on Intelligence, October 30, 1985, https://www.cia.gov/library/readingroom/docs/CIA-RDP86M00191R000300560002-0.pdf.

365. National Counterintelligence Center, *Annual Report to Congress on Foreign Economic Collection and Industrial Espionage, 1995.*

366. Commission to Review Department of Defense Security Policies and Practices, *Keeping the Nation's Secrets.*

367. John Hamre, "Department of Defense Reform Initiative Directive #2: New Defense Security Service," November 25, 1997, https://archive.defense.gov/dodreform/drids/drid2.htm.

368. Government Accountability Office, *Industrial Security: DOD Cannot Ensure Its Oversight of Contractors Under Foreign Influence Is Sufficient* (Washington, DC, 2005), https://www.gao.gov/assets/250/247113.html.

369. National Counterintelligence Executive, *US Industry and Advanced Technology: Annual Report to Congress on Foreign Economic Collection and Industrial Espionage, 2004* (Washington, DC, 2005).

370. US Senate, "Nomination of William R. Evanina to Be the Director of the National Counterintelligence and Security Center," 115th Congress (Washington, DC, 2018), https://www.congress.gov/115/chrg/CHRG-115shrg30120/CHRG-115shrg30120.pdf.

371. "DSS Marks 40th Anniversary" *DSS Access* (Fall 2012), https://www.dcsa.mil/Portals/69/documents/about/err/DSS_ACCESS_Vol_1_Issue_3.pdf.

372. Gregory F. Treverton, *The Next Steps in Reshaping Intelligence* (Santa Monica: RAND, 2005).

373. US Congress, *H.R. 6588: National Intelligence Act of 1980,* before the Permanent Select Committee on Intelligence, House of Representatives, 96th Congress (1980).

374. Treverton, *The Next Steps in Reshaping Intelligence.*

375. Cyber Threat Intelligence Integration Center, https://www.dni.gov/index.php/ctiic-home; Office of the Director of National Intelligence, https://www.dni.gov/index.php.

376. US Senate, *S. 2726 to Improve U.S. Counterintelligence Measures,* before the Select Committee on Intelligence, 101st Congress (Washington, DC, 1990).

377. Defense Personnel Security Research Center, *Foreign Intelligence Threat Awareness Programs.*

378. White House, "Statement by the Press Secretary on United States Counterintelligence Effectiveness," May 3, 1994, https://www.govinfo.gov/content/pkg/PPP-1994-book1/pdf./PPP-1994-book1-doc-pg834.pdf.

379. Defense Personnel Security Research Center, *Foreign Intelligence Threat Awareness Programs.*

380. Richard Gid Powers, *Broken: The Troubled Past and Uncertain Future of the FBI* (New York: Free Press, 2004).

381. "The Presidential Decision Directive on CI-21: Counterintelligence for the 21st Century," fact sheet, January 5, 2000, https://www.hsdl.org/?view&did=447430.

382. Ibid.

383. "Time-Line of CI Milestones," https://www.dni.gov/index.php/ncsc-who-we-are/ncsc-history/ncsc-time-line-of-ci-milestones.

384. "The Presidential Decision Directive on CI-21"; "§ 402b: National Counterintelligence Executive," https://www.govinfo.gov/content/pkg/USCODE-2002-title50/pdf./USCODE-2002-title50-chap15-subchapI-sec402c.pdf.

385. "Time-Line of CI Milestones."

386. Defense Personnel Security Research Center, *Foreign Intelligence Threat Awareness Programs.*

387. National Counterintelligence Center, *Annual Report to Congress on Foreign Economic Collection and Industrial Espionage, 1995.*

388. US Senate, *Current and Projected National Security Threats to the United States,* before the Select Committee on Intelligence, 108th Congress (Washington, DC, 2003), https://www.govinfo.gov/content/pkg/CHRG-108shrg89797/pdf./CHRG-108shrg89797.pdf.

389. Defense Personnel Security Research Center, *Foreign Intelligence Threat Awareness Programs.*

390. General Accounting Office, *Critical Infrastructure Protection: Federal Efforts Require a More Coordinated and Comprehensive Approach for Protecting Information Systems* (Washington, DC, 2002), https://www.gao.gov/new.items/d02474.pdf; Executive Order 13231, "Critical Infrastructure Protection in the Information Age," October 16, 2001, https://www.hsdl.org/?view&did=620.

391. "Additional Questions for Mr. William R. Evanina."

392. "§ 3383: National Counterintelligence and Security Center," https://www.govinfo.gov/content/pkg/USCODE-2017-title50/pdf./USCODE-2017-title50-chap45-subchapV-sec3383.pdf.

393. Congressional Research Service, *The National Counterintelligence and Security Center (NCSC): An Overview* (Washington, DC, 2018), https://crsreports.congress.gov/product/pdf./IF/IF11006.

394. "Additional Questions for Mr. William R. Evanina."

395. Ibid.

396. National Counterintelligence and Security Center, *National CI Strategy 2020–2022* (Washington, DC, 2020), https://www.dni.gov/files/NCSC/documents/features/20200205-National_CI_Strategy_2020_2022.pdf.

4

Countering Proliferation and Terrorism

COUNTERINTELLIGENCE (AND ITS SUBDISCIPLINE OF COUNTERESPIONAGE) focuses on preventing an adversary or competitor from gaining access to an informational advantage. However, counterproliferation and counterterrorism, although they are equally significant national security concerns, differ slightly from counterintelligence in that they emphasize preventing items from reaching end-users (rather than preventing end-users from getting to sensitive information). The concept of counterproliferation sits at the nexus of counterintelligence and counterterrorism. It has elements of counterintelligence—in the transfer of knowledge—but the illicit activity may have a legal beginning. Counterproliferation also crosses over into the field of counterterrorism, since much of the restricted information and technology is dual-use in nature. This means that it can either have a civilian use or become part of something that goes boom in the night (or any other time of day). Foreign actors of a both state and nonstate variety—especially those that cannot confront the United States through symmetrical (military-to-military) operations—may opt to use dual-use items for asymmetrical attacks.

Counterproliferation

Like all facets of intelligence work, counterproliferation is a complex problem. Individual agencies—notably the Federal Bureau of Investigation (FBI) and Department of Homeland Security (DHS) (including Customs, a precursor agency)—have handled much of the work with a nexus to the United States. Disruption of proliferators has included both traditional investigations as well as the occasional covert action. Proliferators have used varied methods that have spanned countries and continents, a complicated reality that has necessitated interagency collaboration, as well as engagement with the private sector, which proliferators target and exploit.

Historical Understanding of the Threat

US security agencies have a long history of combating the threat from entities seeking to put dual-use items as well as weapons into the hands of nefarious actors. As early as 1924, J. Edgar Hoover—who had just become director of the Bureau of Investigation (the term "Federal" would not be appended to the agency's name until approximately a decade later)—advised Congress that during the past year, the Bureau had investigated shipments of arms and ammunition on which the US government had placed export restrictions.[1]

The evolution of counterproliferation reflects changing geopolitical realities. Soviet penetration of the US atomic program has been rehashed in numerous books and other publications. Once the Soviet Union's wartime alliance with the United States dissolved into the Cold War, Moscow continued to pursue restricted items without compunction. In 1947, US authorities removed radiation detection equipment, destined for the Soviet Union sans an export license, from the SS *Murmansk,* which was docked in New York.[2] Nearly two decades later, the Soviets used their presence at the United Nations in another attempt to illicitly acquire technology. In 1965, Vadim Anatolevich Isakov, a Soviet employee of the United Nations International Children's Fund (UNICEF), contacted the operator of a surplus equipment business in Paterson, New Jersey, ostensibly with an inquiry about purchasing items for UNICEF. Although the equipment was not classified, much of it, including certain missile-tracking components, could not legally be shipped to the Soviet Union. In 1966, the State Department requested that UNICEF terminate Isakov's employment.[3]

Amtorg—the Soviet trading company that served as a platform for Soviet intelligence—was also a participant in efforts to illicitly acquire technology from the United States. According to a 1947 FBI assessment, when a given manufacturer would refuse to sell a product for delivery to the Soviet Union, due to export controls, Amtorg would attempt to purchase the item through a "middleman" for delivery to Amtorg, which would in turn arrange for shipment of the item to the Soviet Union.[4] Amtorg provides an example of how a foreign government can use a "front company" to obtain controlled information or technology. According to the US National Counterintelligence Executive, a front company is a firm—whether an actual subsidiary of a foreign entity or a complete sham—that acquires technology on behalf of a foreign power.[5]

The Soviets also used their satellite countries to acquire restricted technology. For instance, in August 1956, Milos Prochazka, an official at the embassy of Czechoslovakia, provided an American contact with specifications for the components of two steel mills that Czechoslovakia wanted to acquire from the United States. To effect this transfer, the mills would be shipped to a Czechoslovakian agent in a Western country, from which they would be shipped to Czechoslovakia.[6] According to a 1981 Central Intelligence Agency

(CIA) assessment, Poland and Hungary functioned as significant Eastern European surrogates for Soviet technology acquisition.[7] In late 1983, the US Department of Justice and US Customs Service, in conjunction with West German authorities, seized a sophisticated US-built computer that could do structural analysis—and also monitor troop and weapon movements—before it was to be shipped to the Soviet Union from Hamburg via Sweden.[8]

Less well known are China's efforts, in the years following the establishment of a communist regime in 1949, to make illicit acquisitions of US technologies. For instance, a 1954 assessment by the FBI, titled *Chinese Communist Intelligence Activities in the United States,* noted the intent of the Pacific Trading Corporation to provide steel to China.[9] In 1962, the Department of Commerce noted the successful disruption of an attempt to ship x-ray tubes to China; the shipment was attempted by a firm that had already been denied export privileges for the same offense.[10]

Discussion of counterproliferation usually focuses on the outward movement of restricted items to end-recipients. It is worth noting that smuggling goes both ways. In 1951, the CIA published a National Intelligence Estimate assessing that the Soviet Union might use part of its atomic stockpile to conduct a clandestine attack against the United States through means including smuggling the device into the United States; bringing a weapon piecemeal into the United States, where Soviet operatives would then assemble it; and even bringing a weapon in among the household effects of Soviet diplomatic representatives.[11] Contemporaneous with these early assessments, the FBI was following up on a lead suggesting that a foreign power had smuggled an atomic device into the United States. An investigation, which began in 1951, predicated on information from São Paulo, Brazil, attempted to determine whether the Soviets had placed an atom bomb in the New York consulate of an unknown country, to be detonated at a time that Moscow deemed expedient.[12] As part of this intelligence effort, the FBI tried to determine through US Customs whether any suspiciously heavy packages, under diplomatic seal, entered the United States, and also sought information from the Atomic Energy Commission about the physical characteristics of a disassembled atomic weapon.[13] In 1952, the FBI placed the case in a closed status but also set forth instructions to periodically solicit information from sources that had knowledge of Soviet and Soviet satellite consulates and forward the information to FBI headquarters under the caption "Smuggling of Atomic Bombs and Parts Thereof and Other Weapons of Mass Destruction into the United States."[14] The FBI continued to track this issue as indicated by a 1958 memorandum stating that communications to field offices about this matter should be furnished to the front office of the Domestic Intelligence Division for approval.[15]

As the Cold War progressed, the Agency's concerns encompassed China after that country acquired a nuclear capability. Even though China lacked diplomatic establishments in the United States, Mexico, and Canada, and therefore could not use diplomatic pouches for the clandestine introduction of nuclear weapon components, the CIA suggested that China could introduce agents under the guise of bona fide immigrants.[16] After 9/11, concern about smuggling of a weapon of mass destruction (WMD) device into the United States rose again. This time, the fear was that a foreign entity could smuggle a WMD into the United States via a cargo container.[17]

The Federal Bureau of Investigation's Response

The development of new technologies—and the ability to weaponize them—drove the evolution of the US approach to combating the movement of restricted information. In late 1953, then-president Dwight D. Eisenhower made the FBI responsible for investigating all violations of the Atomic Energy Act—including the illegal export of fissionable material.[18] The FBI had already established an Atomic Energy Unit as of the mid-1940s.[19] However, counterproliferation is a multifaceted problem and several decades later, in 2006, the FBI's National Security Branch—in response to Mueller's direction in 2005—established the WMD Division.[20] Creation of this division represented an effort to align and consolidate the Bureau's various counterproliferation initiatives, which multiple divisions had managed.[21] The persistent problem of counterproliferation belonging to multiple disciplines prompted the FBI to refine its approach to the problem when, in 2011, the Bureau combined the subject-matter expertise in the WMD Division with operational activities of the Counterintelligence Division, and with the analytical capabilities in the Bureau's Directorate of Intelligence to create its Counterproliferation Center.[22] The Counterproliferation Center, as of 2013, managed all of the FBI's counterproliferation investigations.[23]

Department of Homeland Security's Counterproliferation Activities

The Department of Homeland Security and its predecessors have also been key players in counterproliferation efforts. Customs, which would become part of the DHS's Immigration and Customs Enforcement (ICE), had until 1973 done little investigation of illicit exports, but during the middle to late 1970s began to place additional emphasis on export control enforcement. In 1981, the Customs Service initiated a program of spot-checking advanced technology exports to ensure that they had proper licenses.[24] This program, known as Operation Exodus, had three primary objectives. The first was to stem the illicit flow of critical technology—defined as items that would be technological advances for the Soviet bloc. Additionally, Operation Exodus was meant to disrupt the flow of high technology—defined by items that a country had the capability to manufacture but instead chose to acquire from the United States in order to reduce costs and improve quality—to the Soviet bloc.[25] (This is an interesting revelation—as another Reagan project, the Strategic Defense Initiative, better known as "Star Wars," also drove the Soviet Union to divert resources in an effort to keep pace with the supposed antimissile program of the United States.)[26] Finally, Operation Exodus focused on disrupting the shipment of commodities being exported in violation of sanctions and embargoes.[27]

Operation Exodus evolved throughout its years of operation. The program began as a reactive initiative and evolved into an anticipatory effort that developed intelligence, selective cargo examinations, and investigations.[28] Its longevity illustrated how the threat from proliferators shifted and how the United States followed suit. The program's primary purpose had been to disrupt the Soviets. However, after the dissolution of the Soviet Union, new threats filled the vacuum. In 2002, the director of investigative programs for the US Customs Service noted that the threat was from proliferators such as China, rogue states, international terrorists, and transnational criminal organizations.[29] The actors were not the only change. Technology had shifted dramatically over time and whereas early Exodus wins had included the seizure of computer terminals in Boston, microwave equipment in New York, and microprocessors in Germany, the post-9/11 iteration of Operation Exodus included materials and technologies for weapons of mass destruction.[30]

Covert Action

Traditional investigations are the bread and butter of the FBI and ICE's Homeland Security Investigations, but there is another aspect—covert action—to disrupting foreign entities engaged in proliferation. At times, this has been done in furtherance of a traditional investigative end. On other occasions, agencies have specifically sought an unattributable outcome.

Role in Traditional Investigations

Customs—prior to its absorption into the DHS—conducted several operations that involved sending unexpected merchandise to end-users. In 1982, after determining that a multispectral scanner had been smuggled out of the United States and was going to be routed to East Germany via Mexico and Switzerland, Customs took action. The Customs attaché in Mexico City arranged for the scanner—which was designed for tracking the movement of troops and supplies—to be loaded onto a flight destined for Zurich, with a convenient stopover in Houston. When the flight landed in Texas, the illicit cargo was off-loaded and seized by Customs agents, who swapped in a far less dangerous export: sandbags.[31] Several years later, in 1985, Customs officials, after identifying an attempt to send aircraft replacement parts to Iran, replaced some of the parts with "Love Me Tender Chunks" dog food and then tracked the crates until they reached a transshipment point in Great Britain.[32]

Disinformation and Disruption

There are indications that the FBI has used counterproliferation activities as a conduit for providing disinformation to foreign threats trying to pilfer US technological information. In 1960, the Bureau initiated its Deception Program for Antimissile Defense, which identified sources whom the Bureau could use to pass information about antimissile capabilities on behalf of the Joint Chiefs of Staff.[33] In 1979, then–FBI director William Webster advised Congress that in the field of counterintelligence the Bureau did pursue "legitimate efforts to confuse foreign activities, and to make them consume substantial amounts of their available time wondering about the effectiveness of their operation."[34] This was consistent with an incident that began during the previous year. In 1978, Canadian authorities advised the FBI about a disgruntled US nuclear plant worker who had visited the Soviet embassy in Ottawa. The Bureau doubled-agented this worker and used him to ply the Soviets with what the United States wanted the Soviets to know.[35]

Not quite a decade later the FBI again alluded to its role in deception. Webster, speaking to Congress in 1987, noted that the Bureau used "disinformation" when appropriate to confuse hostile intelligence activities.[36] His statement may have hinted at an intelligence operation in which the FBI had played a role several years prior. In 1981, a KGB officer, Vladimir Vetrov—working in the field of espionage against scientific and technical targets—had volunteered his services to French intelligence and provided approximately 4,000 pages of material regarding Soviet collection against scientific and technical targets.[37] This included information about the goals, achievements, and unfilled objectives of the Soviet Union's collection against Western technology.[38]

The French proved willing to share this intelligence windfall. In mid-1981, French president François Mitterrand provided then–US president Ronald Reagan with what were known as the Farewell (the source's codename) papers at an Ottawa summit meeting.[39] The Farewell material was, as Gus Weiss wrote, the KGB's "shopping list."[40] Based on this, the US intelligence community began making these items—albeit with a few modifications—available to KGB collectors. The Soviets found themselves installing flawed turbines into a gas pipeline and following defective plans that resulted in the disruption of output by chemical plants and a tractor factory. Although the CIA was responsible for the overseas aspect of this campaign, the FBI handled the domestic requirements.[41]

Proliferators

Obfuscating the destination of technology is key to proliferators' success. According to a CIA assessment, diversion of technology to an unauthorized recipient can be accomplished through the illegal diversion of technology from legitimate trade channels to proscribed destinations. This is where front companies fit into proliferation. As discussed by the CIA, foreign governments may use foreign firms willing to engage in profitable impropriety, agents in place within foreign firms, foreign subsidiaries of US firms, country-owned but locally chartered firms (demonstrated by the activities of POLAMCO), and foreign purchasing agents (à la Amtorg).[42] This is a long-recognized problem. A 1952 CIA document noted that the United States required assurance from Western European countries against transshipment or shipment of identical items to the Soviet bloc (including China).[43] According to a 1962 report, transshipment cases (movement of an item to an end-user via a third party) accounted for 70 percent of cases handled by the Office of Export Control (an entity at the US Department of Commerce).[44]

Countries have created entire apparatuses to engage in trade diversion. The Soviet Union established a program, managed by the Ministry of Foreign Trade and its intelligence services (the KGB and GRU), to engage in this activity. Through this program, the Soviet Union pursued microelectronics and fabrication equipment and computers.[45] Soviet satellite countries also made use of transshipment. For instance, East Germany's Ministry for State Security established front companies in Vienna, Austria, for the sole purpose of illicitly acquiring advanced microelectronics.[46]

In the post-Soviet era, the United States continues to be a target for proliferators attempting to acquire restricted technology. According to a 2007 US government assessment, foreign countries increasingly used front companies or middlemen, either in the United States or third countries, to transfer trade secrets and technologies. Canada and the United Kingdom both provide excellent venues for front companies due to the light US export controls on these countries. Free trade ports such as Singapore and the United Arab Emirates are also useful locations, since they facilitate the transfer of goods with little concern for US trade restrictions. The Department of Justice provided some insight into the modern-day instigators of proliferation when, in 2007, it announced that the majority of recent US criminal export prosecutions had involved either China or Iran.[47]

Even after an item is delivered to its end-user, it may still be the object of a proliferation scheme. As a CIA assessment noted, in-place diversion occurs when legally acquired technologies are put to illegal end-uses or are used by unauthorized end-uses.[48] The US government has taken a number of measures to ensure that private sector exports end up with the right people in the right places. According to 1952 CIA document, end-use checks are required in many areas to ascertain that the goods can and will be used as specified in a license application.[49] More recently, in 1990, the US government established the Blue Lantern program.[50] The US Department of State manages this program, which monitors the end-use of defense articles, technical data, services, and brokering activities exported through commercial channels.[51] The Department of Commerce's Bureau of Industry and Security also selectively conducts end-use checks on certain dual-use and munitions exports in order to monitor license condition compliance; monitor compliance of nonlicensed transactions; confirm the end-user; and determine if the company in question is a reliable end-user.[52]

Verifying legitimate end-usage is difficult enough with physical items and even more problematic when it comes to the application of knowledge. A number of countries—especially China—have long histories of seeking out expertise rather than a specific widget. Essentially, the person with the know-how—rather than a piece of technology—and where they apply that knowledge constitute the restricted item. China has zeroed in on this reality and has attempted to attract foreign scientists and researchers to work within the country. In 2006, for instance, China initiated Project 111, with the objective of recruiting 1,000 foreign experts in strategic sectors from the world's top 100 universities and research institutes. In 2008, China launched its Thousand Talents program, which brought foreign expertise to China's scientific laboratories, companies, and research centers. These programs attract participants with research and startup funding.[53] It would be practically impossible to conduct an end-use verification on an individual working in China.

The potential for movement of knowledge—rather than technology—is made even more problematic for the United States due to China's hunting and recruiting of talent. Through its State Administration of Foreign Experts Affairs (SAFEA), China certifies foreign experts to work in the Chinese mainland. It directly manages the China Association for the International Exchange of Personnel (CAIEP), which, among other activities, recruits US scientists, academics, engineers, and other experts to work in China. CAEIP functions as the overseas arm of SAFEA. Its employees in the United States recruit experts from universities, research institutes, and other entities capable of assisting with the Chinese government's technological and economic development.[54] In 2019 the FBI arrested Zhongsan Liu, who operated CAIEP's New York office and who allegedly attempted to obtain an academic visa for an individual who would come to the United States not to conduct research but rather to engage in full-time talent recruitment.[55]

Complicated Problems, Complex Solutions

Effective counterproliferation work requires a far more sprawling intelligence apparatus—one often demanding collaboration across intelligence and security services as well as with the private sector—since it is not only about disrupting the acquisition of material but also about preventing the material from reaching a restricted destination. It is therefore not surprising that the FBI's atomic energy unit also included the Bureau's liaison functions.[56] Several decades later, in 1980, then–FBI director William Webster advised the director of central intelligence that, at the request of the National Security Council, the FBI's Intelligence Division was working with the export control community on ways in which to improve enforcement. According to Webster, the FBI was an active participant in an ad hoc interdepartmental working group that reported its finding to the US National Security Council's ad hoc Technology Transfer Group.[57] In 1996 the FBI created an interagency Counterterrorism Center, under the auspices of the National Security Division.[58] This new center had WMD counterproliferation responsibilities and included representatives from eighteen federal agencies including the CIA and the Defense Intelligence Agency.[59]

After 9/11, the Bureau continued to look for ways to work across agencies on the unwieldy problem of counterproliferation. In 2015, then-director James Comey advised Congress that the FBI had established a Hybrid Threat Center within the Counterintelligence Division, which would support the Department of Commerce's "Entity List" investigations. According to Comey, the Bureau brought together "a lot of elements of the intelligence community and other parts of the US government" to think about the threat that corporations that allowed themselves to be co-opted and act as agents of foreign powers, as well as foreign powers attempting to penetrate the supply chain, posed to the United States. Supposedly, according to Comey, the Hybrid Threat Center had been "well-received" by the intelligence community.[60]

Like the FBI, Customs—through Operation Exodus and its broader efforts—worked in conjunction with other federal agencies. In a 1982 letter to the commissioner of Customs, the director of central intelligence noted that he "had been impressed with the effectiveness of Operation Exodus to date and [he would] continue to encourage the Intelligence Community to further develop a close working relationship" with it.[61]

The creation of the DHS invested ICE with Customs's investigative responsibilities. In 2010, ICE created a Homeland Security Investigations Directorate.[62] One can easily come to the conclusion that this directorate suffers from bureaucratic envy (and overcompensation) when it comes to the FBI, based on the directorate's dramatic claim to be the "nation's leading law enforcement agency" for investigations of US export control, laws under which it claims to have the "broadest" investigative and enforcement authorities among federal law enforcement agencies.[63] It has referred to itself as "an elite law enforcement agency" at least as far as counterproliferation is concerned.[64]

The Homeland Security Investigations Directorate is responsible for several significant counterproliferation entities. Its Counterproliferation

Investigation Centers are located throughout the United States and are supposed to function as regional hubs for manpower, expertise, deconfliction, and undercover operational expertise. These centers are sited based on multiple criteria including their proximity to Department of Defense (DoD) and other US government agencies involved with export enforcement, threat assessments in the areas of responsibility, and significant cases. This arrangement improves the Homeland Security Investigations Directorate's domestic ability to combat illegal exports and illicit procurement networks.[65]

The directorate is also home to the Export Enforcement Coordination Center,[66] a Homeland Security–led interagency body. It serves as the "primary forum within the federal government for executive departments and agencies to coordinate and enhance their export enforcement efforts."[67] According to the executive order that established the center in 2010, its director would be a senior DHS official, and its deputy directors would be from the Department of Commerce and the Department of Justice. It would also include an intelligence community liaison. In addition to serving as a platform for information-sharing and coordination across federal agencies, the Export Enforcement Coordination Center was to be responsible for government-wide statistical tracking capabilities for US criminal and administrative export control enforcement.[68]

Assistance from Industry

The integral involvement of the private sector in activities that can be subverted by proliferators means that government cannot effectively engage in counterproliferation without their assistance. Customs, historically, found ways to develop coverage within the processes that could be exploited for proliferation. During a 1982 congressional hearing, Customs commissioner William von Raab explained that Customs had developed sources among shippers, brokers, and freight forwarders.[69] In addition to sources—or in an endeavor to establish groundwork for developing specific sources—the US national security community has conducted extensive outreach to educate the private sector about the proliferation threat.

Department of Homeland Security

The Department of Homeland Security inherited—and has continued to implement—information-sharing functions. As part of Operation Exodus, the Customs Service (the investigative functions of which became the basis of ICE's counterproliferation work) had engaged in outreach to increase industry awareness of US export controls and to elicit cooperation from the private sector with preventing illegal exports.[70] In a 1982 letter to the commissioner of Customs, then–director of central intelligence William Casey indicated that he believed this outreach "should prove helpful in further alerting the US business community to the administration's concern over the loss of US technology to the Soviet Bloc and other hostile countries."[71]

The merging of the Customs Service into the DHS moved the outreach aspect of Operation Exodus to the newly created ICE. The DNA of Operation Exodus was clearly evident in ICE's Project Shield America. Under this project, ICE agents enlisted cooperation and support from US companies in order to identify suspect orders prior to the sale and exportation of technology.[72] ICE is clearly quite self-satisfied with its work in this field. According to a high-level ICE official, "one of the most effective tools that [ICE uses] is . . . Project Shield America."[73] However, external observers have pointed to a history of inefficacy. In 2004, the DHS's inspector general assessed that ICE's outreach program could be improved to ensure that presentations covered deemed export requirements.[74] Furthermore, in 2011, Congress received testimony from a member of the export control community that DHS needed to "refine and build upon Project Shield America to better inform private industry of export control issues and to more effectively engage the private sector as a full partner."[75]

The DHS has also stumbled over the FBI in the latter's implementation of its own responsibilities vis-à-vis the private sector. According to the 9/11 Review Commission, as of 2015 there was significant overlap between the FBI and the DHS in engagement of critical infrastructure entities. Private sector elements advised the commission that the lack of coordination between the DHS and the FBI was apparent. The commission painted a picture of "FBI agents frustrated to find DHS personnel on their way out as they [were] headed in to meet with an industry partner."[76]

Department of Commerce

The Department of Commerce, especially its Bureau of Industry and Security (BIS), has also engaged industry in furtherance of counterproliferation. Within the BIS, the Office of National Security and Technology Transfer Controls, which governs the export of dual-use technologies (think satellites and semiconductors), works closely with industry to identify items and technologies that warrant enhanced control as well as to identify when flexibility or licensing policy and reduced controls are warranted. Additionally, the BIS conducts outreach to companies that are affected by export controls in order to increase those companies' awareness of export administration regulations and to enhance compliance with those regulations.[77] For instance, in 2002 the BIS's Office of Strategic Trade and Foreign Policy Controls initiated a deemed export outreach program to a variety of entities involved with technologies and equipment that were subject to deemed export controls.[78]

Counterterrorism

The United States is no stranger to attacks by foreign actors. On September 11, 2001, al-Qaeda hijackers struck at symbols of US financial and military power. However, foreign actors—state and nonstate—have long planned to attack more prosaic, but high-impact, critical infrastructure targets. Although such attacks have fortunately been limited, their impact—on energy, water, transportation, or other sectors that keep society functioning—could extend much further than even the September 11 attacks.

Adversaries' Collection Against Critical Infrastructure Targets

During the Cold War, the Soviet Union conducted reconnaissance against multiple critical infrastructure targets. As early as 1960, the FBI noted that Soviet road trips regularly included drives around the perimeters of industrial facilities.[79] Sites of interest had previously included chemical factories and an ordnance plant.[80] In 1961, Soviets had made more than a dozen trips throughout the United States, which included visits to sites of strategic significance including industrial facilities.[81] According to then–FBI director J. Edgar Hoover, Soviet trips included close study of these locations.[82] In 1964, the FBI divulged even more information about these travels. Soviet military intelligence, the GRU, divided the United States into geographical regions. A GRU officer was responsible for each region and, in furtherance of developing Soviet awareness, visited this territory in order to familiarize the officer with locations including industrial facilities.[83] Soviets compiled information gleaned through their roaming into detailed, illustrated reports.[84]

From the early days of the Cold War it was apparent that the Soviets considered sabotage of such facilities to be a tactic on the table. The Communist Party of the United States—which took its direction and funding from the government of the Soviet Union—stood by to do Moscow's bidding. In 1949, one US communist advised that it was the duty of the party to make the United States insecure in the event that hostilities erupted between the United States and the Soviet Union. Two years later, another US communist made the party's intentions even more explicit, offering an explanation that the party was attempting to place trusted members in key positions within machine, tool, and die shops as potential saboteurs.[85] In 1956, Hoover advised Congress that party members sought to secure employment within vital industries in order to be available for sabotage and subversion during a national crisis.[86]

Efforts to place sleeper saboteurs provides context for efforts by the Communist Party of the United States in the labor field. In 1953, the FBI assessed that the party was pushing a policy of "colonization" within selected basic industries.[87] For instance, in New York, the party had established a goal of having 65 percent of its membership working within these fields.[88] However, this policy had limitations, since, according to the Bureau "many, of [the party's members were] simply not qualified by training or experiences for employment" in industrial fields where the party was attempting to make inroads. The party instructed its members who did obtain employment with basic industries to cut all ties with the party and to go "underground."[89] In 1954, the FBI provided Congress one example of this drive to control labor within US industry, stating that the Communist Party of the United States had designated the automobile industry as a prime target because it was one of the most vital for national defense production.[90]

As the Cold War progressed, the Old Left (Communist Party of the United States) and the New Left competed to infiltrate US industry. The Communist Party of the United States, as of 1969, was establishing a youth organization with an emphasis on membership within industry.[91] In 1970 it launched its trade unionist National Rank and File Conference. Although the party dominated this organization, it made strenuous efforts to hide its role. This conference spawned the National Coordinating Committee for Trade Union Action and Advocacy, which the party planned to use in regaining its influence within unions.[92] In 1976, the FBI advised Congress that the Communist Party of the United States continued to pursue influence within the trade union movement, distributing its publications at industrial facilities and directing its members to pursue leadership roles within organized labor.[93] Contemporaneously, the Progressive Labor Party, which aligned itself with the Chinese—rather than Soviet—regime, directed its efforts at exploiting employee problems and working conditions within US industry.[94]

The Soviets did not limit their collection to industry itself but to the critical infrastructure integral to the operation of the US economy. In 1961, the FBI advised Congress that the Soviet Union and its satellites were collecting information regarding transportation and communications systems.[95] One example of such collection occurred more than a decade earlier, in 1946, when a group of ten public utility engineers, all of whom were specialists in municipal planning, visited US cities including New York, Philadelphia, Washington, DC, Chicago, San Francisco, and Los Angeles. During these visits the engineers—who were experts and thus undoubtedly understood what they were examining—studied, photographed, and took copious notes about gas works, water supply infrastructure, and sewage systems. In some instances, the engineers even obtained blueprints of these installations.[96]

According to the FBI, such visits continued under the auspices of the East-West Exchange Program. These exchanges originated with the Geneva Summit Conference of mid-1955 and were formally established in 1956.[97] The purpose was, in part, "to increase the knowledge of the Soviet and satellite people as to the outer world so that their judgements will be based upon fact and not upon Communist fiction."[98] However, by 1959, the FBI noted that the Soviets' "fact"-gathering might be doing more harm than good. A Bureau memo from that year assessed that the "Soviets [were] obviously using the East West cultural exchange programs as a further means of advancing their own objectives" and noted that the individuals whom the Soviet Union had permitted to visit the United States included construction engineers, mining engineers, industrial engineers, railroad engineers, and mechanical and metallurgical engineers.[99] A Bureau assessment noted that "Soviet intelligence regards the visits of East-West Exchange Agreement participants as opportunities to promote intelligence activity."[100] Not surprisingly, the FBI had kept an eye on these visits since 1955, when it initiated its East-West Exchange Program, which was directed at providing "internal safeguards regarding the admission of Soviet and satellite nationals."[101]

Foreign powers have also collected against—and apparently targeted—US energy infrastructure. In 1946 and 1947, Russian engineers visited Tennessee Valley Authority installations at Knoxville and Chattanooga in order to learn about carrier-current communications and protection equipment for power transmission lines. During these visits, the Soviets queried Tennessee Valley Authority personnel about the maintenance and operation of generating plants.[102] More than a decade later, the FBI noted that during reconnaissance trips, Soviet officials had visited, photographed, and made extensive notes about dams and power plants.[103]After the fall of the Soviet Union, the world learned, thanks to KGB archivist Vasili Mitrokhin, what the Soviets had been planning. The KGB had been assessing US targets, including the Flathead Dam in Montana, which it believed to generate the largest power supply system in the world. If it deemed sabotage necessary, the KGB would bring down a series of pylons on a steep mountain slope approximately 3 kilometers downstream from the dam. Furthermore, the KGB had developed plans to disrupt the power supply to the state of New York.[104] It is worth noting, briefly, that the Russians have apparently continued their Soviet predecessors' targeting of the US energy supply. In March 2018, the FBI and the Department of Homeland Security issued a joint alert warning that Russian government entities had targeted multiple critical infrastructure sectors—including energy—since 2016.[105]

Nonstate actors both domestic and foreign in origin have also targeted US critical infrastructure. Early on May 28, 1961, explosions destroyed AT&T microwave relay stations at Wendover, Nevada, and Cedar Mountain, Utah, as well as an underground cable relay station at Knowles, Utah. Because this was a possible act of sabotage, the FBI took responsibility for an investigation that had triggered a nationwide alert out of fear that these incidents presaged additional attacks.[106] The attacks were the work of two individuals—Bernard Jerome Brous and Dale Chris Jensen—who styled themselves as the "American Republican Army." The two fled to Mexico, where they were seized from aboard a 47-foot two-masted sailing vessel that was stocked with machine guns, hand grenades, pistols, and ammunition.[107]

Less than a decade later, another domestic nonstate actor targeted the country's electrical infrastructure. In late January 1969, Cameron David Bishop, an antiwar activist, dynamited four high-powered transmission towers in and around Denver, Colorado, to protest US involvement in Vietnam.[108] Although he was convicted in 1975, the court overturned his case on appeal because he had not received proper notice that his conduct was forbidden by the statute under which he was prosecuted.[109]

Antigovernment types have continued to target critical infrastructure. In 1999, the FBI's Joint Terrorism Task Force in Sacramento arrested several individuals, associated with an antigovernment group, who planned to attack a propane storage facility. When arrested, two of the group's members were in possession of a detonation cord, blasting caps, grenade hulls, and numerous weapons.[110]

A groundswell of extremism does not augur well for any expectation that threats to critical infrastructure from fringe groups will abate. In 2021, the DHS released its first Homeland Threat Assessment. Among its findings was that "racially and ethnically motivated violent extremists—specifically white supremacist extremists—will remain the most persistent and lethal threat in the Homeland."[111] A number of racially motivated extremists have already engaged in acts of violence. The radical left—which lionizes symbols of violence such as Angela Davis and George Jackson—is also culpable in acts of violence. For instance, the Center for Strategic and International Studies discussed how approximately half of what it described as "far left" terrorism instances had used explosives and incendiaries.[112] To quote the Tony-winning musical *Urinetown,* "no one is innocent" and it seems inevitable that at some point an extremist element will decide to attack an element of critical infrastructure with the hope that the resulting disruption will derail what it perceives to be an unacceptable status quo.

Foreign terrorist groups—including homegrown violent extremists who opt in to such groups' ideologies—have also persistently targeted critical infrastructure. In 2005, an individual offered to assist al-Qaeda engage in acts of terrorism by identifying targets and planning attacks.[113] The individual was specifically interested in attacks on the Transcontinental Pipeline and refineries in Wyoming and New Jersey, with the expectation that attacks on these targets would reduce energy reserves, create environmental hazards, and increase anxiety.[114] Additionally, multiple individuals have targeted the New York Stock Exchange and the US financial infrastructure. Between 2000 and 2001, several individuals surveilled the stock exchange. They were subsequently arrested and charged with conspiring to use weapons of mass destruction.[115] Even more recently, al-Qaeda operatives were sentenced for targeting an essential part of US critical financial infrastructure. In 2015, the United States identified Wesam el-Hanafi and Sabirhan Hasanoff as al-Qaeda operatives who had arranged to perform surveillance of the New York Stock Exchange. Hasanoff conducted the surveillance and el-Hanafi transmitted the report documenting Hasanoff's surveillance to terrorist operatives in Yemen.[116]

The Role of Open-Source Information in Planning Attacks

In addition to firsthand observation—or infiltration by proxy—the Soviet Union conducted a significant amount of collection that could inform its targeting of critical infrastructure. Soviet officials, for instance, persistently collected maps.[117] These were probably not for road trips. Sometimes this was done quite openly. In 1954, Soviet officials stationed in Washington, DC, had obtained topographic maps covering North Carolina, Michigan, Illinois, Kentucky, and an area within a 50-mile radius of Washington, DC.[118] Then, in 1962, a map company in New York City identified a Soviet national employed by the United Nations Secretariat as an individual who—over the course of two years, posing as a student—had purchased between 200 and 300 maps. To combat this exploitation of open-source information, the State Department had issued restrictions in 1955 that prohibited Soviets from acquiring maps and charts of a scale of 4 miles to the inch or larger.[119]

The Soviet UN official's masquerading as a student was one of multiple indicators that the Soviets were specifically attempting to conceal their acquisition of maps. Another indicator that the Soviets were attempting to obscure their activities was their efforts to obtain these items far away from their Washington, DC, and New York diplomatic establishments. For instance, in 1953, two Soviets traveled to Missouri and Texas in order to obtain aerial maps of Dallas, Tulsa, Fort Worth, and surrounding areas. Soviets also used cutouts to clandestinely acquire maps. In 1958, a Soviet employee of the United Nations met with an individual in the darkened parking lot of a train station in Scarsdale, New York, to retrieve an aerial map of New York City.[120]

Similar to their collection of maps, Soviet officials relentlessly collected aerial and other photographs. In 1954, the Soviets had purchased aerial photographs of five Long Island communities as well as aerial photographs of Boston, Massachusetts, and Newport, Rhode Island. During that same year, three Soviets traveled to California and ordered $80 worth of aerial photographs covering Los Angeles. Not only did the Soviets purchase aerial photography, but they also commissioned it. In May 1954, an assistant Soviet air attaché stationed in Washington, DC, attempted to enlist a photographer—for $700—to rent an airplane for taking non–commercially available photographs of New York City. The assistant air attaché provided specifics as to the scale to be used and the altitude from which the photographs should be taken. This "brazen abuse of his diplomatic privileges" got the assistant air attaché declared persona non grata by the US government shortly after his photographic foray.[121] In the wake of these instances, the State Department, in 1955, specifically prohibited Soviet citizens from acquiring aerial photographs except when they appeared in or were appendixes to newspapers, periodicals, technical journals, atlases, and books commercially available to the American public.[122] Yet in 1960, the FBI noted that the Soviets had purchased fifteen aerial photographs of Minneapolis and St. Paul, Minnesota.

It is hardly a surprise that the Soviets took steps to cloak their acquisitiveness rather than abstaining from it. Only a month after the restrictions took effect, a Soviet official in Mexico began negotiations with an individual on the West Coast of the United States for aerial photographs of forty-five major US cities, nineteen of which were, probably not coincidentally, located near Strategic Air Command bases. Others were also in vicinity of sensitive locations such as atomic energy installations and important industrial facilities. That dark Scarsdale, New York, train station parking lot again became the site of a clandestine handoff when a Soviet employee of the United Nations received aerial photographs of Chicago, Illinois. As part of this same operation, the Soviets attempted to acquire aerial photographs of Portland, Oregon; Seattle, Washington; and San Diego and San Francisco, California.

Soviet officials also used their road trips to wheedle aerial photographic material from locals far from Washington, DC, and New York. In 1959, two Soviets, including an assistant Soviet military attaché, obtained an aerial photograph of Glasgow Air Force Base in Montana by presenting themselves to the local chamber of commerce. These same officials acquired an aerial photograph of Thermapolis, Wyoming, through a similar ploy.[123]

US Government Efforts to Protect Infrastructure Against Terrorism

The US government (especially in the work of both the FBI and the DHS) has, since the early twentieth century, endeavored—with varying degrees of success—to understand the country's vulnerabilities to sabotage and terrorism. Historically this was the function of the FBI, which—perhaps because of its focus on responding to threat actors, rather than proactive assessment of future threats—never quite managed to successfully implement a program to identify and harden critical infrastructure in the absence of an immediate threat. The DHS, upon its creation, inherited the FBI's efforts in this field and did no better, with far less of an excuse. After all, the DHS's existence is centered on assessing and hardening vulnerabilities and yet it has failed to systemically assess what those vulnerabilities might be.

Since the beginning of the twentieth century, the US government has made efforts to protect privately owned entities—such as critical infrastructure—that are integral to national security. In 1912, the US Army War College (which in 1917 would become officially responsible for espionage and counterespionage) planned—but never implemented—to follow up on reports that the Japanese were obtaining information about the infrastructure (e.g., bridges and tunnels) of railroads leading to the Pacific Coast. The army was concerned about the possibility that the Japanese planned to sabotage this infrastructure in the event of US-Japanese conflict.[124]

Even before the United States entered World War I it experienced foreign sabotage. On July 30, 1916, German saboteurs ignited 2 million tons of war materials that had been stored at the Black Tom railroad yard in Jersey City, New Jersey.[125] The United States was still a neutral nation, but its ability to provide resources for a combatant made it a target. In 1917, Ralph van Deman, a War Department intelligence officer, outlined a counterespionage plan that included thwarting sabotage activities against munitions plants and transportation facilities.[126]

The FBI's Plant Survey Program, established in late 1939, handled aspects of counterterrorism in addition to counterintelligence. Training of agent personnel provided insights about the Bureau's objectives. For instance, the curriculum for agents preparing for plant survey work included instruction, by the FBI's technical laboratory, on subjects including sabotage.[127] Additionally, the personnel record of V. P. Keay, an FBI agent, shows that as of 1939 he had received training in plant protection regarding explosives and arson and outlines for discussion of bombs and explosives with plant guards.[128]

Consistent with the theme of destruction by explosion, the FBI ensured that its Plant Survey Program took into account associated hazards. In 1941, the Bureau advised its offices that some of the plants for which it had responsibility were located adjacent to or nearby large stores of commercial gasoline or were situated on waterways that carried barges loaded with large shipments of commercial gasoline. According to the FBI, offices should take these hazards into consideration during the course of its plant survey work.[129] Offices also received instructions to assess the methods for combating incendiaries. The Bureau directed its agents to account for the adequacy of firefighting equipment, including the capacity of the gravity tank and the proper distribution of sprinkler heads, as well as alternative sources of water.[130]

The FBI's postwar history illustrates that it continued to pay attention to how foreign actors might target US infrastructure. During the late 1950s, for instance, it sought to use its informants to get ahead of terrorist threats. In preparation for a meeting between an informant and Soviet officials, the FBI planned to provide the informant with a number of intelligence objectives, including determining whether the Soviet Union intended to use members of the Communist Party of the United States for sabotage activities.[131] In anticipation of another informant's possible trip to China, the Bureau expanded its collection requirements to include whether Beijing planned to use the party to conduct sabotage.[132]

Despite its experience with plant surveys and its identification of collection and targeting of infrastructure by foreign governments and nonstate actors, the FBI's assessment of vulnerabilities to terrorism was intermittent at best. In 1982, the Bureau advised Congress that it did not have a program specifically directed at assessing the risk or vulnerability of potential US targets to terrorism.[133] Several years later, the Bureau took measures to resolve this issue. In 1985 it created a new unit at FBI headquarters to work with other elements in the US counterterrorism community to identify potential targets.[134]

The Bureau's next step toward systematizing its approach to protecting critical infrastructure was the development of an Infrastructure Vulnerability / Key Asset Program. It initiated this program in 1988.[135] In 1991, the FBI advised Congress that the program was directed at addressing actual and potential acts of terrorism directed against US infrastructure.[136] Unfortunately, the program—according to Congress—atrophied due to a lack of personnel and nonpersonnel resources.[137]

Developments during the early 1990s—such as the 1993 World Trade Center bombing—demonstrated the need for awareness about critical infrastructure vulnerabilities. In 1995, the Clinton administration acknowledged these challenges in the issuance of a presidential decision directive titled "U.S. Policy on Counterterrorism." According to the directive, the FBI was to "reduce vulnerabilities by an expanded program of counterterrorism."[138] The implication of this was that the Bureau needed to revitalize its Infrastructure Vulnerability / Key Asset Program. As of 1997, this program, which included industrial assets, involved the development of a cooperative liaison with the owners and operators of facilities. By gathering information regarding facilities, the FBI would be able to readily determine the significance of the facility and the possible consequences of its destruction.[139] Echoes of the Plant Survey Program resonated.

As part of this push to reduce the vulnerability of critical infrastructure, the FBI created the National Infrastructure Protection Center (NIPC). A presidential decision directive issued in 1998 tasked the Bureau with establishing the center. This new entity was supposed to facilitate information-sharing between member agencies and the private sector.[140] The NIPC, starting in 1998, took responsibility for developing the FBI's Key Asset Initiative, which, in its identification of entities vital to national (including economic) security, seemed very similar to the Infrastructure Vulnerability / Key Asset Program.[141] Under the Key Asset Initiative, the FBI was supposed to identify national, regional, and local infrastructure components that, if disrupted, could cause widespread social and economic consequences.[142] (The term "Key Asset Program" remained in use into the early 2000s.)[143] This effort was part of an initiative to protect infrastructure against both physical and cyber attacks.[144]

The Key Asset Initiative was plagued by several problems. First, the FBI lost sight of a crucial aspect of critical infrastructure protection—collaboration with the owners and operators who had the expertise to assess the implications of an attack or other disruption.[145] According to the presidential directive that informed this effort, protection of critical infrastructure was "a shared responsibility and partnership between owners, operators, and the government."[146] However, the FBI identified more than 5,000 entities as part of the Key Asset Initiative, often without involving the parties who owned or controlled the assets.[147] As the General Accounting Office noted, the result was that "the key assets recorded may not be the ones that infrastructure owners considered to be the most important."[148] Second, the initiative suffered from a lack of information technology. FBI field offices were supposed to place Key Asset Initiative information associated with their respective areas of responsibility into a database.[149] In mid-2002, the FBI assured Congress that the NIPC had "leveraged the [Key Asset Initiative] to undertake an all-agency effort to prepare a comprehensive, centralized database of critical infrastructure assets in the United States."[150] However, according to a 2002 report from the Department of Justice's inspector general, the FBI lacked an adequate database management system and inconsistently characterized the priority of key assets.[151]

The presidential directive that informed this effort established the framework for US government interaction with private industry responsible for critical infrastructure. It designated specific departments to serve as lead agencies for liaison with industry sectors that could be the targets of physical, as well as cyber, attacks. (In addition, several agencies served as leads for specific functions—e.g., Justice/FBI was the lead for law enforcement and internal security.) Lead agencies, in collaboration with the relevant private sector counterparts, would develop and implement a vulnerability awareness and education program for their respective sectors. The presidential directive also strongly encouraged the creation of information sharing and analysis centers (ISACs), described as "the mechanism for gathering, analyzing, appropriately sanitizing and disseminating private sector information to both industry and the NIPC" as well as potentially serving to "gather, analyze and disseminate information from the NIPC for further distribution to the private sector."[152] However, this formulation seemed to be one of stovepipes that made the NIPC a potential point of failure, since information would have to percolate up through one ISAC to the NIPC, which would then need to share this information with all of the relevant ISACs.

The FBI—despite its difficulty with establishing a functioning national-level critical infrastructure protection program, did address this concern at the field office level. In 1982, the Bureau advised Congress that field offices had developed contingency plans to inform their reaction to a terrorist incident in their respective geographic territories. Preparation of these plans included the identification of potential terrorist targets.[153] Once the FBI established field elements that corresponded to the NIPC, it began to develop additional information about key assets. NIPC field squads were supposed to conduct a thorough search for key assets in their regions in each of the eight sectors that the presidential directive had outlined. This sounds like a national security scavenger hunt. Not surprisingly, it proved to be a dysfunctional enterprise. Once agents had collected the information, they were supposed to categorize it as being of national, regional, or local importance. The General Accounting Office determined that the results were scattershot. One agent had omitted assets because the agent believed they were too sensitive to include. Another agent had used a telephone book as the agent's primary source.[154]

Furthermore, the FBI had a long-standing history of engaging local authorities—who arguably knew their respective areas of responsibility better than even the Bureau—on issues of terrorism. In September 1939, President Franklin D. Roosevelt issued a directive that entrusted the FBI with responsibility for investigative work in all matters related to sabotage. In furtherance of this, Roosevelt requested that all law enforcement officers in the US provide the Bureau with any sabotage-related information. In order to facilitate this type of information-sharing, the FBI, in 1939, developed a "law enforcement mobilization plan."[155] Under the auspices of this initiative, the heads of the FBI's field offices held quarterly conferences with the police officials within the field offices' areas of geographic responsibility. The FBI was thinking about the critical infrastructure aspects of this information-sharing from the outset. As Hoover advised Congress in 1942, "it was anticipated that the FBI was going to receive thousands of complaints" about matters including sabotage and that all of these would have to be "handled, analyzed, and investigated."[156] In addition, the Bureau furnished local police forces with training on matters of civilian defense, including how to handle and dispose of incendiary devices.[157] This ethos of collaboration could be seen decades later. In 1997, the FBI advised Congress that the heads of each field office had developed working groups with local officials in order to identify local critical infrastructure vulnerabilities.[158] While these efforts were laudable, they were not organized to take into account the wide-ranging disruption—beyond the arbitrary geographic confines of a field office's area of responsibility—that an attack on critical infrastructure could produce.

Coordination at the federal level was, at times, more problematic. The presidential directive that established the NIPC noted that it would contain representatives from the Department of Defense. However, the directive also mandated the Department of Defense, in conjunction with the Department of Commerce, to offer their expertise to private owners and operators of critical infrastructure to develop "security-related best practice standards."[159] As of 1999, the FBI was supposedly coordinating its aggregation of key asset information with the DoD.[160] Not surprisingly, approximately two years later, Congress heard testimony indicating that the Bureau was not coordinating its Key Asset Initiative with DoD and Commerce.[161] FBI officials had failed to reach an agreement with the DoD. Although NIPC and DoD officials volleyed multiple drafts of a memorandum of understanding regarding coordination between the NIPC and the DoD's Joint Program Office for Special Technology Countermeasures and Infrastructure Assurance to identify vulnerabilities that might have implications for DoD facilities, the officials had reached no agreement by the end of 2000.[162] The FBI was not entirely to blame for a lack of coordination with the DoD. NIPC officials had approached the National Communications System, for which DoD was the executive agent, but learned that telecommunications industry provided National Communications System information only for internal use and to facilitate restoration of priority communications during emergencies.[163] The Bureau encountered a similar snag when it attempted to engage the Department of Commerce's Critical Infrastructure Assurance Office regarding Project Matrix. This initiative led by that office was supposed to identify critical infrastructure components and related interdependencies that could impact government operations. According to an official associated with Project Matrix, information gathered as a result of these efforts belonged to individual federal agencies and was not shareable without the agencies' express permission.[164]

In the wake of the September 11, 2001, terrorist attacks, the US government undertook a significant reorganization that produced the Department of Homeland Security. As part of the Homeland Security Act of 2002, the newly created department took control of the NIPC.[165] In 2002, the Bureau indicated that it was more than happy to relinquish at least one of the NIPC's functions, advising Congress that the Key Asset Initiative would be best performed by the DHS.[166] (The FBI continued to assess certain aspects of critical infrastructure as a part of what it referred to as "know your domain."[167] According to then-director Robert Mueller, knowing the domain meant "understanding every inch of a given community" including its economy and vulnerabilities.)[168] According to the National Strategy for the Physical Protection of Critical Infrastructures and Key Assets, the DHS would "serve as the primary liaison and facilitator for cooperation among federal departments and agencies, state and local governments, and the private sector."[169] The DHS incorporated the NIPC into its Information Analysis and Infrastructure Protection Directorate.[170]

This directorate picked up where the FBI's critical infrastructure protection efforts had left off. According to the Homeland Security Act of 2002, the Information Analysis and Infrastructure Protection Directorate was supposed to recommend measures necessary to protect US critical infrastructure.[171] The DHS created an Office of Infrastructure Protection under the directorate. Within this office, the Infrastructure Coordination Division provided expertise regarding the nation's infrastructure sectors and key assets, monitored the operational status of the sectors and assets, and supported the two-way exchange of critical infrastructure information among the DHS; federal, state, and local partners; and private sector entities. The office's Protective Security Division was responsible for coordinating strategies to protect the nation's physical infrastructure. It also pursued efforts—in conjunction with state, local, and private sector partners—to identify and list critical infrastructure assets.[172] In other words, the Office of Infrastructure Protection assumed a number of the NIPC's functions.

Despite inheriting the NIPC, the Information Analysis and Infrastructure Protection Directorate began developing a database of critical infrastructure and key resources from scratch. As described by the White House in February 2003, critical infrastructures "include human assets and physical and cyber systems that work together in processes that are highly interdependent. They also consist of key nodes that, in turn, are essential to the operation of the critical infrastructures in which they function."[173] The same White House document explained that while key assets "may not be vital to the continuity of critical services at a national level," their degradation may cause "a significant loss of life and property in addition to long-term, adverse public health and safety consequences."[174] Additionally, according to the White House, certain key assets were "symbolically equated with traditional American values and institutions or U.S. political and economic power."[175]

These were definitely broad parameters for the Information Analysis and Infrastructure Protection Directorate. The cataloging and protection efforts began in 2003, as part of Operation Liberty Shield.[176] Tom Ridge, then secretary of homeland security, assessed that this was a "first of its kind initiative," during which the federal government, working with state and local governments, as well as the private sector, "provided an unprecedented level of security" in anticipation of potential hostile terrorist action because of US military involvement in Iraq.[177] The DHS launched this program and used state resources to protect critical pieces of infrastructure during a heightened threat period.[178] According to Ridge, it was the first time that the federal government planned for and worked with and through state and local governments and the private sector to "literally add an overt security presence at critical places around the country."[179]

Liberty Shield began as a relatively small program. It emphasized setting priorities among infrastructure entities based on applied gross consequences and significant economic impact criteria.[180] Based on these factors, the DHS identified 160 nationally critical assets. Congress could not leave well enough alone and, later in 2003, pressured the Office of Infrastructure Protection to produce a prioritized list of assets, which led to the identification of 1,849 entities in sectors including chemical, hazardous material, nuclear, business and finance, electric, oil and natural gas, transportation, commercial, and government facilities. This iteration of the compendium of critical infrastructure information was known as the Protected Measures Target List.[181]

Whether the DHS was any more effective than the NIPC is questionable. As early as 2004, the DHS inspector general noted that "identifying critical infrastructure is a critical step to implementing a national infrastructure protection plan" and that the Protective Security Division was responsible for maintaining a prioritized national list of critical infrastructure and key assets.[182] Unfortunately, the Office of Infrastructure Protection was not able to live up to this expectation. It instead—following on from the establishment of the Protected Measures Target List—created National Threat Vulnerability and Asset Databases.[183] By 2007, the National Asset Database contained nearly 80,000 different entities.[184] However, not all critical infrastructure and key resources assets are created equal. Entities within the National Asset Database ranged from the expected—such as nuclear power plants, dams, and hazardous materials sites—to less-expected, including multiple petting zoos, something called "bean fest," and a popcorn purveyor[185] (the list also included a "Bourbon Festival").[186] In 2006, the DHS advised Congress that the National Asset Database contained a detailed inventory of the nation's critical infrastructure and key resources.[187] However, as the database became increasingly cluttered, the concept of criticality disappeared, and the DHS determined that the database was not actually a list of critical assets but rather a national asset inventory providing a "universe" that contained data to populate lists of critical assets.[188]

The DHS attempted to reconfigure its approach to critical infrastructure and key resources issues amid the reports of its overstuffed database. In September 2006, the department opted to suspend the use of the database.[189] The DHS changed its approach to critical infrastructure and key resources by developing the National Critical Infrastructure Prioritization Program. The Office of Infrastructure Protection, now under the National Protection and Programs Directorate (NPPD), which was the successor to the Information Analysis and Infrastructure Protection Directorate, specifically Office of Infrastructure Protection's Infrastructure Analysis and Strategy Division, managed the program.[190] The primary goals of the National Critical Infrastructure Prioritization Program include identifying infrastructure that, if disrupted or destroyed, could have a significant impact on US public health, safety, economic, or national security, and focusing planning, foster coordination, and support preparedness efforts for incident management response, and restoration activities among federal, state, and private sector partners.[191]

Congress got into the act during the following year. In the Implementing the Recommendations of the 9/11 Commission Act of 2007, Congress directed the DHS to develop a National Asset Database that included entities that—in the event of their interruption, incapacity, or destruction—would have a negative or debilitating effect on the economic security, public health, or safety of the United States, any state, or any local government. Additionally, the legislation required the DHS to develop a list of prioritized critical infrastructure that, if destroyed or disrupted, would produce "national or regional catastrophic effects."[192]

In 2007, a new DHS initiative prompted memories of the countersabotage functions that the FBI's Plant Survey Program had carried out. An explosion at the Texas City Refinery that year killed 15 people and injured 170, illustrating the extent of damage inherent to a chemical facility.[193] Although the Texas City explosion was not a terrorist act, it undoubtedly prompted some thinking about the potential for destruction that could result from terrorists targeting similar facilities. The DHS Appropriations Act for fiscal year 2007 introduced the Chemical Facility Anti-Terrorism Standards (CFATS).[194] The DHS, in late 2007, published a list of 322 chemicals of interest based on potential consequences. These consequences were divided into three categories: (1) release (toxic, flammable, or explosive chemicals that have the potential to create significant adverse consequences for human life or health if intentionally released or detonated); (2) theft/diversion (chemicals that have the potential if stolen or diverted to be used as, or converted into, weapons capable of causing significant adverse consequences for human life); and (3) sabotage/contamination (chemicals that, if mixed with other readily available materials, have the potential to create significant adverse consequences for human life or health).[195]

Facilities with chemicals proceeded through a multistep process in order to determine whether they were subject to CFATS. They first provided an initial submission, known as a top screen, to the Infrastructure Security Compliance Division, within the Office of Infrastructure Protection, which in turn was part of the National Protection and Programs Directorate. (In the DHS shell game, the National Protection and Programs Directorate was the successor to the Information Analysis and Infrastructure Protection Directorate and the predecessor to Cybersecurity and Infrastructure Security Agency.) Based on this information, the DHS determined which facilities were at high risk for terrorist attack or exploitation.[196] In 2009, the DHS indicated that it was attempting to identify facilities that might meet the threshold for CFATS compliance but had not yet registered with the department.[197] Despite this effort, in 2013, an explosion at the West Fertilizer plant in West, Texas, disastrously demonstrated that the DHS was unaware of thousands of facilities across the United States that should be subject to CFATS.[198]

Based on the top screens, the DHS tiered the facilities. Locations that the DHS deemed to be high-risk based on this rubric had to submit a security vulnerability assessment and either a site security plan or alternative security program.[199] These plans and programs describe security measures that a facility will take and how those measures meet specific risk-based performance standards.[200] Although CFATS took effect in 2007, the department did not issue guidance regarding the eighteen risk-based performance standards until mid-2009.[201] Once a facility developed and submitted a site security plan or alternative security program, the NPPD's Infrastructure Security Compliance Division conducted an authorization inspection to review specific details of the plan.[202] As part of the inspection, cyber analysts at the CFATS program office in the compliance division reviewed the plan and highlighted any areas of concern regarding cybersecurity measures.[203] Following the inspection, the compliance division made its final determination of whether the plan satisfied all applicable risk-based performance standards.[204]

In 2013, the DHS inspector general identified significant problems with the progress of CFATS. Among other things, the CFATS tiering system of the standards was created too quickly and, by late 2009, multiple errors in the data and formulas used to tier chemical facilities had been identified.[205] Also during 2013, the Government Accountability Office assessed that, although the Infrastructure Security Compliance Division had assigned approximately 3,500 high-risk chemical facilities to risk-based tiers, the compliance division had not fully assessed the approach it took to implement this process. According to the Government Accountability Office, the compliance division's approach up to this point did not account for all the elements of consequence, threat, and vulnerability associated with a terrorist attack involving specific chemicals.[206]

These reports—along with several legislative and executive developments—prompted the DHS to overhaul its approach. Specifically, the Protecting and Securing Chemical Facilities from Terrorist Attacks Act of 2014 prompted the DHS to develop an enhanced risk assessment and tiering methodology. The DHS informed its revisions with input from external experts in government, academia, and industry. Sandia National Laboratories independently validated the DHS's new approach.[207] The second significant change came in the form of an executive order titled "Improving Chemical Facility Safety and Security"—issued in the wake of the West, Texas, explosion. The order created a Chemical Facility Safety and Security Working Group, cochaired by the secretary of homeland security; the administrator of the Environmental Protection Agency; and the secretary of labor. Among other items, the working group was tasked with enhancing information collection and sharing efforts in order to address the failing that had contributed to the West, Texas, tragedy.[208]

The process of assessing CFATS implementation is not a simple one since it is not—nor should it be—a checklist job. CFATS is a nonprescriptive regulatory framework, one that gives a facility flexibility in its negotiations with the DHS to develop security measures tailored to each facility's unique needs.[209] Cybersecurity vulnerabilities add even more complexity to the implementation of the standards. Analysts at the Chemical Facility Anti-Terrorism Standards program office in the compliance division must conduct an assessment of the site security plan in order to determine the extent to which a facility is reliant on network-connected systems.[210]

In order to accurately assess vulnerabilities and advocate for the interests of the DHS when negotiating a site security plan, the Infrastructure Security Compliance Division must have a work force with acumen in a variety of fields including cyber and physical security and chemistry. Unfortunately, there are indications that the DHS has been unable to attract or retain necessary talent. As of 2012, for instance, the compliance division employed five chemical engineers, one chemist, three general engineers, and seven information technology management specialists.[211] This is hardly a sufficient work force to deal with the approximately 40,000 facilities that had provided the DHS with top screens by 2018.[212] The insufficiency of the work force is indicated by how industry has outpaced the DHS's ability to provide meaningful guidance. For instance, officials from two out of five industry associations assessed that the cybersecurity guidance that the DHS provided was not very relevant for the associations' larger members that had sufficient monetary and human capital to implement cybersecurity guidance that met or exceeded the guidance that the DHS furnished.[213]

Furthermore, the DHS's efforts to expand its CFATS-focused work force did not appear to emphasize the acquisition of expertise. According to the DHS, its chemical security inspectors were supposed to function as "boots on the ground" and serve as the face of the standards in the field. The DHS explicitly stated that these inspectors filled an "important role in helping to identify appropriate security measures during the authorization inspection process."[214] This suggests that the DHS needed a work force possessing specialized knowledge and capable of providing expert advice. However, the CFATS force was originally staffed not by chemistry and cyber experts but by law enforcement detailees from the Federal Protective Service.[215] The primary function of this service is to provide security guards for federal facilities.

Part of the problem was that the DHS had implemented the chemical security inspectors program without actually having a good grasp on what the role of this work force was supposed to be. According to the DHS, it was difficult to know what exactly the role of the inspectors would be. Initially the idea was that these individuals would be a first responder of sorts, capable of responding to a failed security plan or an actual breach, which would benefit from the Federal Protective Service's hazmat technician force.[216] The Infrastructure Security Compliance Division even considered arranging the inspector work force to have law enforcement status.[217]

However, after conferring with industry and reviewing the setup of other compliance and regulatory agencies, the compliance division decided to go a different direction.[218]

Once it had acknowledged that its initial conceptualization of the inspector role was not the one it wanted to implement, the DHS decided that it would grow its own talent. As part of its initial idea for the inspectors, the National Protection and Programs Directorate's Infrastructure Security Compliance Division planned to establish an academy to train inspectors who would enforce CFATS regulations.[219] However, although inspectors received training, several of the inspectors that the DHS's inspector general cited claimed that the course was "a waste of time because it did not address the issues CFATS was experiencing."[220] Furthermore, even under this grow-your-own philosophy, the Infrastructure Security Compliance Division lagged. According to a 2013 DHS inspector general report, the compliance division had failed to develop training for employees at the division's headquarters. This was problematic, since an uneducated headquarters element was arguably ill-suited to oversee the CFATS program. The 2013 report—issued approximately five years after the CFATS program commenced operation—stated that the Infrastructure Security Compliance Division needed to develop training plans that would "prepare employees to perform assigned duties."[221]

The cyber component of the CFATS program posed specific problems. As of 2013, the DHS advised Congress that the NPPD's Office of Cybersecurity and Communications had developed training materials for inspectors to assist facilities with integrating cybersecurity into the overall security posture.[222] However, as of 2019, the DHS advised Congress that inspectors received "a kind of base level of training on cybersecurity" and that approximately half of the inspector work force had gone through an addition two-week advanced cybersecurity course.[223] The Government Accountability Office, in 2020, presented the training of personnel in cybersecurity as far less robust than the DHS claimed. For instance, it cited a supervisory inspector who stated that the inspectors were not required to take additional cybersecurity-related training and were not expected to have cybersecurity certifications. Not only was expertise not encouraged, but also its development was inconsistently facilitated. The same Government Accountability Office report noted that training was not guaranteed, and pointed to the case of a CFATS inspector who actually used his own funds to pay for an external course—as well as associated travel expenses—because the program denied his training request.[224] If an agency is unable to hire true expertise, then it should at least ensure that it supports the development of such in a way that exceeds lip service.

Without significant expertise, there is little to separate the inspectors, in reality, from the DHS's protective security advisers. The adviser program works with the private sectors to communicate threat advisories and specific warning information, as well as provide guidance on appropriate security measures and countermeasures for the communities they serve.[225] According to the DHS, the advisers function as the DHS's on-site critical infrastructure and vulnerability assessment specialists.[226] The private sector—with which the advisers work—has cited the adviser program as an effective initiative that the DHS should use to inform its CFATS program.[227]

CFATS provides an example of how the federal government must leverage external expertise not only in the development of human capital but also for the successful implementation of its responsibilities. First, the Infrastructure Security Compliance Division has had to leverage private sector organizations in order to disseminate information. For instance, the division used contact with the ISAC sector coordinating councils to engage the chemical, oil and gas, and other fields of industry. Additionally, the division has used industry-organized functions as venues of information-sharing.[228] The DHS has also used industry associations such as the Fertilizer Institute and the Agricultural Retailers Association as channels through which to reach state-level agricultural association executives.[229]

However, the Infrastructure Security Compliance Division failed to refine this relationship, a step that would have closed the public-private feedback loop and made the DHS more effective in its work. The compliance division did not require its personnel to solicit feedback during interactions with the private sector regarding the effectiveness of outreach. The Government Accountability Office found that industry was not impressed by the DHS's outreach to industry, with seven of eleven trade associations describing their experience with the compliance division's outreach as either mixed or negative.[230]

The DHS, as of this writing, is reconsidering the CFATS program. According to its fiscal year 2021 budget, the department eliminated funding for the CFATS program, which is now under the auspices of the NPPD's successor, the Cybersecurity and Infrastructure Security Agency. Consistent with the de facto similarity, in terms of expertise, between inspectors and advisers, the DHS's budget significantly increased the funding for the adviser program. This is a decision that also aligns with the private sector's assessment of the adviser program as an effective tool for engagement, an important consideration if the DHS is going to successfully engage industry. By emphasizing the adviser program, the Cybersecurity and Infrastructure Security Agency will be able to provide voluntary support to chemical facilities and put that sector on par with other critical infrastructure sectors.[231] Of course the diminishment of expertise is never something to be encouraged, but if the DHS is going to emphasize the lowest common denominator, it might as well not create bureaucratic complications through the fiction of specialization.

Preventing Insider Threats at Chemical Facilities

Employees-gone-bad (the "insider threat") can exploit chemical facilities' vulnerabilities to do damage in a variety of ways. Terrorism is the most obvious concern. For instance, in 1998 the FBI advised Congress that a white supremacist organization in Dallas, Texas, planned to bomb a natural gas refinery, as a diversion from their planned robbery of an armored car. The attack would have released a deadly cloud of hydrogen sulfide.[232]

Additionally, individuals seeking to gain information about proprietary chemicals pose an economic espionage threat. For instance, during the late 2000s, Hong Meng, an employee of the DuPont Corporation who had obtained a faculty position at China's Peking University, College of Engineering, stole a protected chemical process from DuPont by emailing it to his Peking University account.[233] Meng also obtained 109 samples of DuPont intermediate chemical compounds and enlisted a colleague to send them to Meng once he was in China.[234] Furthermore, in 2012 the United States sentenced Wen Chyu Liu, a former research scientist for Dow Chemical, to sixty months in prison for stealing trade secrets and attempting to peddle these to Chinese companies.[235] Incidents such as these could be facilitated by individuals working within the locations manufacturing the compounds in which foreign threat actors have an interest.

CFATS included measures to mitigate the insider threat. As of 2013, the DHS indicated that, in order to implement the personnel surety aspect of CFATS, it would require facilities to perform background checks on personnel and visitors and to check for terrorist ties by comparing employee information with the federal government's consolidated terrorist watch list.[236] As of 2019, facilities—under the risk-based performance standards—were required to implement measures to validate identity; check criminal history; validate legal authorization to work in the United States; and identify people with terrorist ties.[237]

Imagery and Geospatial Intelligence and Critical Infrastructure Protection

Imagery intelligence (IMINT) and geospatial intelligence (GEOINT) have a long-standing—if sometimes controversial—role in the protection of critical infrastructure. The Office of the Director of National Intelligence defines IMINT as "representations of objects reproduced electronically or by optical means on film, electronic display devices, or other media."[238] According to the National Geospatial Intelligence Agency, GEOINT is the "exploitation and analysis of imagery and geospatial information to describe, assess, and visually depict physical features and geographically referenced activities on Earth."[239] In other words, IMINT is the raw information, and GEOINT is value derived from that raw information.

The unique perspectives that IMINT and GEOINT provide can help the United States to better understand physical vulnerabilities of critical infrastructure. (Certainly the Soviets had figured this out, as indicated by their Cold War collection of aerial photography.) However, multiple factors—ranging from classification concerns, finite resources, and political opposition, on civil liberties grounds—have impeded Washington from fully exploiting its IMINT and GEOINT resources in furtherance of protecting critical infrastructure.

Overhead Imagery for Domestic Uses

The United States has a long—albeit conflicted—history with the use of overhead imaging from "national technical means" (such as satellites) for domestic missions. In 1965, the US Bureau of the Budget directed the Department of Defense to examine the potential for using National Reconnaissance Program (NRP) satellite photography—which facilitated mapping of foreign locations—to improve and expedite US civil mapping.[240] (The NRP was established as a national effort to address the US government's foreign intelligence needs through overhead reconnaissance satellites. NRP data fulfills national, political, economic, strategic and tactical military, and scientific and technological intelligence needs.)[241] On November 7 of the following year, the US government—facilitated by the special assistant to the president for science and technology—initiated a study to evaluate satellite photography from the perspective of civilian agencies' needs in the areas of economic, social, and material resource surveys.[242] This project—known initially as "peaceful uses of high level aircraft and satellite photograph"—received the name ARGO in mid-1967.[243] Subsequent to receipt of this new name, the ARGO Project formally launched in July 1967 as a cooperative venture among multiple agencies including the Department of Defense and the Central Intelligence Agency.[244] In early 1968, the participants completed their study.[245]

Once the report was finished, the next step was the creation of an ARGO steering committee. Participating agencies agreed that the committee's functions would include the collection and consolidation of civilian agencies' data needs from existing reconnaissance satellite systems and communicating these needs to the US intelligence community—especially the National Reconnaissance Office, which was responsible for satellite coverage. This committee—suggested by the special assistant to the president for science and technology and approved by the director of central intelligence—convened for the first time on June 10, 1968.[246] The steering committee's charter was not developed and approved until 1970. In addition to the functions of coordinating and communicating civilian agencies' needs, the charter also spelled out the role of the committee in exploring the intelligence community's new systems and techniques for possible application to the needs of ARGO members.[247]

The development of an ARGO study group on national disasters is of particular significance to the role of domestic imagery in the protection of critical infrastructure. The ARGO National Disaster Support Group began meeting regularly in February 1970.[248] This study group was responsible for providing the ARGO committee with a plan for generating data using methods such as flying photography; tasking classified systems; processing data, including integration of information in a single database and redaction of the data; and analysis of the data, including mapping and definition of problem areas. The study group would provide recommendations regarding plans for action by other government agencies, as well as by state and local governments.[249]

Satellite coverage of domestic areas was disrupted by governmental reorganization and chilled by government inquiry. In 1973, the Office of the President's Scientific Adviser—which had chaired ARGO—was disbanded, which spelled the end of the committee.[250] Furthermore, there was growing concern within the intelligence community about government inquiry. Director of Central Intelligence James Schlesinger had ordered a review of the CIA's activities in order to identify any questionable or illegal activities. This resulted in the production of 693 pages of information, which the CIA's executive management committee received for review on May 17, 1973.[251] Through the ARGO committee, the intelligence community had effectively separated itself from the appearance of domestic meddling by preventing the community from passing judgment on the needs that civil agencies articulated or on how they used the information. However, there were still hints of consternation within the intelligence community. As an early 1975 memo noted, "possible criticism could be levied on the intelligence community for being involved in a program which takes pictures of domestic areas from intelligence platforms."[252] However, in mid-June, when the Commission on CIA Activities within the United States issued its report, it determined that there was "no impropriety" in the continued use of classified overhead photography by civilian agencies.[253]

The commission did usher in a new era for civilian use of intelligence community imagery. Among its recommendations was the establishment of a Civil Applications Committee (officially known as the Committee for Civil Applications of Classified Overhead Photography of the United States) to receive, evaluate, consolidate, standardize, prioritize, and transmit members' requests to the director of central intelligence and assist committee members with knowledgably formulating their requests for information.[254] The US government established the committee in late 1975 under the direction of the Department of the Interior, and the committee held its first meeting in early 1976.[255] (The US Department of the Interior's US Geological Survey had opened a classified facility in 1968 that became a national depository of classified information.)[256] However, it was not until November 1976 that the committee finally approved a policy statement that outlined the proper use of classified imagery and derived information by civilian agencies.[257]

Following the September 11, 2001, attacks on the United States, various entities within the US government began to consider the use of imagery in a context that would become known as homeland security. For instance, in May 2002, the Senate Select Committee on Intelligence acknowledged the homeland security role that the National Imagery Mapping Agency, which would evolve into the National Geospatial Intelligence Agency, could play.[258] Similar to the Bureau of Budget's 1965 request for a DoD assessment in order to determine new uses for capabilities, the director of national intelligence and the director of the US Geological Survey commissioned a study in 2005 to review the Civil Applications Committee's current and future role and to examine whether the intelligence community was making adequate use of national technical means for homeland security.[259] The resulting final report of an independent study group, issued in 2005, found that policies governing the use of intelligence capabilities to support domestic users reflected the pre-9/11 era and that the nation was missing opportunities to protect itself.[260]

One of the report's key points was that the current organization—rather than just the policies—for sharing domestic imagery needed reform. The DHS, according to the report, should establish a Domestic Applications Office.[261] This, as then–DHS secretary Michael Chertoff explained to Congress in 2007, would take the form of a National Applications Office, which would include the functions of the Civil Applications Committee.[262] The functions of the National Applications Office were not unlike the responsibilities of ARGO and the Civil Applications Committee. They included receiving, evaluating, consolidating, and prioritizing National Applications Office participants' requests for intelligence community capabilities; promoting effective and efficient use of community assets in part by facilitating access to unique community capabilities; and educating National Applications Office participants about the community's capabilities.[263] As Chertoff noted, the DHS was "not proposing to expand the uses to which satellites [were] put" but rather to "rationalize and, in a more orderly way, control the way satellites are used domestically."[264]

The National Applications Office would consist of three components: the Civil Applications Domain Working Group, which would continue the efforts of the Civil Applications Committee; the Homeland Security Domain Working Group, which would include the government agencies involved in the prevention and mitigation of, preparation for, response to, and recovery from disasters including terrorism and other threats to the homeland; and finally—and perhaps most controversially—a Law Enforcement Domain Working Group, whose customers would be federal, state, local, and tribal entities. This working group would focus on activities in support of the enforcement of both criminal and civil laws.[265] According to the DHS, the National Applications Office would not accept any requests from the Law Enforcement Domain until all associated, legal, privacy, civil rights, civil liberties, and policy issues had been resolved.[266]

Despite these assurances, the National Applications Office drew fire from elected officials and civil liberties advocates. Congress had initially been amenable to the establishment of the office and had provided funding to initiate operations.[267] However, on August 22, 2007, Congressman Bennie Thompson fired off a letter to Chertoff expressing concern about the DHS's "failure to vet [the office] with the Privacy and Civil Liberties Oversight Board" and referred to "the failure to consult the Board on a matter as controversial as using spy satellites for domestic homeland security law enforcement purposes" as "particularly worrisome."[268] On September 6, 2007, Thompson, along with Jane Harman—who once referred to the National Applications Office as "a homeland security intelligence program gone wrong," and Christopher Carney, wrote to Charles Allen, the DHS's chief intelligence officer, advising that the signers were calling for a "moratorium" on the National Applications Office until "constitutional, legal, and organizational questions" were answered.[269] As expected, groups such as the Center for National Security Studies and the American Civil Liberties Union (ACLU) hopped aboard the bandwagon of naysayers. The ACLU put the worst possible spin on the National Applications Office, claiming that the government was using "spy satellites to monitor its own people," and specifically insisted that the intent was not for monitoring "national infrastructure."[270]

Histrionics helped to scuttle the National Applications Office. After some initial noises in 2009 about the DHS working with state homeland security advisers and other key stakeholders to "determine how the [office] might meet their homeland security needs to protect lives and property," the DHS ultimately decided, in June 2009, to shut down the office.[271] Then–DHS secretary Janet Napolitano assessed that there were other programs that would "better meet the needs of law enforcement, protect the civil liberties and privacy of all Americans, and make [the United States] more secure."[272]

The termination of the National Applications Office has not completely stopped US agencies responsible for intelligence within the domestic setting from using IMINT in furtherance of their missions. First, the Civil Applications Committee continued to operate.[273] Congress, following termination of the National Applications Office, attempted to prevent the Civil Applications Committee from engaging in law enforcement functions.[274] However, as of 2014, according to the Civil Applications Committee's executive steering committee, it could support law enforcement missions (although it could not target US persons). The DHS, as of 2016—years after the demise of the National Applications Office—was an associate member of the Civil Applications Committee.[275] Furthermore, the DHS, as a member of the US intelligence community, has other ways to obtain IMINT and derive GEOINT. Specifically, the National Geospatial Intelligence Agency maintains a liaison presence at DHS headquarters.[276] (This was not a post–National Applications Office workaround; the liaison presence had been a fixture at DHS headquarters since at least 2005, when a study noted that the National Geospatial Intelligence Agency was one of the agencies working to understand and respond to identified needs and bringing to the DHS's attention information that might warrant exploration).[277] The National Geospatial Intelligence Agency presence "engages with DHS and DHS Components on all requests for satellite imagery, map-based intelligence and geospatial information."[278] In other words, derailing the National Applications Office did not deny the DHS access to IMINT and GEOINT; it simply kept the processes for information-sharing disjointed and, potentially, less transparent.

The Role of IMINT and GEOINT in Protecting Critical National Infrastructure

Inherent to the history of domestic uses of satellite imagery is the potential for using this imagery to assess and mitigate vulnerabilities in critical infrastructure. As early as 1967, the interagency Committee on Overhead Reconnaissance identified several domestic entities for IMINT coverage by the KH-4 satellite.[279] Several of these sites fit within what would come to be known as critical infrastructure. They included the Thiokol Corporation, a New Jersey–based chemical company; the Wyandotte Chemical Corporation; and the Alcoa Aluminum Plant. Additionally, the committee indicated interest in satellite imagery of US atomic energy sites including those at Savannah River and Oak Ridge.[280]

Multiple projects under the auspices of ARGO demonstrated the role that IMINT/GEOINT could play in securing critical infrastructure. An example of this was the mapping of energy-related entities. The US Geological Survey conducted an analysis of a route from the Alaskan North Slope, through Fairbanks, to Prudhoe Bay on the south Alaskan shore as part of planning a new pipeline.[281] (Use of traditional methods—rather than KH-4 imagery—would have been a very costly endeavor, requiring at least three summer field sessions.)[282]

Similar efforts informed the planning of transportation infrastructure. In 1969, the US Army's Chief of Engineers used satellite photography to assess the feasibility of constructing a Lake Michigan–Wabash River Barge Canal in northeastern Indiana. KH-4 photography provided information about the multiple variables involved in such a project, including transportation facilities, populations and land use patterns, vegetation, and soil conditions along the proposed alignment.[283] The photo-mosaic facilitated by KH-4 coverage identified that there were six problems that made the construction and use of the proposed canal impractical.[284]

From both of these planning scenarios, one can extrapolate the national security use of IMINT/GEOINT for ensuring the soundness of critical infrastructure against sabotage. Through the ability to identify, in advance, the vulnerabilities inherent to planned projects that threat actors could exploit to cripple the United States (e.g., cutting off energy or transportation of essential goods), planners—informed by the appropriate subject-matter experts—could mitigate the shortcomings through a variety of means (e.g., reconsideration of routing, and hardening particularly at-risk elements of the project).

ARGO's work also illustrated how imagery could be used to assess existing infrastructure. The Army Corps of Engineers (Civil Works) conducted a project to determine the location and size of dammed water bodies of over a capacity of 50 acre feet in North Carolina.[285] As with planning of critical infrastructure, the ability to assess existing critical infrastructure within a broader geographic context could help to identify and mitigate points of vulnerability. The Office of Emergency Planning, seeing value in the synoptic imagery that KH-4 satellites could provide, levied requirements for imagery of 115 US cities in furtherance of "precontingency" photo coverage.[286] More recently, the DHS has demonstrated the value of such imagery. For instance, the DHS and federal law enforcement agencies have used imagery to identify the vulnerabilities of facilities that host high-profile events such as the Super Bowl.[287]

Preparation for natural disasters has also driven domestic customers' desire for IMINT/GEOINT. ARGO's National Disaster Support Task Group had, from its outset, recognized the need for a pre-disaster photographic database. The group recognized a need—and requested acquisition planning efforts—for coverage of the Gulf Coast in advance of hurricane season.[288] Material of this nature was useful decades later, following Hurricane Katrina, when the National Geospatial Intelligence Agency provided graphics of relief efforts that depicted the locations of major airports, police and fire stations, emergency operations centers, highways, and schools, as well as potential dangers such as hazardous materials.[289]

Impediments to IMINT

There are real impediments—beyond agitated civil libertarians—to the use of IMINT and GEOINT in furtherance of protecting critical infrastructure. Classification has been a problem from the outset of such endeavors, and the expertise to process the information has been another. On the other end of the equation, ARGO and the Civil Applications Committee have had a long-standing difficulty in educating customers about the value of IMINT and GEOINT data to their respective missions.

Classification. The sensitivity of imagery data has posed a problem for non-clearance-holding customers since the intelligence community began exploring wider uses of imagery data. As the ARGO committee noted, even when agencies had the appropriate cleared personnel, the decisions made based on classified information could not be substantiated to uncleared planners and engineers. This was uniquely problematic in the field of policymaking, where "Congressmen, who by nature require proof of information derogatory to their pet projects, prove to be the [the committee's] worst enemy."[290] ARGO and its successor, the Civil Applications Committee, both grappled with the question of classification. As the ARGO steering committee's charter noted, one of the committee's functions would be to provide a forum for discussion of problems and procedures related to the security classification of reconnaissance technologies and products.[291] The director of central intelligence would designate a representative to the Civil Applications Committee who would serve in an ex officio capacity in part to address classification issues. In the establishment of the committee, it was acknowledged that "civil usage of data is enhanced if unclassified" and that the director's representative would facilitate this when appropriate.[292]

Resources. There has been a tension between the intelligence community and civilian uses of national technical means–derived information in relation to the prioritization of resources. As early as 1968, the National Photographic Interpretation Center (NPIC) was reviewing KH-4 domestic target material.[293] (The center was an element of the CIA until 1996, when it was merged into the National Imagery and Mapping Agency, a predecessor to the National Geospatial Intelligence Agency.)[294] However, the NPIC was not entirely comfortable with this arrangement. According to a 1970 NPIC memo, civilian uses of satellite photography had reached a point that caused concern about the involvement of intelligence resources and "it [was] time to encourage civil agencies to limit their dependence upon [the NPIC] to security aspects, to essential technical support, and to minor supplementary funding support."[295] The NPIC indicated concern that "an enlarging pattern of dependence upon [it] and other components of the satellite reconnaissance community" would result in a "distraction of intelligence resources" and work against civilian agencies' development of independent capabilities.[296]

Availability. From early on it was clear to those involved with promoting the use of IMINT and GEOINT by civilian agencies that those agencies often could not exploit the data to its full potential, as they did not have knowledge of what was available. In 1970, the President's Office of Science and Technology observed that "if one is candid it would have to be said that though some 'tasking' of [National Reconnaissance Office] resources has taken palace, the ARGO Committee has not stressed the capabilities that are available."[297] The Civil Applications Committee was supposed to "stimulate and facilitate the operational use of classified systems" in an apparent effort to ensure that customers knew what was available.[298]

In furtherance of this, the committee provided information about the availability and potential of classified satellite imagery to civilian agencies that had not yet used the imagery but that might find it beneficial to their work.[299] Nearly three decades later, a Civil Applications Committee study determined that one of the reasons for the failure of domestically oriented agencies to use intelligence community capabilities was that "they didn't know what existed."[300] One of the National Applications Office's functions was to remedy this by helping customers learn about intelligence community remote sensing capabilities, including the benefits and limitations of those capabilities.[301]

Notes

1. US Congress, *Appropriations, Department of Justice, 1926,* before a subcommittee of House Committee on Appropriations, 68th Congress, December (Washington, DC, 1924).

2. Federal Bureau of Investigation, *Soviet Intelligence Targets in the United States, 1946–1953* (Washington, DC, 1953), https://www.governmentattic.org/2docs/FBI_Monograph_Soviet-Targets-US_1953.pdf.

3. Federal Bureau of Investigation, *FBI Annual Report 1966* (Washington, DC), https://ia801005.us.archive.org/29/items/FBIAnnualReport1966/FBI%20Annual%20Report%201966.pdf.

4. Federal Bureau of Investigation, *Soviet Intelligence Targets in the United States.*

5. National Counterintelligence Executive, *A Counterintelligence Reader,* vol. 4 (Washington, DC, 2011), https://www.hsdl.org/?view&did=449820.

6. US Senate, *Exposé of Soviet Espionage 1960* (Washington, DC), https://www.cia.gov/library/readingroom/docs/CIA-RDP65B00383R000200040033-2.pdf.

7. Central Intelligence Agency, *Summary Report on Technology Transfer to Communist Countries and the Intelligence Community's Role and Effectiveness* (Langley, 1981), https://www.cia.gov/library/readingroom/docs/CIA-RDP85T00176R000900020001-5.pdf.

8. Jon Zonderman, "Policing High Tech Exports," *New York Times,* November 27, 1983, https://www.nytimes.com/1983/11/27/magazine/policing-high-tech-exports.html.?searchResultPosition=5.

9. Federal Bureau of Investigation, *Chinese Communist Intelligence Activities in the United States* (Washington, DC, 1954), https://ia801908.us.archive.org/35/items/FBIPRCSpying/fbi-prc-spying.pdf.

10. US Congress, *Investigation and Study of the Administration, Operation, and Enforcement of the Export Control Act of 1949 and Related Acts,* report of the Select Committee on Export Control, House of Representatives, 87th Congress (Washington, DC, 1962), https://books.google.com/books?id=iNQz97wiQbgC&pg=RA105-PA59&lpg=RA105-PA59&dq=secretary+of+commerce+revoked+soviet+license+1961&source=bl&ots=hZNjDZORfi&sig=ACfU3U1AaiQ2_7nTvwGC0s2EWNCd_Ag8Ow&hl=en&sa=X&ved=2ahUKEwiE1OPKoqPqAhV3mHIEHeGAAdEQ6AEwCXoECA0QAQ#v=onepage&q=secretary%20of%20commerce%20revoked%20soviet%20license%201961&f=false.

11. Central Intelligence Agency, *National Intelligence Estimate: Soviet Capabilities for Clandestine Attack Against the US with Weapons of Mass Destruction and the Vulnerability of the US to Such Attack, Mid 1951 to Mid 1952* (Langley,

September 4, 1951); Federal Bureau of Investigation, "Atomic Bomb in Unknown Consulate, New York City, Internal Security," November 7, 1951.

12. Ibid.

13. Ibid.

14. Ibid.

15. Federal Bureau of Investigation, A. H. Belmont, memorandum to H. L. Edwards, Domestic Intelligence Division/Inspection, October 14, 1958 (William Cleveland personnel file FOIA release).

16. Central Intelligence Agency, *National Intelligence Estimate* no. 4-68 (Langley, June 18, 1968).

17. General Accounting Office, *Container Security: A Flexible Staffing Model and Minimum Equipment Requirements Would Improve Overseas Targeting and Inspection Efforts* (Washington, DC, 2005), https://www.gao.gov/new.items/d05557.pdf.

18. US Congress, *Domestic Intelligence Operations for Internal Security Purposes,* pt. 1, before the Committee on Internal Security, House of Representatives, 93rd Congress (Washington, DC, 1974).

19. Federal Bureau of Investigation, Victor P. Keay, memorandum (Keay personnel file FOIA release), https://archive.org/details/VictorP.Keay/Keay%2C%20Victor%20P.%20-1.

20. Federal Bureau of Investigation, *FY 2015 Budget Justification: US Senate Ten Years After 9/11, 2011,* before the Committee on Homeland Security and Governmental Affairs, US Senate, 112th Congress (Washington, DC, 2012).

21. US Congress, *Commerce, Justice, Science, and Related Agencies Appropriations for 2010,* pt. 1, before a subcommittee of the Committee on Appropriations, House of Representatives, 111th Congress (Washington, DC, 2009).

22. US Congress, *Commerce, Justice, Science, and Related Agencies Appropriations for 2016,* pt. 2B, before a subcommittee of the Committee on Appropriations, House of Representatives, 114th Congress (Washington, DC, 2015).

23. US Congress, *Commerce, Justice, Science, and Related Agencies Appropriations for 2014,* pt. 2B, before a subcommittee of the Committee on Appropriations, House of Representatives, 113th Congress (Washington, DC, 2013).

24. Zonderman, "Policing High Tech Exports."

25. US Senate, *Transfer of United States High Technology to the Soviet Union and Soviet Bloc Nations*, before the Committee on Governmental Affairs, 97th Congress (Washington, DC, 1982), https://www.cia.gov/library/readingroom/docs/CIA-RDP85M00364R001001520004-2.pdf.

26. Tim Weiner, "Lies and Rigged 'Star Wars' Test Fooled the Kremlin, and Congress," *New York Times,* August 18, 1993, https://www.nytimes.com/1993/08/18/us/lies-and-rigged-star-wars-test-fooled-the-kremlin-and-congress.html.

27. US Senate, *Transfer of United States High Technology to the Soviet Union and Soviet Bloc Nations.*

28. Bruce Wejrauch, "Operation Exodus: The United States Government's Program to Intercept Illegal Exports of High Technology," *Computer Law Journal* 7, no. 2 (Fall 1986), https://repository.jmls.edu/cgi/viewcontent.cgi?article=1478&context=jitpl.

29. US-China Commission, *Export Controls and China* (Washington, DC, 2002), https://www.uscc.gov/sites/default/files/transcripts/1.17.02HT.pdf.

30. Ibid.; Jim McGee, "Cuban Fumbles Try at Spying, Was Too Straightforward, U.S. Says," *Philadelphia Inquirer,* July 20, 1982, https://www.cia.gov/library/readingroom/docs/CIA-RDP90-00965R000201090074-2.pdf.

31. US Senate, *Transfer of United States High Technology.*

32. Mary Thornton, "Customs Fights KGB on High-Tech Thefts," *Washington Post,* February 5, 1986, https://www.cia.gov/library/readingroom/docs/CIA-RDP90-00965R000706710007-7.pdf.

33. Federal Bureau of Investigation, W. A. Brangian, memorandum to W. C. Sullivan, "DEPAND (Deception Program for Antimissile Defense)," November 6, 1961, https://ia800305.us.archive.org/29/items/FBI_Confidential_Files-HQ-1/FBI_Confidential_Files-HQ-1.pdf.

34. US Senate, *FBI Statutory Charter,* pt. 1, before the Committee on the Judiciary, 95th Congress (Washington, DC, 1979).

35. Central Intelligence Agency, director of security to deputy director of central intelligence, "Subject: Newsweek Article Entitled 'The Soviets' Dirty Tricks Squad' of 23 November 1981," https://www.cia.gov/library/readingroom/docs/CIA-RDP87S00869R000200250003-5.pdf.

36. US Senate, *DEA and FBI,* pt. 3, before the Committee on the Judiciary, 100th Congress (Washington, DC, 1987).

37. Bill Gertz, "Three in Congress once with KGB, Says Author," *Washington Times,* February 24, 1986, https://www.cia.gov/library/readingroom/docs/CIA-RDP90-00965R000302320060-9.pdf.CIA.

38. Gus Weiss, "The Farewell Dossier: Duping the Soviets," 2007, https://www.cia.gov/library/center-for-the-study-of-intelligence/csi-publications/csi-studies/studies/96unclass/farewell.htm.

39. John Krige, "Regulating International Knowledge Exchange: The National Security State and the American Research University from the 1950s to Today," *Technology and Culture* 60 (January 2019): 252–277.

40. Weiss, "The Farewell Dossier."

41. Ibid.

42. Central Intelligence Agency, *The Technology Acquisition Efforts of the Soviet Intelligence Services* (Langley, 1982), https://www.cia.gov/library/readingroom/docs/CIA-RDP82M00786R000104810001-5.pdf.

43. Central Intelligence Agency, *US Mechanisms for the Control of Exports and of Transshipment of U.S. Exports to Communist China* (Langley, 1952), https://www.cia.gov/library/readingroom/docs/CIA-RDP62-00647A000200070005-9.pdf.

44. US Congress, *Investigation and Study of the Administration, Operation, and Enforcement of the Export Control Act of 1949 and Related Acts.*

45. US Senate, *Meeting the Espionage Challenge: A Review of United States Counterintelligence and Security Programs,* report of the Select Committee on Intelligence (Washington, DC, 1986), https://www.cia.gov/library/readingroom/docs/CIA-RDP90-00530R000300620021-3.pdf.

46. Central Intelligence Agency, *The Technology Acquisition Efforts of the Soviet Intelligence Services.*

47. Department of Justice, "Justice Department and Partner Agencies Launch National Counter-Proliferation Initiative," October 11, 2007, https://www.justice.gov/archive/opa/pr/2007/October/07_nsd_806.html.

48. Central Intelligence Agency, *The Technology Acquisition Efforts of the Soviet Intelligence Services.*

49. Central Intelligence Agency, *U.S. Mechanisms for the Control of Exports and of Transshipments.*

50. Thomas D. Little, "End Use Monitoring Is the Key to Success in Foreign Military Sales," August 28, 2017, https://www.army.mil/article/192447/end_use_monitoring_is_the_key_to_success_in_foreign_military_sales.

51. "End-Use Monitoring of Defense Articles and Defense Services Commercial Exports, FY 2018," https://www.pmddtc.state.gov/sys_attachment.do?sysparm_referring_url=tear_off&view=true&sys_id=d53a84efdb9177045564ff1e0f961910.

52. Kevin J. Kurland, "End Use Monitoring and Effective Export Compliance" (Washington, DC: Department of Commerce, 2016), https://www.bis.doc.gov/index.php/documents/pdf.s/1593-end-user-verification-kurland/file.

53. US-China Economic and Security Review Commission, *How Chinese Companies Facilitate Technology Transfer from the United States* (Washington, DC, 2019), https://www.uscc.gov/sites/default/files/Research/How%20Chinese%20Companies%20Facilitate%20Tech%20Transfer%20from%20the%20US.pdf.

54. *United States of America v. Zhongsan Liu,* https://www.justice.gov/opa/press-release/file/1202996/download.

55. Department of Justice, "Chinese Government Employee Charged in Manhattan Federal Court with Participating in Conspiracy to Fraudulently Obtain U.S. Visas," September 16, 2019, https://www.justice.gov/opa/pr/chinese-government-employee-charged-manhattan-federal-court-participating-conspiracy.

56. Federal Bureau of Investigation, H. B. Fletcher, memorandum to D. M. Ladd, "Personnel Advancement," December 14, 1948 (Keay personnel file FOIA release).

57. William H. Webster, memorandum to the director of central intelligence, "Report for the Administration," December 1, 1980, https://www.cia.gov/library/readingroom/docs/CIA-RDP05T00644R000601780002-1.pdf.

58. US Senate, *Current and Projected National Security Threats to the United States,* before the Select Committee on Intelligence, 105th Congress (Washington, DC, 1998).

59. US Senate, *Biological Weapons: The Threat Posed by Terrorists,* before the Subcommittee on Technology, Terrorism, and Government Information of the Committee on the Judiciary, 105th Congress (Washington, DC, 1998); US Senate, *Foreign Terrorists in America: Five Years After the World Trade Center,* before the Subcommittee on Technology, Terrorism, and Government Information of the Committee on the Judiciary, 105th Congress (Washington, DC, 1998); US Senate, *Counterterrorism: Evaluating the 5 Year Plan,* before a subcommittee of the Committee on Appropriations, 105th Congress (Washington, DC, 1998).

60. James B. Comey, director of the Federal Bureau of Investigation, statement before the Committee on the Judiciary, US Senate, December 9, 2015.

61. William J. Casey, memorandum to William von Raab, July 20, 1982, https://www.cia.gov/library/readingroom/docs/CIA-RDP83M00914R002200190006-6.pdf.

62. Jerome P. Bjelopera, *Homeland Security Investigations, a Directorate Within U.S. Immigration and Customs Enforcement: In Brief* (Washington, DC: Congressional Research Service, 2015).

63. US Senate, *The Homeland Security Department's Budget Submission for Fiscal Year 2012,* before the Committee on Homeland Security and Governmental Affairs, 112th Congress (Washington, DC, 2011); Bjelopera, *Homeland Security Investigations.*

64. US Senate, *S. Doc. 112-196: The Homeland Security Department's Budget Submission for Fiscal Year 2012,* before the Committee on Homeland Security and Governmental Affairs, 112th Congress (Washington, DC, 2011).

65. US Congress, *Economic Espionage: A Foreign Intelligence Threat to American Jobs and Homeland Security,* before the Committee on Homeland Security, House of Representatives, 112th Congress (Washington, DC, 2012), https://www.govinfo.gov/content/pkg/CHRG-112hhrg79843/html./CHRG-112hhrg79843.htm.

66. Ibid.

67. US Immigration and Customs Enforcement, Export Enforcement Coordination Center, https://www.ice.gov/eecc.

68. Executive Order 13558, "Export Coordination Enforcement Center," https://obamawhitehouse.archives.gov/the-press-office/2010/11/09/executive-order-13558-export-coordination-enforcement-center.

69. US Senate, *Transfer of United States High Technology.*

70. General Accounting Office, *Export Controls: Actions Needed to Improve Enforcement* (Washington, DC, 1993), https://www.gao.gov/assets/160/154080.pdf.

71. Central Intelligence Agency, William J. Casey, memorandum to William von Raab, July 20, 1982, https://www.cia.gov/library/readingroom/docs/CIA-RDP83M00914R002200190006-6.pdf.

72. Department of Homeland Security, *Audit of Export Controls for Activities Related to China* (Washington, DC, 2006), https://www.governmentattic.org/2docs/4DHS-OIG_Reports_2005-2007.pdf.

73. US Congress, *Economic Espionage.*

74. Department of Homeland Security, *Review of Deemed Exports* (Washington, DC, 2004), https://www.governmentattic.org/5docs/DHS-OIG-DeemedExportsReview_2004.pdf.

75. US Congress, *Homeland Security Investigations: Examining DHS's Efforts to Protect American Jobs and Secure the Homeland,* before the Subcommittee on Oversight, Investigations, and Management of the Committee on Homeland Security, House of Representatives, 112th Congress (Washington, DC, 2011), https://www.govinfo.gov/content/pkg/CHRG-112hhrg72254/pdf./CHRG-112hhrg72254.pdf.

76. 9/11 Review Commission, *The FBI: Protecting the Homeland in the 21st Century* (Washington, DC, 2015), https://ucr.fbi.gov/stats-services/publications/protecting-the-homeland-in-the-21st-century.

77. Department of Commerce, Bureau of Industry and Security, *Fiscal Year 2019 President's Submission,* http://www.osec.doc.gov/bmi/budget/FY19CBJ/BIS_FY19_President%27s_Budget_FINAL.pdf.

78. Department of Commerce, Bureau of Industry and Security, *Deemed Export Controls May Not Stop the Transfer of Sensitive Technology to Foreign Nations in the U.S.* (Washington, DC, 2004).

79. US Senate, *Exposé of Soviet Espionage.*

80. Federal Bureau of Investigation, A. H. Belmont, memorandum to the director, "Director's Brief for President Eisenhower on Khrushchev," September 5, 1959, https://ia801308.us.archive.org/4/items/KHRUVIS19571959/KHRUVIS_1957-1959.pdf.

81. US Congress, *Departments of State, Justice, and Commerce, the Judiciary, and Related Agencies Appropriations for 1963,* before a subcommittee of the Committee on Appropriations, House of Representatives, 87th Congress (Washington, DC, 1962).

82. US Congress, *Departments of State, Justice, and Commerce, the Judiciary, and Related Agencies Appropriations for 1964,* before a subcommittee of the Committee on Appropriations, House of Representatives, 88th Congress (Washington, DC, 1963).

83. US Congress, *Departments of State, Justice, and Commerce, the Judiciary, and Related Agencies Appropriations for 1965,* before a subcommittee of the Committee on Appropriations, House of Representatives, 88th Congress (Washington, DC, 1964).

84. US Congress, *Departments of State, Justice, and Commerce, the Judiciary, and Related Agencies Appropriations for 1966,* before a subcommittee of the Committee on Appropriations, House of Representatives, 89th Congress (Washington, DC, 1965).

85. Federal Bureau of Investigation, *The Sabotage Plans and Potential of the Communist Party, USA* (Washington, DC, 1953), https://ia800809.us.archive.org/2/items/TheSabotagePlansAndPotentialOfTheCPUSA/The%20Sabotage%20Plans%20and%20Potential%20of%20the%20Communist%20Party%2C%20U.S.A%20%281953%29%20%5B1953-10%5D.pdf.

86. US Congress, *Departments of State and Justice, the Judiciary, and Related Agencies Appropriations for 1957,* before a subcommittee of the Committee on Appropriations, House of Representatives, 84th Congress (Washington, DC, 1956).

87. Federal Bureau of Investigation, *The Sabotage Plans and Potential of the Communist Party, USA.*

88. US Congress, *Departments of State, Justice, and Commerce Appropriations for 1955,* before a subcommittee of the Committee on Appropriations, House of Representatives, 83rd Congress (Washington, DC, 1954).

89. Federal Bureau of Investigation, *The Sabotage Plans and Potential of the Communist Party, USA.*

90. US Congress, *Departments of State, Justice, and Commerce Appropriations for 1955,* before a subcommittee of the Committee on Appropriations, House of Representatives, 83rd Congress (Washington, DC, 1954).

91. Federal Bureau of Investigation, *Annual Report 1969,* https://ia803109.us.archive.org/20/items/FBIAnnualReport1969/FBI%20Annual%20Report%201969.pdf.

92. US Senate, *Departments of State, Justice, and Commerce, the Judiciary, and Related Agencies Appropriations for Fiscal Year 1972,* before a subcommittee of the Committee on Appropriations, 92nd Congress (Washington, DC, 1971).

93. US Congress, *Departments of State, Justice, and Commerce, the Judiciary and Related Agencies Appropriations for 1977,* pt. 4, before a subcommittee of the Committee on Appropriations, House of Representatives, 94th Congress (Washington, DC, 1976).

94. Federal Bureau of Investigation, *Annual Report 1971,* https://ia803106.us.archive.org/23/items/FBIAnnualReport1971/FBI%20Annual%20Report%201971.pdf.

95. US Senate, *Departments of State, Justice, the Judiciary, and Related Agencies Appropriations, 1962,* before a subcommittee of the Committee on Appropriations, 87th Congress (Washington, DC, 1961).

96. Federal Bureau of Investigation, *The Sabotage Plans and Potential of the Communist Party, USA.*

97. Central Intelligence Agency, *Informal History: US Intelligence Involvement in the East-West Exchange Program* (1965), https://www.cia.gov/library/readingroom/docs/DOC_0001495225.pdf.

98. NSC 5607, "Statement of Policy on East-West Exchanges," June 29, 1956, https://history.state.gov/historicaldocuments/frus1955-57v24/d104.

99. Federal Bureau of Investigation, A. H. Belmont, memorandum to the director, "Director's Brief for President Eisenhower on Khrushchev."

100. Federal Bureau of Investigation, *Inspection: Domestic Intelligence Division* (Langley, 1971), https://www.archives.gov/files/research/jfk/releases/docid-32989638.pdf.

101. Ibid.

102. Federal Bureau of Investigation, *The Sabotage Plans and Potential of the Communist Party, USA.*

103. Federal Bureau of Investigation, A. H. Belmont, memorandum to the director, "Director's Brief for President Eisenhower on Khrushchev."

104. Christopher Andrew and Vasili Mitrokhin, *The Sword and the Shield: The Mitrokhin Archive and the Secret History of the KGB* (New York: Basic, 1999).

105. US-CERT, "Alert (TA18-074!): Russian Government Cyber Activity Targeting Critical Infrastructure Sectors," March 15, 2018, https://us-cert.cisa.gov/ncas/alerts/TA18-074A.

106. Federal Bureau of Investigation, W. C. Sullivan, memorandum to A. H. Belmont, "Sabotage," June 21, 1961, https://ia801800.us.archive.org/3/items/foia_Belmont_Alan_7/Belmont_Alan_7.pdf.

107. Robert F. Whitney, "2 Seized in Mexico in Sabotage Blasts of Towers in U.S.," *New York Times,* June 19, 1961, https://timesmachine.nytimes.com/timesmachine/1961/06/19/97674007.pdf.?pdf._redirect=true&ip=0.

108. Federal Bureau of Investigation, *Annual Report 1969.*

109. "Conviction Reversed for Antiwar Activist in Power Line Blast," *New York Times,* May 7, 1977, https://www.nytimes.com/1977/05/07/archives/conviction-reversed-for-antiwar-activist-in-power-line-blast.html.?searchResultPosition=2.

110. Federal Bureau of Investigation, *Terrorism 2002–2005* (undated), https://www.fbi.gov/stats-services/publications/terrorism-2002-2005.

111. Department of Homeland Security, *Homeland Threat Assessment* (Washington, DC, 2020), https://www.dhs.gov/sites/default/files/publications/2020_10_06_homeland-threat-assessment.pdf.

112. Seth G. Jones, Catrina Doxsee, Nicholas Harrington, Grace Hwang, and James Suber, *The War Comes Home: The Evolution of Domestic Terrorism in the United States* (Washington, DC: Center for Strategic and International Studies, 2020), https://csis-website-prod.s3.amazonaws.com/s3fs-public/publication/201021_Jones_War_Comes_Home_v2.pdf.

113. Federal Bureau of Investigation, *Terrorism 2002–2005.*

114. Ibid.; Steven Bucci, James Carafano, and Jessica Zuckerman, *60 Terrorist Plots Since 9/11: Continued Lessons in Domestic Counterterrorism* (Washington, DC: Heritage Foundation, 2013), https://www.heritage.org/terrorism/report/60-terrorist-plots-911-continued-lessons-domestic-counterterrorism.

115. Department of Justice, "Three British Nationals Indicted on Charges of Conspiring to Use Weapons of Mass Destruction, Providing Material Support to Terrorists," April 12, 2005, https://www.justice.gov/archive/opa/pr/2005/April/05_crm_180.htm.

116. Department of Justice, "Brooklyn Man Sentenced in Manhattan Federal Court to 15 Years in Prison for Providing Material Support to Al Qaeda," January 20, 2015.

117. US Congress, *Departments of State, Justice, and Commerce, the Judiciary, and Related Agencies Appropriations for 1963;* US Congress, *Departments of State, Justice, and Commerce, the Judiciary, and Related Agencies Appropriations for 1965.*

118. US Senate, *Exposé of Soviet Espionage 1960.*

119. US Congress, *Departments of State, Justice, and Commerce, the Judiciary, and Related Agencies Appropriations for 1964.*

120. US Senate, *Exposé of Soviet Espionage 1960.*

121. Ibid.

122. Ibid.

123. Ibid.

124. Joan M. Jensen, *Army Surveillance in America, 1775–1980* (New Haven: Yale University Press, 1991).

125. Federal Bureau of Investigation, "Black Tom 1916 Bombing," https://www.fbi.gov/history/famous-cases/black-tom-1916-bombing.

126. Jensen, *Army Surveillance in America.*

127. Federal Bureau of Investigation, "Plant Protection Program," July 31, 1940.

128. Federal Bureau of Investigation, memorandum to the director, "In-Service Training Received for V. P. Keay," November 22, 1939 (Keay FOIA release).

129. Federal Bureau of Investigation, Bulletin no. 9: First Series 1941, March 1, 1941, https://ia802702.us.archive.org/2/items/foia_FBI_Confidential_Informants-HQ-1a/FBI_Confidential_Informants-HQ-1a.pdf.

130. Federal Bureau of Investigation, Bulletin no. 17: (D) Plant Surveys, May 15, 1941, https://ia802702.us.archive.org/2/items/foia_FBI_Confidential_Informants-HQ-1a/FBI_Confidential_Informants-HQ-1a.pdf.

131. Federal Bureau of Investigation, director to SAC Chicago, "SOLO: Internal Security," July 29, 1959, https://vault.fbi.gov/solo/solo-part-13-14-of/view.

132. Federal Bureau of Investigation, director to SAC New York, "SOLO: Internal Security," December 23, 1959, https://vault.fbi.gov/solo/solo-part-17-18-of/view.

133. US Congress, *Departments of Commerce, Justice, and State, the Judiciary, and Related Agencies Appropriations for 1983,* pt. 7, before a subcommittee of the Committee on Appropriations, House of Representatives, 97th Congress (Washington, DC, 1982).

134. US Senate, *FBI Budget and Oversight for Fiscal Year 1987,* before the Subcommittee on Security and Terrorism of the Committee on the Judiciary, 99th Congress (Washington, DC, 1986).

135. US Congress, *Departments of Commerce, Justice, and State, the Judiciary, and Related Agencies Appropriations for 1998,* before a subcommittee of the Committee on Appropriations, House of Representatives, 105th Congress (Washington, DC, 1997); Department of Justice, *A Review of the Federal Bureau of Investigation's Counterterrorism Program: Threat Assessment, Strategic Planning, and Resource Management* (Washington, DC, 2002), https://fas.org/irp/agency/doj/oig/fbi02sum.html.

136. US Congress, *Departments of Commerce, Justice, and State, the Judiciary, and Related Agencies Appropriations for 1992,* pt. 2, before a subcommittee of the Committee on Appropriations, House of Representatives, 102nd Congress (Washington, DC, 1991).

137. US Congress, *Departments of Commerce, Justice, and State, the Judiciary, and Related Agencies Appropriations for 1998.*

138. Presidential Decision Directive 39, "U.S. Policy on Counterterrorism," June 21, 1995, https://www.hsdl.org/?view&did=462942.

139. US Congress, *Departments of Commerce, Justice, and State, the Judiciary, and Related Agencies Appropriations for 1998.*

140. Presidential Decision Directive 63, "Critical Infrastructure Protection," May 22, 1998, https://irp.fas.org/offdocs/pdd/pdd-63.htm.

141. US Senate, *Securing Our Infrastructure: Private/Public Information Sharing,* before the Committee on Governmental Affairs, 107th Congress (Washington, DC, 2002).

142. General Accounting Office, *Critical Infrastructure Protection: Significant Challenges in Developing National Capabilities* (Washington, DC, 2001), https://www.gao.gov/assets/160/157052.pdf.

143. Jeffrey J. Berkin, assistant special-agent-in-charge, Milwaukee Division, Federal Bureau of Investigation, testimony before the House Committee on Government Reform Subcommittee on Government Efficiency, Financial Management, and Intergovernmental Relations, July 10, 2002, https://archives.fbi.gov/archives/news/testimony/milwaukee-division-counterterrorism-initiatives.

144. US Senate, *Improving Our Ability to Fight Cybercrime: Oversight of the National Infrastructure Protection Center,* before the Subcommittee on Technology,

Terrorism, and Government Information of the Committee on the Judiciary, 107th Congress (Washington, DC, 2001).

145. Ibid.

146. Presidential Decision Directive 63.

147. US Senate, *Improving Our Ability to Fight Cybercrime.*

148. General Accounting Office, *Critical Infrastructure Protection: Significant Challenges in Developing Analysis, Warning, and Response Capabilities* (Washington, DC, 2001), https://www.gao.gov/assets/110/108944.pdf.

149. US Senate, *Improving Our Ability to Fight Cybercrime.*

150. Larry Mefford, assistant director, Cyber Division, Federal Bureau of Investigation, testimony before the House of Representatives Committee on Government Reform, June 11, 2002, https://archives.fbi.gov/archives/news/testimony/nipcs-role-in-the-new-department-of-homeland-security.

151. Presidential Decision Directive 63.

152. Ibid.

153. US Senate, *FBI Oversight Hearing,* before the Subcommittee on Security and Terrorism of the Committee on the Judiciary, 97th Congress (Washington, DC, 1982).

154. General Accounting Office, *Critical Infrastructure Protection.*

155. US Congress, *Department of Justice Appropriation Bill for 1943,* before a subcommittee of the Committee on Appropriations, House of Representatives, 76th Congress (Washington, DC, 1942).

156. Ibid.

157. Ibid.

158. US Congress, *The Activities of the Federal Bureau of Investigation,* pt. 2, before the Subcommittee on Crime of the Committee on the Judiciary, House of Representatives, 105th Congress (Washington, DC, 1997).

159. Presidential Decision Directive 63.

160. US Senate, *Department of Defense Authorization for Appropriations for Fiscal Year 2000 and the Future Years Defense Program,* pt. 5, before the Committee on Armed Services, 106th Congress (Washington, DC, 1999).

161. US Senate, *Improving Our Ability to Fight Cybercrime.*

162. General Accounting Office, *Critical Infrastructure Protection.*

163. Ibid.; National Communications System, https://www.hsdl.org/?view&did=13856.

164. General Accounting Office, *Critical Infrastructure Protection.*

165. Public Law 107-296, "Homeland Security Act of 2002," https://www.dhs.gov/sites/default/files/publications/hr_5005_enr.pdf.

166. US Senate, *Counterterrorism,* before the Committee on the Judiciary, 107th Congress (Washington, DC, 2002).

167. Robert S. Mueller III, director, November 10, 2008, https://archives.fbi.gov/archives/news/speeches/using-intelligence-to-protect-our-communities. For the record, the archived version of this page refers to Mueller as the "Director, Central Intelligence Agency"—and it is on an FBI website.

168. Ibid.

169. White House, "The National Strategy for the Physical Protection of Critical Infrastructures and Key Assets" (Washington, DC, 2003), https://www.dhs.gov/xlibrary/assets/Physical_Strategy.pdf.

170. US Congress, *How Is America Safer? A Progress Report on the Department of Homeland Security,* before the Select Committee on Homeland Security, House of Representatives, 108th Congress (Washington, DC, 2003).

171. US Senate, *Department of Homeland Security's Information Analysis and Infrastructure Protection Budget Proposal for Fiscal Year 2005,* before the Select Committee on Homeland Security, 108th Congress (Washington, DC, 2004).

172. Department of Homeland Security, *Survey of the Information Analysis and Infrastructure Protection Directorate* (Washington, DC, 2004), https://www.oig.dhs.gov/sites/default/files/assets/Mgmt/OIG_SurveyIAIP_0204.pdf.

173. White House, "The National Strategy."

174. Ibid.

175. Ibid.

176. Department of Homeland Security, *Progress in Developing the National Asset Database* (Washington, DC, 2006).

177. US Senate, *Department of Homeland Security Appropriations for Fiscal Year 2004,* before the Committee on Appropriations, 108th Congress (Washington, DC, 2003), https://www.govinfo.gov/content/pkg/CHRG-108shrg2910448/html/CHRG-108shrg2910448.htm.

178. US Congress, *How Is America Safer? A Progress Report on the Department of Homeland Security*, before the Select Committee on Homeland Security, House of Representatives, 108th Congress (2003), https://www.govinfo.gov/content/pkg/CHRG-108hhrg96366/pdf./CHRG-108hhrg96366.pdf.

179. US Senate, *Department of Homeland Security Appropriations for Fiscal Year 2004.*

180. US Congress, *How Is America Safer?*; Department of Homeland Security, *Progress in Developing the National Asset Database.*

181. Department of Homeland Security, *Progress in Developing the National Asset Database.*

182. Department of Homeland Security, *Survey of the Information Analysis and Infrastructure Protection Directorate* (Washington, DC, 2004), https://www.oig.dhs.gov/sites/default/files/assets/Mgmt/OIG_SurveyIAIP_0204.pdf.

183. US Senate, *Department of Homeland Security's Information Analysis and Infrastructure Protection Budget Proposal for Fiscal Year 2005.*

184. US Senate, *Department of Homeland Security Status Report: Assessing Challenges and Measuring Progress,* before the Committee on Homeland Security and Governmental Affairs, 110th Congress (Washington, DC, 2007).

185. Congressional Research Service, *Critical Infrastructure: The National Asset Database* (Washington, DC, 2007), https://fas.org/sgp/crs/homesec/RL33648.pdf; Department of Homeland Security, *Progress in Developing the National Asset Database.*

186. Ibid.

187. US Senate, *Department of Homeland Security Appropriations for Fiscal Year 2007,* pt. 2, before the Committee on Appropriations, 109th Congress (Washington, DC, 2006).

188. Congressional Research Service, *Critical Infrastructure.*

189. Department of Homeland Security, *Efforts to Identify Critical Infrastructure Assets and Systems* (Washington, DC, 2009), https://www.oig.dhs.gov/sites/default/files/assets/Mgmt/OIG_09-86_Jun09.pdf.

190. Government Accountability Office, *DHS List of Priority Assets Needs to Be Validated and Reported to Congress* (Washington, DC, 2013), https://www.gao.gov/assets/660/653300.pdf.

191. Ibid.

192. Public Law 110-53, "Implementing Recommendation of the 9/11 Commission Act of 2007," https://www.congress.gov/110/plaws/publ53/PLAW-110publ53.htm.

193. Department of Justice, *The Accomplishments of the U.S. Department of Justice, 2001–2009* (Washington, DC, undated), https://www.justice.gov/sites/default/files/opa/legacy/2010/03/08/doj-accomplishments.pdf.

194. Government Accountability Office, *Critical Infrastructure: Preliminary Observations on DHS Efforts to Assess Chemical Security Risk and Gather Feedback on Facility Outreach* (Washington, DC, 2013), https://www.gao.gov/assets/660/653022.pdf.

195. US Congress, *Evaluating Internal Operation and Implementation of the Chemical Facility Anti-Terrorism Standards Program (CFATS) by the Department of Homeland Security,* before the Committee on Energy and Commerce, House of Representatives, 112th Congress (Washington, DC, 2012), https://www.govinfo.gov/content/pkg/CHRG-112hhrg75573/pdf./CHRG-112hhrg75573.pdf.

196. Acting deputy assistant secretary for infrastructure protection, testimony before the House Committee on Energy and Commerce, June 14, 2018, https://energycommerce.house.gov/sites/democrats.energycommerce.house.gov/files/documents/Testimony-Wulf-EE-Hrg-on-CFATS-A-Progress-Report-2018-06-14.pdf.

197. US Congress, *H.R. 2868: The Chemical Facility Anti-Terrorism Act of 2009,* before the Committee on Homeland Security, House of Representatives, 111th Congress (Washington, DC, 2009), https://www.govinfo.gov/content/pkg/CHRG-111hhrg51493/pdf./CHRG-111hhrg51493.pdf.

198. US Congress, *West Fertilizer, Off the Grid: The Problem of Unidentified Chemical Facilities,* before the Committee on Homeland Security, House of Representatives, 113th Congress (Washington, DC, 2013), https://www.govinfo.gov/content/pkg/CHRG-113hhrg86244/pdf./CHRG-113hhrg86244.pdf.

199. Acting deputy assistant secretary for infrastructure protection, testimony before the House Committee on Energy and Commerce, June 14, 2018.

200. Government Accountability Office, *Critical Infrastructure Protection.*

201. US Congress, *Evaluating Internal Operation and Implementation of the Chemical Facility Anti-Terrorism Standards Program.*

202. US Congress, *Chemical Facility Anti-Terrorism Standards (CFATS) Program: A Progress Update,* before the Committee on Energy and Commerce, House of Representatives, 113th Congress (Washington, DC, 2013), https://www.govinfo.gov/content/pkg/CHRG-113hhrg80377/pdf./CHRG-113hhrg80377.pdf.

203. Government Accountability Office, *Critical Infrastructure Protection.*

204. US Congress, *Evaluating Internal Operation and Implementation of the Chemical Facility Anti-Terrorism Standards Program.*

205. US Congress, *H.R. 4007: The Chemical Facility Anti-Terrorism Standards Authorization and Accountability Act of 2014,* before the Committee on Homeland Security, House of Representatives, 113th Congress (Washington, DC, 2014), https://www.govinfo.gov/content/pkg/CHRG-113hhrg88171/pdf./CHRG-113hhrg88171.pdf.

206. US Congress, *West Fertilizer.*

207. Acting deputy assistant secretary for infrastructure protection, testimony before the House Committee on Energy and Commerce, June 14, 2018.

208. Executive Order 13650, "Improving Chemical Facility Safety and Security," August 1, 2013, https://obamawhitehouse.archives.gov/the-press-office/2013/08/01/executive-order-improving-chemical-facility-safety-and-security.

209. Acting deputy assistant secretary for infrastructure protection, testimony before the House Committee on Energy and Commerce, June 14, 2018.

210. Government Accountability Office, *Critical Infrastructure Protection.*

211. US Congress, *The Chemical Facilities Anti-Terrorism Standards Program: Addressing Its Challenges and Finding a Way Forward,* before the Committee on

Homeland Security, 112th Congress (Washington, DC, 2012), https://www.govinfo.gov/content/pkg/CHRG-112hhrg76601/pdf./CHRG-112hhrg76601.pdf.

212. Acting deputy assistant secretary for infrastructure protection, testimony before the House Committee on Energy and Commerce, June 14, 2018.

213. Government Accountability Office, *Critical Infrastructure Protection.*

214. US Congress, *H.R. 4007.*

215. US Congress, *The Chemical Facilities Anti-Terrorism Standards Program: Addressing Its Challenges.*

216. US Congress, *Department of Homeland Security Appropriations for 2013,* pt. 5, before the Committee on Appropriations, House of Representatives, 112th Congress (Washington, DC, 2012), https://www.govinfo.gov/content/pkg/CHRG-112hhrg77757/pdf./CHRG-112hhrg77757.pdf.

217. US Congress, *The Chemical Facilities Anti-Terrorism Standards Program: Addressing Its Challenges.*

218. US Congress, *Department of Homeland Security Appropriations for 2013,* pt. 5.

219. US Congress, *H.R. 4007.*

220. Department of Homeland Security, *Effectiveness of the Infrastructure Security Compliance Division's Management Practices to Implement the Chemical Facility Anti-Terrorism Standards Program* (Washington, DC, 2013), https://www.oig.dhs.gov/assets/Mgmt/2013/OIG_13-55_Mar13.pdf.

221. Ibid.

222. US Congress, *Chemical Facility Anti-Terrorism Standards (CFATS) Program: A Progress Update.*

223. US Congress, *Securing Our Nation's Chemical Facilities: Building on the Progress of the CFATS Program,* before the Committee on Homeland Security, House of Representatives, 116th Congress (Washington, DC, 2019), https://www.govinfo.gov/content/pkg/CHRG-116hhrg35379/pdf./CHRG-116hhrg35379.pdf.

224. Government Accountability Office, *Critical Infrastructure Protection.*

225. US Senate, *Department of Homeland Security Appropriations for Fiscal Year 2007,* pt. 2.

226. US Congress, *Department of Homeland Security, Appropriations for 2012,* pt. 4, before a subcommittee of the Committee on Appropriations, House of Representatives, 112th Congress (Washington, DC, 2012).

227. US Congress, *H.R. 4007.*

228. US Congress, *Chemical Facility Anti-Terrorism Standards (CFATS) Program: A Progress Update.*

229. US Congress, *West Fertilizer.*

230. Ibid.

231. Christopher Krebs, director, Cybersecurity and Infrastructure Security Agency, testimony for the hearing "CISA Fiscal Year 2021 President's Budget," before the House Committee on Homeland Security, March 11, 2020, https://homeland.house.gov/imo/media/doc/Testimony%20-%20Krebs.pdf.

232. US Senate, *Current and Projected National Security Threats to the United States,* before the Select Committee on Intelligence, 105th Congress (1998), https://www.govinfo.gov/content/pkg/CHRG-105shrg51954/pdf./CHRG-105shrg51954.pdf.

233. Office of the National Counterintelligence Executive, *Foreign Spies Stealing US Economic Secrets in Cyberspace* (Washington, DC, 2011); Department of Justice, "Summary of Major U.S. Export Enforcement, Economic Espionage, Trade Secret, and Embargo-Related Criminal Cases," January 2008–present (October 2014), https://www.hsdl.org/?view&did=825549.

234. Department of Justice, "Summary of Major U.S. Export Enforcement, Economic Espionage, Trade Secret, and Embargo-Related Criminal Cases."

235. Department of Justice, "Former Dow Research Scientist Sentenced to 60 Months in Prison for Stealing Trade Secrets and Perjury," January 13, 2012, https://www.justice.gov/opa/pr/former-dow-research-scientist-sentenced-60-months-prison-stealing-trade-secrets-and-perjury.

236. Government Accountability Office, *Critical Infrastructure Protection.*

237. David Wulf, *The Chemical Facility Anti-Terrorism Standards Program,* before the Committee on Homeland Security, House of Representatives. (2019), https://homeland.house.gov/imo/media/doc/Testimony-Wulf.pdf.

238. Office of the Director of National Intelligence, "What Is Intelligence?" https://www.dni.gov/index.php/what-we-do/what-is-intelligence.

239. National Geospatial Intelligence Agency, *Geospatial Intelligence (GEOINT) Basic Doctrine,* publication 1.0 (Washington, DC, 2006), https://fas.org/irp/agency/nga/doctrine.pdf.

240. Central Intelligence Agency, memorandum to Honorable Daniel K. Inouye, "Transmittal of Requested Report on Non-Military Uses of Intelligence Assets," October 13, 1977, https://www.cia.gov/library/readingroom/docs/CIA-RDP83M00171R001200190001-4.pdf; "Background Paper for Information of CIA Oversight Committees on the Partial Use of NRP Assets for Civil Applications," undated, https://www.cia.gov/library/readingroom/document/cia-rdp80t01137a000300020001-5.

241. Authority for the National Reconnaissance Program, "Domestic Satellite Reconnaissance Activities," https://nsarchive2.gwu.edu/NSAEBB/NSAEBB229/23.pdf.

242. "Project ARGO: Comments on Draft of Memo to Accompany ARGO Final Report," February 13, 1968, https://www.cia.gov/library/readingroom/docs/CIA-RDP80T01137A000300030007-8.pdf; "ARGO Chronology," https://www.cia.gov/library/readingroom/docs/CIA-RDP80T01137A000200060038-2.pdf.

243. Memorandum for the record, "Peaceful Uses," April 27, 1967, https://www.cia.gov/library/readingroom/docs/CIA-RDP80T01137A000600010014-9.pdf.

244. "Project Argo," https://www.cia.gov/library/readingroom/docs/CIA-RDP80T01137A000200060044-5.pdf.

245. "Project ARGO: Comments on Draft of Memo to Accompany ARGO Final Report."

246. Minutes of the ARGO Steering Committee meeting, June 10, 1968, https://www.cia.gov/library/readingroom/docs/CIA-RDP80T01137A000200060034-6.pdf. (Project Argo, September 23, 1968), https://www.cia.gov/library/readingroom/docs/CIA-RDP80T01137A000200060025-6.pdf.

247. ARGO Steering Committee Charter, October 28, 1970, https://www.cia.gov/library/readingroom/docs/CIA-RDP80T01137A000300030012-2.pdf.

248. Memorandum to chairman, Committee for Imagery Requirements and Exploitation (COMIREX), "KH-4 Coverage of Selected Areas of the United States," via chairman of the ARGO Steering Group, March 30, 1970, https://www.cia.gov/library/readingroom/docs/CIA-RDP80T01137A000200030016-9.pdf.

249. Memorandum for the record, "Argo Meeting, 28 January 1970," February 3, 1970, https://www.cia.gov/library/readingroom/docs/CIA-RDP80T01137A000200030041-1.pdf.

250. "Note for A/DDS&T," January 23, 1975, https://www.cia.gov/library/readingroom/docs/CIA-RDP80T01137A000300040001-3.pdf.

251. John Prados, *The Family Jewels: The CIA, Secrecy, and Presidential Power* (Austin: University of Texas Press, 2013).

252. "Note for A/DDS&T."

253. "Implementation of the Recommendation of the Commission on CIA Activities Within the United States Pertaining to Civilian Agencies' Use of Classified Overhead Photography," August 19, 1975, https://www.cia.gov/library/readingroom/docs/CIA-RDP80M01133A000900130007-9.pdf.

254. Ibid.

255. Richard A. Best Jr. and Jennifer K. Elsea, *Satellite Surveillance: Domestic Issues* (Washington, DC, 2011), https://fas.org/sgp/crs/intel/RL34421.pdf.

256. Memorandum to Honorable Daniel K. Inouye, "Transmittal of Requested Report on Non-Military Uses of Intelligence Assets."

257. Memorandum to the deputy director of central intelligence, "Current Guidelines on Photography of US," December 14, 1976, https://www.cia.gov/library/readingroom/docs/CIA-RDP79M00467A002400090001-6.pdf.

258. Best and Elsea, *Satellite Surveillance.*

259. US Congress, *Turning Spy Satellites on the Homeland: The Privacy and Civil Liberties Implications of the National Applications Office,* before the Committee on Homeland Security, House of Representatives, 110th Congress (Washington, DC, 2007), https://www.govinfo.gov/content/pkg/CHRG-110hhrg48963/pdf./CHRG-110hhrg48963.pdf.

260. Civil Applications Committee (CAC) Blue Ribbon Study, *Independent Study Group Final Report* (2005), https://nsarchive2.gwu.edu//NSAEBB/NSAEBB229/40.pdf.

261. Ibid.

262. US Senate, *Confronting the Terrorist Threat to the Homeland: Six Years After 9/11,* before the Committee on Homeland Security and Governmental Affairs, 110th Congress (Washington, DC, 2007), https://www.govinfo.gov/content/pkg/CHRG-110shrg38842/pdf./CHRG-110shrg38842.pdf.

263. "Charter: National Applications Office," https://nsarchive2.gwu.edu//NSAEBB/NSAEBB229/48.pdf.

264. US Congress, *The President's Fiscal Year 2009 Budget Request for the Department of Homeland Security,* before the Committee on Homeland Security, House of Representatives, 110th Congress (Washington, DC, 2008), https://www.govinfo.gov/content/pkg/CHRG-110hhrg44512/pdf./CHRG-110hhrg44512.pdf.

265. "Fact Sheet: National Applications Office," August 15, 2007, https://nsarchive2.gwu.edu//NSAEBB/NSAEBB229/43.pdf.

266. "Charter: National Applications Office."

267. "Fact Sheet: National Applications Office."

268. Bennie G. Thompson, memorandum to Michael Chertoff, August 22, 2007, https://fas.org/irp/congress/2007_cr/thompson082207.pdf.

269. Bennie G. Thompson, Jane Harman, and Christopher P. Carney, memorandum to Michael Chertoff and Charles Allen, September 6, 2007, https://www.democraticleader.gov/newsroom/homeland-security-chairs-call-for-moratorium-on-spy-satellite-program; US Congress, *Homeland Security Intelligence: Its Relevance and Limitations,* before the Committee on Homeland Security, House of Representatives, 111th Congress (Washington, DC, 2009), https://www.govinfo.gov/content/pkg/CHRG-111hhrg49943/pdf./CHRG-111hhrg49943.pdf.

270. US Congress, *Turning Spy Satellites on the Homeland.*

271. US Congress, *Fiscal Year 2010 Budget for the Office of Intelligence and Analysis of the Department of Homeland Security,* before the Committee on Homeland Security, House of Representatives, 111th Congress (Washington, DC, 2009), https://www.govinfo.gov/content/pkg/CHRG-111hhrg51634/pdf./CHRG-111hhrg51634.pdf; Best and Elsea, *Satellite Surveillance.*

272. Best and Elsea, *Satellite Surveillance.*

273. US Congress, *Spending Priorities and Missions of the USGS Survey and the President's FY 2012 Budget Proposal,* before the Committee on Natural Resources, House of Representatives, 112th Congress (Washington, DC, 2011), https://www.govinfo.gov/content/pkg/CHRG-112hhrg65119/pdf./CHRG-112hhrg65119.pdf; US Group on Earth Observations, *2019 National Plan for Civil Earth Observations,* https://www.whitehouse.gov/wp-content/uploads/2019/12/Natl-Plan-for-Civil-Earth-Obs.pdf.

274. *Department of the Interior, Environment, and Related Agencies Appropriations Bill, 2010,* report together with minority views to accompany H.R. 2996, https://www.govinfo.gov/content/pkg/CRPT-111hrpt180/pdf./CRPT-111hrpt180.pdf.

275. Civil Applications Committee and National Civil Applications Center, Overview Briefing to NSTC's Subcommittee on Disaster Response (undated), https://www.sdr.gov/pdfs/Presentations/2017/2017-0105%20-%20Paul%20Young-Dan%20Opstal%20(USGS)%20-%20CAC-NCAC%20Overview.pdf.

276. "Memorandum of Understanding Between the Department of Defense and the Department of Homeland Security Regarding the Non-Reimbursable Exchange of Liaisons Within the National Capital Region," 2013, https://www.jcs.mil/Portals/36/Documents/Doctrine/Interorganizational_Documents/doj_mou_liaisons_cap_region2013.pdf.

277. Civil Applications Committee (CAC) Blue Ribbon Study, *Independent Study Group Final Report.*

278. "Memorandum of Understanding Between the Department of Defense and the Department of Homeland Security."

279. Memorandum to COMOR Photo Working Group, "Revised List of Domestic Targets for KH-4," April 28, 1967, https://www.cia.gov/library/readingroom/docs/CIA-RDP79B01709A003000050004-1.pdf; memorandum to Committee on Overhead Reconnaissance, "Addition of Communications Requirement to COMOR-D-69/32," May 25, 1967, https://www.cia.gov/library/readingroom/docs/DOC_0001041094.pdf.

280. Memorandum to COMOR Photo Working Group, "Revised List of Domestic Targets."

281. Planning, Programming, and Budgeting Staff, National Photographic Interpretation Center, memorandum for the record, "ARGO Meeting, Room 208, EOB, December 9, 1969," December 17, 1969, https://www.cia.gov/library/readingroom/docs/CIA-RDP80T01137A000200060002-1.pdf.

282. "Civil Applications Following ARGO Study" (1973), https://www.cia.gov/library/readingroom/docs/CIA-RDP80T01137A000300040008-6.pdf.

283. Ibid.

284. Memorandum for the record, "Steering Committee / ARGO Meeting, 9 July 1969," August 25, 1969, https://www.cia.gov/library/readingroom/docs/CIA-RDP80T01137A000200060011-1.pdf.

285. "Civil Applications Following ARGO Study."

286. Ibid.; memorandum to Imagery Collection Requirements Subcommittee, "Photographic Coverage Requirements, Office of Emergency Planning," October 9, 1968, https://www.cia.gov/library/readingroom/docs/CIA-RDP79B01709A003300040009-4.pdf.

287. US Congress, *Turning Spy Satellites on the Homeland.*

288. Memorandum to chairman, Committee for Imagery Requirements and Exploitation, "KH-4 Coverage of Selected Areas of the United States."

289. Best and Elsea, *Satellite Surveillance.*

290. Memorandum for the record, "Steering Committee / ARGO Meeting, 9 July 1969."

291. ARGO Steering Committee Charter, October 28, 1970, https://www.cia.gov/library/readingroom/docs/CIA-RDP80T01137A000300030012-2.pdf.

292. "Establishment of the Committee for Civil Applications of Classified Overhead Phtography of the United States," October 3, 1975, https://www.cia.gov/library/readingroom/docs/CIA-RDP79M00467A002400090002-5.pdf.

293. National Photographic Interpretation Center observer, memorandum to chairman, Imagery Collection Requirements Subcommittee, COMIREX, "ARGO Domestic Targets Requirement," October 26, 1968, https://www.cia.gov/library/readingroom/docs/CIA-RDP80T01137A000200060023-8.pdf.

294. Jeffrey T. Richelson, *The US Intelligence Community,* 6th ed. (Boulder: Westview, 2012), p. 25.

295. Director, National Photographic Interpretation Center, memorandum to deputy director of intelligence, "ARGO and Intelligence Resources," May 8, 1970, https://www.cia.gov/library/readingroom/docs/CIA-RDP80T01137A000300030013-1.pdf.

296. Ibid.

297. Executive Office of the President, Office of Science and Technology, memorandum, "ARGO," May 4, 1970, https://www.cia.gov/library/readingroom/docs/CIA-RDP80T01137A000200030013-2.pdf.

298. "Implementation of the Recommendation of the Commission on CIA Activities."

299. Memorandum to Honorable Daniel K. Inouye, "Transmittal of Requested Report on Non-Military Uses of Intelligence Assets."

300. Civil Applications Committee (CAC) Blue Ribbon Study, *Independent Study Group Final Report.*

301. US Congress, *Turning Spy Satellites on the Homeland.*

5

Securing the Cyber Realm

THE CYBER DOMAIN IS NOT IN ITSELF A THREAT BUT RATHER AN ENVIronment where threat actors can pursue intelligence collection and terrorist activities in new ways. As then–FBI director James Comey put it in 2013, "cyber is sort of an evil layer cake."[1] It is a critical domain, because in addition to facilitating threat activities, the interconnectedness of the cyber environment amplifies these activities. Involvement of private industry in combating the exploitation of this space is essential for multiple reasons. As with other aspects of critical infrastructure, private industry plays a significant role in developing and maintaining the networks on which the country relies for secure storage and communication of data. Furthermore, industry is at the cutting edge of cyber technology development, which means it is also positioned to have an early understanding of the vulnerabilities inherent to that technology.

The US government's establishment of a cybersecurity interface with the private sector has produced an alphabet soup of agencies that have been formed, shuffled, and renamed into the current (until the next reorganization occurs) bureaucratic architecture. Although describing the ceaseless cavalcade of component comings and goings may seem to be an exercise in meaningless minutiae, it is important to understand the evolution of government's capability (or lack thereof) to address the cyber threat environment in order to avoid re-creating the wheel or, worse, making the same mistakes again (a not unheard of government pastime). For those who would rather not delve into this particular chapter of history, the important players at present are the Department of Homeland Security's Cybersecurity and Infrastructure Security Agency (CISA) and the Federal Bureau of Investigation's Cyber Division as well as the Bureau's InfraGard program.

Old Targets, New Weapons

Physical manifestation of intelligence collection, counterproliferation, and terrorism all have cyber corollaries. The National Counterintelligence Executive (now the National Counterintelligence and Security Center [NCSC]) noted, in a 2011 report, that cyberspace provides an alluring environment for intelligence collection, since there is less of an opportunity for victims to identify the perpetrator; it is a cost-effective tactic; and it can be done from outside of the United States.[2] In 2014, the US Department of Justice highlighted this reality when it indicted five Chinese military hackers for targeting the US nuclear power, metals, and solar products industries. The conspirators pilfered trade secrets that would have been especially useful to Chinese companies.[3]

Proliferation—transmitting sensitive technologies to problematic end-users—has also gone online. In 2019, the US Department of Justice indicted the Chinese telecommunication company Huawei for a wide-ranging effort to steal trade secrets. As part of the investigation that led to the indictments, the Federal Bureau of Investigation (FBI) discovered that Huawei provided employees of T-Mobile with an encrypted email address to which those employees could provide stolen information.[4]

Finally, terrorism via the internet is a very real possibility. For a number of years, high-level US government officials and informed commentators have warned about the possibility of a "cyber Pearl Harbor."[5] The potential for physical and economic catastrophe has become evident through real-world events. In 2013, for example, Iranian hackers infiltrated the control systems of a dam less than 20 miles from New York City.[6] During the following year, North Korea deployed destructive malware against Sony Pictures Entertainment and rendered thousands of the company's computers inoperable.[7] The use of the internet for sabotage is particularly troubling in its empowerment of otherwise inconsequential actors. James Clapper noted in 2013 that isolated state or nonstate actors could deploy unsophisticated cyber attacks as either retaliation or provocation. However, Clapper warned that these types of attacks could have "significant outcomes due to unexpected system configurations and mistakes, or that the vulnerability of one node might spill over and contaminate other parts of a networked system."[8]

Connectivity of critical infrastructure to the cyber environment increases its vulnerability to terrorist attacks. The electric power grid is one such tempting target. In March 2005, security consultants within the electric industry indicated that hackers had targeted the grid.[9] In 2013, the Department of Homeland Security (DHS) reported that actors probing the grid included Iran. Not to be outdone, North Korean hackers, as of 2017, had targeted US electric companies with a spear-phishing campaign directed at probing the utilities' defenses. Russian hackers—in 2017—had compromised US electric utilities' networks and, according to the DHS, had positioned themselves to cause blackouts.[10]

China has also used cyber intrusions to collect information against critical infrastructure. In 2013, Chinese hackers compromised the Army Corps of Engineers' National Inventory of Dams.[11] Interestingly, in 2014 the US Department of Justice indicted Xiafen "Sherry" Chen—a Chinese-born, naturalized US citizen employed by the National Weather Service—for having downloaded sensitive files from the National Inventory of Dams.[12] Chen had previously received an inquiry from Jiao Yong, the vice minister of China's Ministry of Water Resources, regarding how the United States funded repairs of its reservoir system.[13] The human intelligence (HUMINT) and signal intelligence (SIGINT) targeting of the dam database seems unlikely to be a coincidence. Regardless of whether Chen was innocent or not, the episode highlights how a pernicious and persistent threat actor might use a variety of vectors to develop a detailed picture of US vulnerabilities.

Evolution of the Threat

The evolution of the cyber-enabled threat has two aspects. First, a threat actor must gain access to technology. This includes both the adversary's acquisition of equipment as well as its ability to access the systems that it is targeting. Second, an adversary must continually develop knowledge about how to use the tools it has acquired as the cyber environment changes over time. Exploiting a bank computer during the 1970s is, after all, very different than finding ways to compromise 5G networks or the Internet of Things.

Gaining Access

Much of the early threat to what is now in the cyber category was Soviet-sponsored. As early as 1961, the FBI warned Congress about Soviet and satellite agents, many of whom were operating under diplomatic cover, against communications systems.[14] *Politico,* in 2016, had obtained similar information regarding odd behavior by Russian diplomats that led US intelligence officials to believe that the Russian spy games[15] were directed at mapping US telecommunications infrastructure.[16] In addition to critical infrastructure, the Soviets targeted US computer technology on US soil. By the mid-1980s, the US government had determined that Soviet trade representatives regularly traveled—as parts of delegations—to California and demonstrated an interest in the work going on in Silicon Valley.[17] Additionally, the Soviets maintained a consulate in San Francisco, which gave Moscow a human footprint useful for targeting high-technology developments. In the early 1980s, US officials noted that there were approximately a hundred Soviet agents, many of whom had technology and engineering backgrounds, assigned at the Soviet consulate in San Francisco.[18] Additionally, the Soviet's San Francisco consulate gave Moscow a SIGINT base near a region burgeoning with new technology. As the head of the FBI's

San Francisco field office told the *Los Angeles Times* in the mid-1980s, "the gear on the building isn't there for picking up TV programs."[19] According to the National Security Agency (NSA), a 1971 KGB directive ordered the strengthening of communications interceptions against scientific and technical targets—including IBM and GE.[20]

Developing Capabilities

Intelligence collection has, not surprisingly, historically included acquisition of technology that would enable threat actors to field cyber capabilities against the United States. For instance, by the mid-1980s, the Soviets had acquired more than 300 different types of US and other Western computer hardware and software that a US government assessment concluded "enabled them to develop the technical ability to penetrate at least some US automated systems."[21] This type of collection about the technical specifications of systems has continued. For instance, a 2003 report stated that, according to the National Security Agency, potential adversaries were developing a body of knowledge about US systems and how to attack those systems.[22]

US intelligence officials in the early 2000s drove home a sobering reality: it was an increasingly simple matter for threat actors to acquire the knowledge and technical capability for cyber-facilitated attacks. In 2000, then–director of central intelligence George Tenet observed that a surprising number of information warfare–related tools and weapons were available on the open market at relatively little cost.[23] Several years later, in 2003, then–FBI director Robert Mueller III painted a similarly gloomy picture. According to Mueller, it was easier than ever for malicious actors to acquire capabilities for cyber attacks due to the "prevalence of publicly available hacker tools."[24]

Threat actors increasingly do not even need to acquire technical capabilities themselves; they can simply recruit individuals who have knowledge. In 2005, the US intelligence community assessed that terrorist organizations had expanded their recruitment efforts to target individuals studying mathematics, computer science, and engineering as the organizations attempted to develop capabilities for attacks against US technical systems.[25] Threat actors have even been able to engage the skills of former US intelligence personnel who have intimate knowledge of communications networks. The United Arab Emirates, for instance, simply hired former NSA personnel.[26]

Early Investigations

The FBI's early experiences with investigating cyber-related threat activities began with computer crimes. In 1974, for example, the Bureau informed Congress about how a bank employee who had perpetrated a fraud after discovering that the bank's computer only read the magnetically printed account number at the bottom of deposit slips and not the account number at the top of the form. The enterprising employee replaced all of the deposit slips on the lobby desks in the bank with his own electronically coded deposit forms. In another computer-aided fraud, a bank employee who directed the bank's system to issue dividend checks in the names of former shareholders sent the checks to an accomplice, and then erased any record of the checks having been issued.[27]

A focus on white-collar crime continued to characterize the FBI's handling of cyber issues into the 1980s. In 1975, then-director Clarence Kelley advised Congress that what the FBI was seeing was not in the area of illicit intrusions into computer systems but rather "the machinations" of those with legitimate access to computers who used them to perpetrate fraud.[28] According to then-director William Webster, speaking in 1983, "computer crime falls generally within the white-collar crime program of the Bureau. . . . The computer is used primarily in fraud cases, in the fraudulent electronic transfer of funds."[29] The following year, Webster alluded to a "newly recognized vulnerability" in modern electronic means of communication.[30]

The Bureau's conceptualization of cyber threats markedly shifted once the US government created the National Security Threat List. In 1992, the attorney general introduced the list as a framework for addressing counterintelligence issues. In 1995, the US government added the issue of targeting the national information infrastructure to it. As defined by the National Security Threat List, the issue covered foreign power–sponsored or foreign coordinated intelligence activity targeting the facilities, personnel, information, or communications systems that compose or are associated with the national information infrastructure. Specific proscribed activities included denial or disruption of computer, cable, satellite, or telecommunications services; unauthorized monitoring of computer, cable, satellite, or telecommunications systems; unauthorized disclosure of proprietary information stored within or communicated through computer, cable, satellite, or telecommunication systems; unauthorized modification or destruction of computer programming codes of information, computer network databases, stored information, or computer capabilities; and manipulation of computer, cable, satellite, or telecommunications services resulting in fraud, financial loss, or other federal criminal violations.[31] In other words, foreign powers should rethink meddling with US cyber networks.

In an early effort to organize its response to address cyber vulnerabilities, the FBI developed a field office–based capability. It announced, in 1992, that it was establishing an entire squad at the Bureau's Washington, DC, metropolitan field office that would address national and international computer crimes. This squad would function as a liaison with all of the FBI's field offices in order to tackle cyber issues, which were not bounded by the same arbitrary, geographic divisions for which individual field offices were responsible, and instead could be national in scope.[32] By 1996, San Francisco had added a similar squad. (It is interesting—and perhaps indicative of an out-of-touch FBI—that the San Francisco office, the office closest to Silicon Valley, was not the first office to have a squad directed at combating computer crime.) The Bureau also planned to establish a similar squad in its New York field office.[33]

As it developed its capability for responding to cyber-related issues, the FBI grappled with how this challenge fit within its organizational concept. Speaking in 1999, then-director Louis Freeh acknowledged that historically the Bureau had built up its offices by program (e.g., a bank robbery squad, a terrorism squad). However, cyber was not an issue to be worked but rather a tool that introduced greater complexity to how threat actors—both state and nonstate—operated. Freeh seemed to recognize this when he stated that the FBI needed "an interdisciplinary squad."[34] The FBI would continue to grapple with how cyber fit within an organizational structure built around threat actors.

Computer Investigations and Infrastructure Threat Assessment Center

The FBI, in addition to its field components, developed a centralized approach to countering threat actors' use of the cyber environment. An executive order issued in July 1996 established an FBI-chaired Infrastructure Protection Task Force (IPTF). The IPTF had multiple ramifications for the relationship between the federal government and private industry in the field of national security. First, the IPTF was responsible for engaging with the private sector in the process of identifying and coordinating expertise within and outside of the federal government to "provide, or facilitate and coordinate the provision of expert guidance to critical infrastructure to detect, prevent, halt, or confine an attack and to recover and restore service."[35] Additionally, the IPTF had a responsibility for information-sharing, as it was expected to "issue threat and warning notices in the event advance information is obtained about a threat."[36] Finally—in terms of responsibilities to the private sector—the IPTF would "provide training and education on methods of reducing vulnerabilities and responding to attacks on critical infrastructure."[37]

It was against this backdrop that the FBI developed its approach to cyber issues. In July 1996 it established the Computer Investigations and Infrastructure Threat Assessment Center (CITAC).[38] According to the budget justification, CITAC was envisioned as "one of the most ambitious projects ever undertaken by the Federal government to counter the threats posed by foreign powers, terrorist groups, criminal organizations, and nonstate entities to the [national information infrastructure]."[39] Calling something "one of the most ambitious projects ever undertaken" is akin to the phrase "unsinkable ship." CITAC was supposed to serve as the point of coordination for criminal investigative, counterterrorism, and counterintelligence responsibilities with a cyber nexus.[40] Its functions included initiating activities to identify and counter the activities of entities intruding into the national information infrastructure.[41] (According to a 1993 US government definition, the national information infrastructure comprises "a wide range and ever-expanding range of equipment including cameras, scanners, keyboards, telephones, fax machines, computers, switches, compact disks, video and audio tape, cable, wire, satellites, optical fiber transmission lines, microwave nets, switches, televisions, monitors, printers, and much more."[42] By 2020 the "much more" included such developments as smartphones and the Internet of Things.)

As the Bureau noted in 1998, national information infrastructure intruders' intentions were not usually readily apparent at the outset of events, and, in order to develop an appropriate response, the US government needed to work with private sector owners and operators.[43] Consistent with the executive order that established the IPTF, CITAC endeavored to develop relationships with the private sector in order to glean its knowledge about the nature and extent of foreign involvement in attacks on commercial, financial, and other sectors.[44] Additionally, it leveraged expertise from across the government (presaging the creation of the National Infrastructure Protection Center [NIPC]).[45] CITAC's deliverable from this engagement was to be an analysis of vulnerabilities in US infrastructure and development of recommendations about how to harden those vulnerabilities.[46]

This headquarters-based entity grew out of the expanding cyber-oriented investigative infrastructure that was developing in the field. According to 1998 congressional testimony, the FBI established CITAC specifically to support the existing network of field-based squads by providing in-house support to criminal and national security investigations. With the creation of CITAC, the FBI reorganized its apparatus for addressing issues with a cyber nexus by staffing up computer investigation and infrastructure threat assessment squads in the field. At least the names of the field and headquarters programs now aligned. In addition to the squads in Washington, DC, New York, and San Francisco, the FBI announced its intention, in 1998, to create such squads in Chicago, Dallas, and Los Angeles.[47]

National Infrastructure Protection Center

Remember the touting of CITAC as one of the most "ambitious projects"? It apparently was not ambitious enough, since in 1997 the Department of Justice and the FBI began developing plans for the National Infrastructure Protection Center, which would supplant CITAC.[48] This new entity was established, in February 1998, as a focal point for the federal government's efforts to protect critical infrastructure.[49] A presidential decision directive officially established the NIPC in 1998.[50] On October 2, 1998, the NIPC became part of the FBI's National Security Division, and subsequently moved to the Bureau's newly created Counterterrorism Division in 1999.[51] The NIPC had an expansive mandate, with a mission of "providing timely warnings of international threats, comprehensive analysis and law enforcement investigation and response."[52] Furthermore, although led by the FBI, the NIPC had a significant interagency aspect. Agencies represented at the NIPC included the Central Intelligence Agency (CIA), the Department of Defense, the US Postal Service, the General Services Administration, and the Department of Energy.[53]

Organization of the NIPC

The NIPC was organized—for the benefit of bureaucracy junkies—into three sections.[54] The NIPC's operational and response arm was known as the Computer Investigations and Operations Section, through which the NIPC managed the FBI field offices' investigation of cyber intrusions. The NIPC's Analysis and Warning Section served as the indications and warning element. It reviewed multiple sources including government and private sector databases and media for information relevant to any aspect of the NIPC's mission, including indicators of a possible attack.[55] (An example of this activity was the NIPC's Project La Resistance,[56] which was an effort to gather information from disparate sources to identify linkages and commonalities among incidents and perpetrators.)[57] Finally, the Training, Administration, and Outreach Section coordinated the training and education of cyber investigators. It also had an essential role in coordinating the NIPC's outreach to the private sector, one of the NIPC's primary constituents. The section was responsible for collecting and cataloging the data for the FBI's inconsistently implemented "key asset" initiative.[58]

The NIPC integrated aspects of multiple FBI divisions and missions. It worked in conjunction with the FBI's counterterrorism efforts to neutralize and penetrate terrorist groups that engaged in cyber crime. Additionally, the NIPC had counterintelligence-related responsibilities, in its efforts directed at state-sponsored threats to infrastructure. Finally, in conjunction with the Criminal Investigative Division, the NIPC looked at the threats to infrastructure from criminal entities.[59] In addition to programmatic integration, the NIPC had responsibilities for oversight and coordination of FBI computer-related investigations. It monitored all of the relevant investigations—as well as the monthly statistical reports—on which field offices were working.[60] Beyond its oversight role, the NIPC served as a hub for coordinating complex investigations of computer-related crimes that impacted the areas of responsibility of multiple field offices and legal attaché (the FBI's liaisons abroad).[61] Finally, the NIPC also attempted to provide resources to field investigations, in the form of technical support, analysis, expert assistance for interviews, tools for mitigation of computer-based attacks, and administrative support.[62]

Consistent with the NIPC's interagency aspirations, its leadership included billets for multiple federal entities. The NIPC deputy director was a two-star navy rear admiral; the executive director was detailed from the Air Force Office of Special Investigations; the head of the unit responsible for analysis and information-sharing came from the National Security Agency; the section chief for analysis and warning was from the CIA; and the US Secret Service provided the assistant section chief for computer investigations and operations.[63] This might seem to have made for one friendly federal family, but that was not really the case. The NIPC began operations less-than-auspiciously, in 1998, with no permanent staff.[64] Then, in 1999, Congress prohibited the FBI from reimbursing other agencies for detailees to the NIPC, which, as the Bureau observed, "made it somewhat more difficult for other agencies to devote scarce resources to our common mission at the NIPC."[65] (As of 2001, the position of chief of the Analysis and Warning Section—which was to have been filled by the CIA—had been vacant for approximately half of the NIPC's existence.[66] Furthermore, the NSA billet was vacant for seventeen months between May 1998 and April 2000.)[67]

Money was not the only impediment to achieving interagency cooperation. The FBI dominated what was supposed to be a cross-governmental enterprise.[68] Detailees complained about poor treatment during their time with the center.[69] Troublingly, detailees felt that they were not afforded the same level of respect and support as were FBI personnel.[70] The FBI also felt stymied by a lack of de facto government cooperation. In November 2000, then-director Louis Freeh complained that other federal agencies did not recognize the NIPC's mission and, without intervention by the National Security Council, would not be able to provide analysis and warning functions.[71]

The NIPC in the Field

The FBI, as it had with CITAC, aligned its field-based cyber investigative capabilities with the headquarters-based NIPC. In October 1998, the FBI established a National Infrastructure Protection and Computer Intrusion (NIPCI) program, managed by the NIPC, at all of its field offices.[72] As of 1999, the Washington, DC, New York, San Francisco, Chicago, Dallas, Los Angeles, Atlanta, Charlotte, Boston, and Seattle field offices had full NIPCI squads, while the rest of the offices had smaller teams.[73] Each squad consisted of approximately eight FBI agents and the teams consisted of between one and five agents.[74] In 1999, the FBI stated its aspiration was to create a full NIPCI squad in each field office.[75] By the end of 2000, sixteen field offices had NIPCI squads. Additionally, by the end of the year the FBI had established a regional task force to address computer-crime cases.[76] However, like at NIPC, the work in the field was less a program and more a jumble of responsibilities. Field personnel assigned to the NIPCI program had to conduct computer intrusion investigations, respond to threats, and gather information for the key asset initiative.[77]

The NIPC's role was one of oversight and assistance. It monitored the open investigations in all of the FBI field offices. Furthermore, it coordinated investigations across FBI field offices—giving a desperately needed national perspective to cyber cases, which by their nature did not fit neatly within the offices' arbitrary geographic boundaries. The NIPC also furnished support in the form of expert assistance for interviews and in the form of technology for analyzing and mitigating computer-based attacks.[78] Additionally, the NIPC handled training for field personnel. Through training, the FBI hoped to keep pace with rapid changes in technology and, in 1999, trained nearly 400 FBI agents, state and local law enforcement representatives, and representatives from other government agencies on topics including computer intrusions and network analysis as well as the workings of the energy and telecommunications key assets.[79]

Further undermining the field NIPCI program was the NIPC's inability to direct resources. It did not have its own agents in the field.[80] Squads were under the control of field office supervisors rather than the NIPC. Although an incident might have an NIPC nexus, individual field officers determined whether it was necessary to open a case.[81] The best that the NIPC could do was to advise the field about the importance of infrastructure protection and to engage in outreach activities in order to solicit cooperation from entities including those in the private sector. However, the field offices in control of resources were often overworked and tended to prioritize ongoing investigations rather than the development of critical infrastructure awareness.[82]

Failure of the NIPC and Creation of the DHS

Where the NIPC fundamentally failed was in its ability to serve one of its primary customers: the private sector. This was due to three reasons. The first was the inability to understand the environment of private industry. It is nearly impossible to effectively protect unknown territory. The second issue was the NIPC's ability to remain ahead of the threat. Arguably, industry knew more about its own threat environment than the FBI did, and many of the NIPC's efforts to warn about threats amounted to too little, too late. Finally, the NIPC was unable to gain the private sector's confidence necessary for collaboration.

Lack of Knowledge

The personnel who remained at the NIPC struggled to advance the organization's mission. Many of the NIPC's problems were in the area of analysis. (This is not surprising given the difficulty that the Bureau encountered in developing a competent analytical cadre both before and after 9/11.) As of 2001, the NIPC had assessed that it needed twenty-four analysts but was operating with only thirteen.[83] (This was at odds with a statement to Congress by then-director Louis Freeh two years earlier claiming that "the NIPC is nearly fully staffed and has been . . . a very, very successful endeavor."[84] Someone clearly got their wires crossed—always a bad thing but especially not great, even from a metaphorical perspective, when talking about computers.) These thirteen analysts were not a gallant band of cyber sleuths with exceptional intellectual firepower ready to combat all threats. Instead, as the General Accounting Office assessed, during the same year, the NIPC lacked expertise.[85] According to NIPC officials, most of the FBI employees assigned to the Analysis and Information-Sharing Unit (part of Analysis and Warning Section) lacked the necessary skills for the functions for which they were responsible. Senior NIPC officials acknowledged that the unit had to rely on detailees from other agencies—not a great position given how difficult it was for the NIPC to attract detailees—to supplement the FBI staff since, as the Bureau acknowledged, it lacked a cadre with experience in critical infrastructure operations.[86]

Subpar analytic capabilities harmed the NIPC (and are a warning to the FBI as it continues to refine its analytic program). Several officials within the intelligence community viewed the NIPC as a second-tier entity that did not generate original analytical products. This was a stinging assessment but not incorrect. The Analysis and Information-Sharing Unit, according to the General Accounting Office, noted that most of the products that the unit issued were merely compilations of information that other entities had previously reported.[87]

The difficulties with developing awareness of key assets undermined the NIPC's ability to carry out its mission. (To paraphrase *The Music Man,* the NIPC didn't know the territory.) There was a definite disconnect between what the NIPC was telling Congress and the reality of its situation. In 1999, Congress received testimony regarding an NIPC initiative to incorporate information technology industry executives into the working of the NIPC in order to assist the center with understanding the communications infrastructure.[88] However, this approach seemed to do little good. In 2001, the General Accounting Office noted that the NIPC lacked industry-specific data on factors such as critical system components, known vulnerabilities, and interdependencies.[89] Even the NIPC's then-director, Ronald Dick, acknowledged the difficulty of getting private sector experts to share their knowledge with the government.[90]

Lack of Information-Sharing

Although one of the NIPC's significant functions was the provision of warning, it found itself behind the curve in this area. As described in the presidential directive that created the center, the NIPC was tasked with sanitizing law enforcement and intelligence information, which it would then provide to relevant critical infrastructure owners and operators.[91] A General Accounting Office assessment of the NIPC's effectiveness in this area was less than laudatory. Between 1998 and 2000, the NIPC issued eighty-one alerts, advisories, and assessments. However, most of these pertained to attacks that were already in progress.[92] (The lag between the NIPC learning of a problem and providing information to the victims of the attack sometimes was weeks or even months.)[93] These alerts and the like gave recipients little or no informational advantage that could enable them to counter their attackers. The FBI, in typical government euphemism-speak, addressed this state of affairs by claiming that "as companies continue to gain experience in dealing with the NIPC and the FBI field offices, as we continue to provide them with important and useful threat information, and as companies recognize that cyber crime requires a joint effort by industry and Government together, we will continue to make real progress in the area."[94] (Translation: We'll get it right someday and by the way it's the other guy's fault.)

Another element of the problem with timely warning was traceable to the NIPC's underpowered analytic apparatus. The Watch and Warning Unit's warnings were based on analyses that the Analysis and Information-Sharing Unit developed.[95] Four years after the creation of the NIPC, it was still struggling to disseminate actionable information to its private sector partners. According to its then-director, Ronald Dick, in 2002, the National Photographic Interpretation Center was still in need of an enhancement that would allow it to "share with the private sector what actionable things that they can do to prevent them from becoming victims."[96]

Part of the warning issue was methodological. Just because the NIPC had an Analysis and Warning Section—which was supposed to be the "indications and warning" element of the NIPC—this did not mean that there was a functional indications and warning component.[97] The section included two units: the Analysis and Information-Sharing Unit and the Watch and Warning Unit. These elements, consistent with the NIPC writ large, were minimally staffed.[98] (This might have been an "indication" that all was not well with the NIPC's indications and warning function.) In 1999, the NIPC noted that one of its long-term goals was the development of a comprehensive indications and warning system.[99] In 1999 the Analysis and Warning Section established an Indications, Analysis, and Warning pilot program, which included agreed-upon criteria for attack indicators specific to the electric power industry.[100] However, the NIPC's lack of two-way dialogue with most other elements of industry meant that its success with the energy sector was not replicable.

Lack of Trust from Private Sector Partners

Industry was leery of working with the NIPC for several reasons. Corporate security officers were nervous about the NIPC's confidential website, since the site used a potentially intrusive system that companies feared could compromise their proprietary information.[101] An additional, significant concern was the role of the FBI—an agency that was supposed to elicit confidence even as it continued to pursue investigations. Because the NIPC was part of the FBI, some saw it as a law enforcement entity.[102] Individuals in both the corporate and academic sectors specifically cited concerns—however unfounded—about the possibility that the Bureau could use its software to spy on corporate networks.[103] US policymakers amplified these concerns. For instance, Senator Bob Bennett noted, in 2000, that giving the coordination function to the FBI "immediately raise[d] suspicions on the part of industry."[104]

Beyond a fear of nosy government, industry had market-driven concerns. Working with law enforcement, some feared, would create a loss of confidence among shareholders and have a negative impact on companies' market share.[105] Elements within the private sector were concerned that the increased focus on critical infrastructure protection could lead to the re-regulation of industries.[106] Although the NIPC was not in the business of policymaking, this overarching fear probably did little to improve the standoffish public-private relationship.

The NIPC's ham-handedness soured the public-private relationship. For instance, the NIPC furnished a questionnaire to telecommunications carriers throughout the United States that prompted confusion and controversy. Not surprisingly, the survey returned almost no useful information, while at the same time leaving a bad taste in the mouth of the industry partners whose assistance the government needed. The survey also alienated a number of key players in the telecommunications industry who had been working on national security and emergency preparedness issues. These figures resented not being consulted prior to the release of the questionnaire.[107] Not only was this incident harmful to an essential relationship, but it also demonstrated the government's inability to understand—and work with established experts in—the environment that it was trying to protect.

Dismantling the NIPC

The NIPC never achieved the monumental role at the nexus of government and industry to which it aspired. In 1998, the year that the NIPC was established, a congressional staff report suggested that the NIPC had already become a "catch-all" for infrastructure protection activities without a clear understanding of what was necessary to succeed at this work.[108] By 2000, there was discussion under way about excising forensic and technical support functions from the NIPC.[109] Clearly the NIPC was a rapidly growing bureaucratic dumpster fire—fueled by an expansive mission and burning through resources (e.g., detailees, relationships). By early 2002, then–FBI director Robert Mueller III was ready to break up the NIPC.[110] The eagerness to divest the Bureau of this lumbering white elephant was apparent in the statement of Larry Mefford, the then–assistant director of the FBI's newly created Cyber Division, who advised Congress in 2002 that "the NIPC will play an important role in the new Department of Homeland Security."[111] What was one more problem for DHS, which, from its inception, was bureaucratically equivalent to Frankenstein's monster? In 2002 the FBI established a new Cyber Division, which retained the NIPC's Computer Investigations and Operations Section.[112]

The National Security Agency and the Private Sector

The National Security Agency, which furnished leadership to an aspect of the NIPC, is the primary signals intelligence collection organization in the United States. It came into existence in 1952, as a successor to the Armed Forces Security Agency.[113] The NSA primarily collects foreign intelligence (although it has had a nexus to the domestic setting through the Cold War–era Shamrock and Minaret programs). It does, however, have a role in protecting industries and their customers from the technical measures used by threat actors.

US SIGINT has contributed to the protection of industry through the development of countermeasures to unwelcome interception. The problem of radiation emanations—which provide opportunities for eavesdropping—is, according to a declassified NSA document, known as Tempest[114] (the term had no special meaning; it was selected from a cover-name list in the early 1950s).[115] In 1960, the NSA briefed the US Communications Security Board on the existing US Tempest vulnerabilities.[116] According to the NSA, it currently oversees the Certified Tempest Manufacturer program, which enables industry to develop, produce, and sell products that meet the national Tempest standard for use by the United States and North Atlantic Treaty Organization nations, as well as their contractors, to process classified information.[117]

The NSA, during the 1980s, inched toward providing a greater informational advantage to the US private sector. According to a 1986 Senate report, the NSA planned to license essential techniques for use in equipment marketed to the public, a step that the report called "unprecedented."[118] Subsequently, the Computer Security Act of 1987 established the NSA's role as one of providing technical advice to the National Institute of Standards and Technology—part of the Department of Commerce—which in turn assists the private sector, as well as the government, with securing unclassified but sensitive computer data.[119] As an example of this relationship, the NSA, in 2005, disseminated a set of recommendations, via the institute, about how to secure a communications network by using publicly available cryptography.[120]

The NSA has developed the capability to push information regarding vulnerabilities directly to the private sector. In late 2019, it established a new Cybersecurity Directorate.[121] According to the directorate, it will work to "prevent and eradicate threats to national security systems and critical infrastructure."[122] Although this sounds similar to the role of CISA at the DHS, the NSA Cybersecurity Directorate's emphasis is on the defense industrial base.[123]

The Department of Homeland Security

Creation of the Department of Homeland Security had significant implications for how the government interacted with the private sector on cyber-related issues. First, moving certain cyber-related responsibilities from the FBI to the DHS shifted—at least in theory—the approach to securing critical infrastructure from an after-the-fact approach (which the FBI, as a law enforcement–oriented agency, especially during the Clinton years, brought to the task) to one of securing vulnerabilities before threat actors could exploit them. However, a second implication was the increasing fragmentation of government–private industry collaboration on national security. The DHS established new channels for contact while the FBI maintained its own InfraGard and counterintelligence outreach efforts. Finally, the DHS's regular reorganization of its cyber element has arguably amounted to a shell game that degrades efficacy by making oversight more difficult and by disrupting program continuity.

The Homeland Security Act of 2002 includes several key concepts that provide context to cybersecurity activities. In establishing the Department of Homeland Security—a department whose mission centered around critical infrastructure—the law enshrined the definition of critical infrastructure formulated in the Patriot Act, which included "means, systems and assets, whether physical or virtual, so vital to the United States that the incapacity or destruction of such systems and assets would have a debilitating impact on security, national economic security, national public health or safety, or any combination of those matters."[124] That word brick gives an agency with critical infrastructure responsibilities an almost limitless reach (or an almost impossible mission). It also means that the United States must continue to reassess what constitutes critical infrastructure as new technologies (and their effects on existing technical and social systems) emerge. For instance, in mid-2020, several MITRE Corporation executives broached the question of whether social media should be designated as critical infrastructure.[125]

Additionally, the Homeland Security Act made clear that the DHS's responsibility was one of monitoring (i.e., intelligence collection) and securing critical infrastructure rather than investigating malignant actors who sought to exploit it. Specifically, the department's primary mission includes reducing the vulnerability of the United States to terrorism and minimizing the damage from terrorist attacks that do occur within the United States.[126]

Old Agencies, New Names

The DHS cyber center of gravity was initially located in the Information Analysis and Infrastructure Protection Directorate. As part of the government reorganization that produced the DHS, this directorate took responsibility for a large portion of the FBI's NIPC functions. Additionally, this directorate absorbed the Critical Infrastructure Assurance Office.[127] This office, part of the Department of Commerce's Bureau of Export Administration, had its origins in the presidential directive that established the NIPC.[128] As part of the directive, the US government created a National Plan Coordinator for Security Infrastructure Protection and Counter-Terrorism to facilitate implementation of the directive, whose staff were responsible for developing a national infrastructure assurance plan; coordinating analyses of the US government's critical infrastructure; and contributing to a national education and awareness program.[129] By 1999 this entity became known as the Critical Infrastructure Assurance Office and, although under Commerce, remained interagency in nature.[130] With the creation of the DHS, the Critical Infrastructure Assurance Office became defunct.

The Information Analysis and Infrastructure Protection Directorate also inherited the National Infrastructure Simulation and Analysis Center, which had been a Department of Energy entity. It was created in 2001 as part of the Critical Infrastructures Protection Act and operated as a joint program between the Department of Energy's Sandia and Los Alamos national laboratories.[131] Its function was to provide computer modeling, simulation, and analysis of the nation's infrastructure, with an emphasis on interdependencies among the various infrastructures.[132] More than a decade later, the DHS noted that the center looked at the consequences of disruptions across the sixteen critical infrastructure sectors at the national, regional, and local levels.[133]

Private industry was an essential Information Analysis and Infrastructure Protection Directorate partner. This was consistent with the directorate's role as the lead agency for critical infrastructure protection activities in the information and telecommunications sector.[134] (The directorate inherited this function from the Critical Infrastructure Assurance Office, as the latter's parent agency, the Department of Commerce, had been responsible for coordination with the information and communications sector.)[135] Part of the overarching mission of the Information Analysis and Infrastructure Protection Directorate included strengthening a national cyberspace security readiness system that would include a public-private architecture capable of facilitating rapid response and dissemination of information in the event of a national-level cyber incident. In furtherance of this effort, the directorate recognized the need for an ongoing relationship with the private sector. In 2004, the DHS advised Congress that the directorate had responsibilities that included cultivating an environment conducive to public-private partnerships as well as coordinating and supporting the development of partnerships with entities including private industry.[136]

To further its responsibilities in the field of critical infrastructure, specific Information Analysis and Infrastructure Protection Directorate elements focused on this work. Its National Cyber Security Division (NCSD), established in 2003, was developed—in response to President George W. Bush's National Strategy to Secure Cyberspace—to serve as the nation's focal point for cybersecurity issues.[137] (The DHS created the NCSD from the remnants of the FBI's National Infrastructure Protection Center.) Multiple components within the division worked to secure the private sector in a variety of ways. The NCSD's Law Enforcement and Intelligence Branch, among other functions, furnished a mechanism to coordinate information-sharing between national security (i.e., intelligence and law enforcement) government entities and the private sector. An essential aspect of this—especially as noncleared private sector entities increasingly impacted US national security concerns—was "cleaning" classified information of sensitive content so that private sector partners could review it (and, hopefully, act accordingly). The NCSD's Outreach and Awareness Branch also engaged private industry, but rather than providing specific information on which recipients might be able to act, it emphasized messaging about cybersecurity awareness. Within the branch, the Coordination Team engaged the private sector—as well as other relevant players—in order to collaborate on events and activities.[138] Although such efforts might have minimal immediate impact, they arguably forge a culture that—in the longer term—helps government and the private sector start from the same point of reference when discussing how to secure US infrastructure.

In addition to information-sharing of varying degrees, the NCSD had more operationally oriented components under the auspices of its Strategic Initiatives Branch. The branch's Critical Infrastructure Protection Cybersecurity Team participated in identifying critical assets and vulnerabilities, mapping interdependencies, and promoting cyber awareness for the information technology sector. Additionally, the branch's Control Systems Team established an assessment capability directed at identifying vulnerabilities within control systems, facilitating control system incident management, and providing recommendations for the future use of control systems and security products.[139]

Finally, the NCSD was home to the US Computer Emergency Readiness Team (US-CERT). Established in September 2003, US-CERT is a partnership between the DHS and the private, as well as the public, sectors.[140] US-CERT's purpose is to make cyber security a national effort, increase public awareness of cyber threats and vulnerabilities, and improve computer security preparedness and response to cyber attacks.[141] (At least that is what the DHS's budget submission claimed.) US-CERT was another legacy from the FBI's National Infrastructure Protection Center. The Computer Emergency Response Team / Coordination Center, at Carnegie Mellon University, had a contractual relationship with the NIPC. It provided the NIPC with advance notice about its advisories as well as about cyber intrusion activities. The NIPC provided it with information that the NIPC obtained through investigations and other sources, which the team/center disseminated to security professionals in the private sector.[142] US-CERT was a direct evolution of this partnership.[143]

One aspect of US-CERT is its involvement in warning functions. It operates an around-the-clock cyber watch, warning, and incident response center, which responds to cyber incidents and operates the National Cyber Alert System, which provides timely, actionable information to the public.[144] In 2004, working with the Department of Energy, the National Cyber Security Division established the US-CERT Control Systems Security Center, which coordinates control system incident management.[145] US-CERT issues warnings to the private sector through multiple channels. The US-CERT Portal is an internet-based collaborative system that enables the sharing of sensitive cyber-related information with industry. Additionally, US-CERT maintains a public website, which also provides the private sector with information that will assist with protecting information systems and infrastructures.[146]

US-CERT also collects and analyzes information. Its Internet Health Services allowed it to gather information from the private sector about vulnerabilities, network attacks, and malicious code activity.[147] It is able to accept voluntarily provided incident reports from private entities.[148] On a voluntary basis, US-CERT engages with the private sector in order to provide remote and onsite incident detection. US-CERT also provides analytic assistance to the private sector. Additionally, also on a voluntary basis, US-CERT helps private industry to assess threats and vulnerabilities.[149]

In 2004, the DHS announced the first of several subsequent reorganizations. US-CERT would move from the Information Analysis and Infrastructure Protection Directorate's National Cybersecurity Division into the National Infrastructure Coordinating Center (NICC). The NICC would also become responsible for the National Coordinating Center for Telecommunications. Located within the Information Analysis and Infrastructure Protection Directorate, the NICC was responsible for maintaining operational awareness of critical infrastructure and key resources and strove to establish a shared capability for information-sharing and coordination among government and private sector entities including critical infrastructure owners and operators.[150]

Several years after the US-CERT's move to the NICC, the DHS bureaucratically relocated it again. In October 2009, the DHS established the

National Cybersecurity and Communications Integration Center (NCCIC).[151] DHS situated this new element in the National Protection and Programs Directorate (NPPD), which DHS had established in 2007 as a successor to the Information Analysis and Infrastructure Protection Directorate.[152] The NCCIC functions as a national-level cyber and communications operations center that fuses information from a variety of civilian, law enforcement, intelligence, and state and local government agencies, with data from the private sector.[153] Along with US-CERT, the NCCIC consolidated the National Cybersecurity Center and the National Coordinating Center for Telecommunications (the operational arm of the National Communications System). Both the National Cybersecurity Center and the National Coordinating Center for Telecommunications conducted—and brought to the NCCIC—functions that engaged the private sector. The National Cybersecurity Center coordinated the work of the six largest federal cyber centers with private sector partners.[154] (In an illustration of how the DHS reorganizations could look like a fast-moving shell-game, the National Cybersecurity Center was only approximately a year old when it was brought under the auspices of the NCCIC.[155] Its creation was the result of a national security presidential directive on cybersecurity policy, issued in early 2008, stating that the DHS would establish a National Cybersecurity Center to "coordinate and integrate information to secure US cyber networks and systems.")[156] The National Communications System—which became part of the DHS under the Homeland Security Act of 2002—had started as a Department of Defense entity in 1963 to ensure national telecommunications survivability.[157] This, as one might surmise, required the cooperation of private industry.

Approximately a decade after this move, the DHS took steps to reorganize US-CERT. The department advised Congress that in 2017 it would divide US-CERT into US-CERT Incident Response and US-CERT Detection and Analysis. Both of these elements would have responsibilities for interaction with the private sector. The former division would be responsible for providing assistance to a variety of entities—including private sector partners—that had been victims of significant cybersecurity compromises. In furtherance of this mission, it would conduct either on-site or remote diagnosis of the compromise, remove the adversary from the victim's network, and help the victim to establish a secure state of operations. The latter division would develop and disseminate cybersecurity threat, vulnerability, and mitigation information to customers, including those in the private sector.[158]

The National Cybersecurity and Communications Integration Center—including US-CERT—changed hands within the DHS approximately a decade after its creation. In 2018, the Cybersecurity and Infrastructure Security Act redesignated the NPPD as the Cybersecurity and Infrastructure Security Agency.[159] In addition to the NCCIC, CISA is also home to the National

Risk Management Center—a planning, analysis, and collaboration entity engaged in identifying the most significant risks to US infrastructure.[160] The management center now houses the National Infrastructure Simulation and Analysis Center.[161] The management center is a reconfiguration of the NPPD's Office of Cyber and Infrastructure Analysis, which the NPPD had established in 2014.[162] The office was meant to be the first step in integrating risk assessment activity—especially the interdependencies across the physical and cyber domains.[163]

Despite having these resources, CISA may not be up to the job. According to the Cyberspace Solarium Commission, CISA's mission is to be the US government's primary coordinating body responsible for developing public-private collaboration in cybersecurity. (The commission was established by the 2019 National Defense Authorization Act in order to "develop a consensus on a strategic approach to defending the United States in cyberspace against cyber attacks of significant consequences.") However, the commission indicated concern that CISA was institutionally limited by inadequate facilities, insufficient resources, lack of buy-in from other federal entities, congressional ambiguity regarding the agency's role, and inconsistent support and integration with the private sector.[164] Even CISA's own director, Chris Krebs, seemed to be stalling for time when, in 2018, shortly after the launch of CISA, he claimed that he saw it as "a 14 year startup organization."[165] At the time of Krebs's comment, fourteen years was nearly as long as the DHS had been in existence.

For historical reference, plans for the creation of a Cyber and Infrastructure Protection Agency at the DHS preceded what ultimately became CISA. That agency—like CISA—would have focused on reorganizing the NPPD into an operational component by consolidating the NCCIC, Infrastructure Security, and Federal Protective Service.[166] Then–secretary of homeland security Jeh Johnson explained, in 2016, that this agency was supposed to "streamline and strengthen existing functions."[167]

The DHS: New Entities, Old Challenges

In addition to inheriting functions with bureaucratic DNA that predates the DHS, the department has established new entities in order to analyze threats pertinent to the private sector. Additionally, the DHS has had to define new relationships with existing entities such as information sharing and analysis centers (ISACs). The DHS's abilities to establish and maintain relationships with the private sector are essential, since it has significant responsibilities in the fields of countering terrorist threats to critical infrastructure and in counterproliferation. Unfortunately, it has had difficulty with providing relevant nongovernment customers with the information that they need in order to secure the assets for which they are responsible.

Protecting Federal Networks to Protect the Public

Although this book focuses on the government–private sector security relationship, it is important to note that programs to enhance government information technology ultimately help the private sector, which is plugged into the networks that compromised government systems can impact. Breaches into multiple government networks, including those at the Departments of Commerce and State, the Office of Personnel Management—as well as the Homeland Security Information Network—illustrate how vulnerable these infrastructures—and those who connect to them—are.[168]

The DHS is responsible for a significant aspect of combating threats to federal networks and, by extension, to the entities that connect to those networks. According to the Cybersecurity Act of 2015, the department is the government's central hub for automated cyber threat indicator-sharing.[169] The DHS's National Cybersecurity Protection System is located within the NCCIC.[170] This system is an integrated system-of-systems that protects the traffic flowing out of—as well as into—federal networks. Additionally, the NCCIC conducts "cyber hygiene" scans to identify vulnerabilities in agencies' internet-accessible devices.[171]

Office of Intelligence and Analysis

The DHS's analytical component, the Office of Intelligence and Analysis (OIA), plays a role (albeit sometimes not as significantly as warranted) in the department's securing of the private sector. It has the primary responsibility within the US intelligence community for analysis, evaluation, and dissemination of information regarding threats to homeland critical infrastructure.[172] The origins of the OIA are in the Information Analysis and Infrastructure Protection Directorate, where it was the Office of Information Analysis. Part of the latter's function was the provision of intelligence support to the DHS's external partners. That latter office's Risk Assessment Division was authorized to establish a two-way exchange of information with the private sector.[173] As a result of a 2005 review of the DHS that led to the dissolution of the Information Analysis and Infrastructure Protection Directorate and the creation of the NPPD, the Office of Information Analysis, in 2007, became the Office of Intelligence and Analysis as a standalone entity that reports to the undersecretary for information and analysis.[174]

The OIA supports multiple DHS entities that conduct work with implications for private industry. These include the Homeland Infrastructure Threat and Risk Analysis Center (HITRAC) and the NCCIC (including US-CERT).[175] Beyond the OIA's support to DHS entities, it also provides assistance to other US government agencies that address the increasingly technical aspects of private industry. For instance, the OIA provides support to the adjudication process of the Committee on Foreign Investment in the United States.[176]

The role—if not the efficacy of—the OIA in analyzing cyber issues with implications for private sector critical infrastructure is a well-established one. Like other government agencies, the DHS has a tendency to entertain grandiose aspirations, as reflected in its claim that it would "evolve towards dynamic real-time situational awareness capabilities, like 'weather maps' for cyberspace." According to the DHS, this situational awareness would "support cyber infrastructure that—much like the human immune system—will be smart enough to detect, adapt to, and defend against new threats."[177] In 2010, the DHS advised Congress that the OIA was providing "substantial and growing" support to the department in this area.[178] Four years later, the DHS notified Congress that the OIA would provide all-source analysis of cyber threats to critical infrastructure networks and systems to assist owners and operators in protecting the cyber infrastructure.[179]

The OIA and HITRAC. The Homeland Infrastructure Threat and Risk Analysis Center is an important element in the DHS's engagement with the private sector. HITRAC is the DHS's infrastructure-intelligence fusion center.[180] According to the DHS, its purpose was to merge DHS infrastructure expertise with intelligence analysts to map terrorist threats to infrastructure vulnerabilities.[181] In 2006, the Information Analysis and Infrastructure Protection Directorate characterized HITRAC as the "linchpin in [its] support to the efforts of the Department and the private sector to determine the risk of attack against key infrastructure as well as the protective measures that may be taken."[182] It was jointly managed by the OIA and the Office of Infrastructure Protection—an element within the NPPD.[183] HITRAC is currently under the auspices of CISA.[184] Within HITRAC, the Risk Analysis Division does, as its title suggests, analysis of risk to infrastructure. Through its Threat Analysis Division, HITRAC conducts critical infrastructure threat analysis, cyber threat analysis, and regional threat assessments.[185] For instance, as of 2005, HITRAC, in conjunction with the NCSD, had developed the draft of a Domestic Cyber Risk Estimate, which evaluated threats emanating from inside the United States.[186] HITRAC products—notably infrastructure intelligence notes—provide the private sector with a perspective on events, activities, and information of importance to security planning in their respective industry areas of responsibility.[187] OIA personnel with assignments at HITRAC have provided regular and incident-specific briefings to a variety of customers including the private sector critical infrastructure protection community.[188]

The OIA and NCCIC. Much of what the OIA does is in conjunction with US-CERT and the NCCIC. The OIA established a national intelligence analytical framework to support customers including the NCCIC and US-CERT.[189] Furthermore, it has provided tactical intelligence support—including situational awareness and warnings of cyber threats—to the NCCIC.[190] The DHS intelligence enterprise also derives raw information from the department's cyber activities. For instance, as of 2010, it was making an effort to examine CERT-derived data in an effort to conduct predictive analysis and attribution of threat activities.[191] (Interestingly, this may have been a response to earlier criticism. The Government Accountability Office, in 2008, had criticized US-CERT for not integrating its work into predictive analyses of broader implications or potential future attacks. Furthermore, the Government Accountability Office assessed that US-CERT lacked the analytic—as well as technical resources—to respond to simultaneous cyber incidents.)[192]

Information Sharing and Analysis Centers

A presidential decision directive on critical infrastructure protection issued in 1998 urged industry to take a larger role in securing the cyber and other vulnerabilities of critical national infrastructure. In furtherance of this it strongly encouraged the development of a private sector information sharing and analysis center. The ISAC, according to aforementioned presidential directive, "could serve as the mechanism for gathering, analyzing, appropriately sanitizing and disseminating private sector information to both industry and the NIPC. The center could also gather, analyze and disseminate information from the NIPC for further distribution to the private sector."[193] The directive seemed to envision ISAC as a one-to-one counterpart to the NIPC. However, ISACs began to develop around individual infrastructure sectors. For instance, by 2000, the financial services industry, the telecommunications industry, and the electric power industry all had their own ISACs.[194]

Despite the presidential directive's intertwining of the NIPC and ISAC concepts, the relationship between the two entities was tenuous at best. The nature of the NIPC's relationship with industry was unclear almost from the outset. According to a 1998 congressional hearing, NIPC outreach did not focus on developing overall protection strategies for sectors but rather on making an operational connection to the NIPC.[195]As of 2002, the NIPC had established an ISAC Support and Development Unit.[196] This unit was supposed to enhance private sector cooperation and trust in order to facilitate two-way information-sharing. A better mechanism for information-sharing was sorely needed. Congressional testimony from the previous year noted that the NIPC had established only one two-way information-sharing partnership with an ISAC.[197] (This partnership, with the North American Electric Reliability Council, had a unique historical precedent. In the late 1980s, the council's board of trustees resolved that each electric utility should develop a close working relationship with FBI field offices. Additionally, the council's staff were supposed to establish and maintain a working relationship with the FBI at the national level.[198] In other words, the electricity sector was a low-hanging fruit and there was no reason to think that the NIPC could replicate this relationship.)

However, the NIPC was not alone in responsibility for the breakdown in communication. ISACs were, at times, unwilling to commit to a two-way exchange of information. The National Communications System ISAC received information from the NIPC, which the ISAC distributed to its membership, but had not shared any incident reports with the NIPC. Similarly, the Financial Services ISAC was willing to accept information from the government and law enforcement sources, but the transfer of information was unidirectional.[199]

ISACs and the DHS

The framework of lead agencies established by the 1998 presidential decision directive on critical infrastructure protection has continued to inform twenty-first-century policy regarding private sector critical infrastructure. The directive established the concept of lead agencies. This arrangement designated specific US government agencies to serve as liaisons with infrastructure sectors that were vulnerable to physical or cyber attacks.[200] In 2003, a homeland security presidential directive acknowledged that "each infrastructure sector possesses its own unique characteristics and operating models" and established sector-specific agencies.[201] The DHS became the sector-specific agency for the chemical sector; the commercial facilities sector; the communications sector; the critical manufacturing sector; the dams sector; the emergency services sector; the information technology sector; and the nuclear reactors, materials, and waste sector. Additionally, the DHS is the co-sector-specific agency for the government facilities sector (along with the General Services Administration) and for the transportation systems sector (along with the Department of Transportation).[202] Each of the sixteen critical infrastructure sectors has both a government coordinating council and a sector coordinating council.[203] The sector councils provide a conduit for private sector input to homeland security. According to the 2013 National Infrastructure Protection Plan, these are "self-organized, self-run, and self-governed private sector councils consisting of owners and operators and their representatives" that serve as "principal collaboration points between the government and private sector owners and operators for critical infrastructure security and resilience policy coordination and planning."[204]

When the DHS assumed most of the NIPC's functions, it took on responsibilities for engaging with the ISACs. According to the National Council of ISACs, there were twenty-five ISACs as of 2020.[205] (In some instances multiple ISACs align with a single critical infrastructure sector or do not neatly align with any of the sectors.) The DHS's role vis-à-vis ISACs is twofold. First, it has integrated representatives of ISACs into DHS operations in order to share information. Additionally, DHS components have become sector-specific agencies for what seems to be a continuously expanding list of critical infrastructure sectors (eight in 1998, sixteen as of 2020), most of which have corresponding ISACs.

Multiple DHS components include ISAC representatives. In 2010, the DHS reached an agreement with the Information Technology–Information Sharing and Analysis Center to embed a full-time analyst at the NCCIC.[206] As of 2019, CISA—which inherited the NCCIC from the NPPD—operated the communications sector ISAC, which included more than sixty private sector communications and information technology companies.[207] New threats have prompted the creation of new ISACs. For instance, CISA funded creation of an election infrastructure ISAC, which co-located representatives with the NCCIC.[208] (In advance of the 2018 midterm elections, election infrastructure ISAC threat alerts were shared with all fifty states, more than 1,400 local and territorial election offices, six election associations, and twelve election vendors.)[209]

Refining the DHS's Relationships with Private Sector Customers

The Department of Homeland Security has encountered several difficulties with ensuring that private industry receives the information it needs to do its part in safeguarding the United States. Getting the right information, to the right people, in time for it to be effective is one challenge. A second challenge—and one less discussed—is how to ensure that the DHS is acquiring the kind of information that its private sector partners will be able to use in furtherance of activities consistent with national security.

Private sector partners have complained about information that the DHS has been able to provide them. One theme seems to be too little, too late. In 2008, the Government Accountability Office noted that the warnings that US-CERT provided were not consistently actionable or timely.[210] Several years later, there was little sign of improvement. According to the Government Accountability Office, in 2010, the review and revision process for US-CERT products could add days to a product's release if it was necessary to remove classified or law enforcement information.[211] Unfortunately, cyber threats operate on time frames of seconds, not days. In an effort to address this, an executive order directed the attorney general, the secretary of homeland security, and the director of national intelligence to "ensure the timely production of unclassified reports of cyber threats to the U.S. homeland that identify a specific targeted entity."[212] Furthermore, the order mandated the establishment of a process that "rapidly disseminates" these reports.[213]

The second problem in the DHS's provision of information is an inability to give private sector counterparts what they need. Part of that is simply unavoidable. For instance, US-CERT is not allowed to provide individualized treatment to one private sector entity over another private sector entity. This makes it difficult to share specific information when entities are directly impacted by a cyber threat.[214] However, the DHS has failed to identify and prioritize customers. The OIA has the primary responsibility within the US intelligence community to analyze, evaluate, and disseminate analysis on threats to homeland critical infrastructure.[215] As of 2014, private critical infrastructure sectors ranked fifth on a list of five OIA customer groups.[216] In other words, private industry ranked dead last. Furthermore, industry indicated that the OIA was unfamiliar with its private sector audience's needs. For instance, a number of private critical infrastructure representatives assessed the OIA's products to be more strategic than what they needed.[217]

Even worse is that the US intelligence community—which the OIA prioritized above the private sector as a customer—did not perceive particular value in the OIA's work. Officials in the Office of the Director of National Intelligence assessed that the OIA did not tailor its work to intelligence community elements and that whereas the intelligence community had an international focus, the OIA focused on the homeland.[218] Although the DHS failed to identify where it could add the greatest value, remarks by officials in the Office of the Director of National Intelligence also indicate trouble within the intelligence community. After all, developments within the homeland—such as activities by foreign government officials, intelligence officers, nonstate terrorists, and criminal actors—can add to the intelligence community's awareness of worldwide trends. (For instance, HITRAC's Domestic Cyber Risk Estimate complemented the intelligence community's international threat assessments and helped to complete a global picture of a problem that is unconstrained by geographic boundaries.)[219]

In order to fulfill its commitment to private sector customers, the DHS must ensure that it—as a member of the intelligence community—is obtaining information that will help those customers to work as effective partners in national security. In order to leverage US government resources in furtherance of obtaining relevant information, the DHS must formulate intelligence requirements. The OIA has been largely responsible for developing intelligence requirements for the DHS intelligence enterprise. As of 2004, the OIA's predecessor, the Office of Information Analysis, facilitated the creation of requirements for DHS components, the intelligence community, and law enforcement entities.[220] This process included specific cyber-related input from the NCSD.[221] In 2010, the DHS advised Congress that the OIA had recently completed a comprehensive set of "standing information needs" that uniformly documented the intelligence and information needs of the entire department.[222] However, representatives of private sector critical infrastructure sectors have stated that the OIA did not understand the needs of their industries.[223] Therefore, it is unlikely that the OIA could formulate requirements that

Industry and Fusion Centers

The DHS, since the mid-2000s, has been the lead agency for engaging with state and local fusion centers. In mid-2006, the DHS directed its Office of Intelligence and Analysis to manage the department's support for fusion centers; this resulted in the OIA's establishment of the State and Local Program Office as a focal point for this function. Then, in 2007, the Implementing Recommendations of the 9/11 Commission Act formalized this arrangement by investing the secretary of homeland security with the responsibility for establishing a state, local, and regional fusion center initiative within the Department of Homeland Security.[224]

The DHS has attempted to use fusion centers to more effectively engage its private industry partners. In 2010, it announced that it would launch the Cybersecurity Partners Local Access Plan. According to the DHS's National Cyber Security Division, this program would help to establish relationships between fusion centers and the critical infrastructure and key resource partners within their areas of responsibility.[225] Through this program, the owners and operators of critical infrastructure and key resources could access secret-level cybersecurity information via their local fusion centers.[226] Owners and operators could also access video teleconference calls.[227] This seemed to be a direct response to concerns that private sector entities had provided to the NCSD about reduced travel budgets.[228]

might elicit information that would be of value to the private sector. Given the DHS's aspiration to a two-way relationship with private sector critical infrastructure, industry should have an opportunity to provide its sector-specific needs to the OIA's requirements.

Public-Private Collaboration

Information-sharing platforms require private sector entities to pull information, whereas collaborative efforts with the government ensure that information is pushed to them. As with counterintelligence and counterterrorism, the US government has developed cyber-specific initiatives to assist private industry. There are not clear lines in the road, though, since cyber is a domain in which actors operate, rather than an actor in and of itself.

The FBI added to its counterintelligence and counterterrorism awareness efforts with a similar cyber initiative. Cyber, of course, cuts across these two areas—as well as criminal investigations—and so this program, known as InfraGard, is complementary rather than parallel to efforts such as the Development of Counterintelligence Awareness (DECA) program and the Awareness of National Security Incidents and Response (ANSIR) program. In 1996, the FBI's Cleveland field office—joined by the Cincinnati and Indianapolis field offices—developed InfraGard in order to obtain expert perspectives on cybersecurity from external sources including private sector information technology firms.[229] The FBI—in describing the Cleveland field office's pilot project—explained that the name referred to "guarding the information infrastructure."[230] Through InfraGard the FBI can provide information regarding vulnerabilities to other InfraGard members.[231] In 1998, the FBI explained that the NIPC planned to turn InfraGard into a national-level program in 1999.[232] By 2002, all of the FBI's field offices had active InfraGard chapters.[233] The InfraGard program is organized around chapters within the jurisdiction of FBI field offices through which private sector owners and operators can share information regarding cyber intrusions and vulnerabilities.[234]

Once the NIPC became responsible for InfraGard, the information collected through the program could be used to develop an understanding of threats beyond the field office level. According to the FBI, the NIPC could analyze information in conjunction with law enforcement intelligence, open-source information, and industry data to determine whether an intrusion was part of a broader attack.[235] The National Plan for Information Systems Protection Version 1.0, issued by the White House in 2000, incorporated InfraGard into its vision. According to the plan, InfraGard would provide its members with prompt, value-added threat advisories, alerts, and warnings; increase the quantity and quality of infrastructure threat information and incident reports provided to local FBI field offices (for coordination, investigation, and follow-up) and to the NIPC (for national-level analysis and warning); and increase interaction and information-sharing among InfraGard members, and their associated local FBI field offices, and the NIPC, on infrastructure threats, vulnerabilities, and interdependencies.[236]

Despite the broadened scope of InfraGard, the FBI acknowledged that it remained difficult to secure the cooperation of industry in combating cyber-enabled threats. In 1998, the Bureau advised Congress that despite its outreach efforts, the government did not receive notification by victims about cyber malfeasance.[237]

The FBI's handling of InfraGard, especially in the early years of the program, was erratic. After elevating the Cleveland pilot project to a national-level program, the NIPC boasted that it had "taken [InfraGard] from its humble roots of a few dozen members in just two states" and turned it into "the largest government / private sector joint partnership for infrastructure protection in the world."[238] The head of the NIPC made this chest-thumping claim in 2001. Yet the General Accounting Office, during that same year, noted the NIPC's lethargy in recruiting InfraGard member companies.[239] World-conquering apparently notwithstanding, the FBI attempted to pawn off InfraGard to the DHS during the following year.[240] The Bureau advised Congress that the DHS was best suited to carry out the program.

According to the Bureau, InfraGard was better positioned to support the DHS's critical infrastructure protection and warning missions than it was to supporting the FBI's investigative mission.[241] However, in 2003 the FBI incorporated InfraGard into the newly created Cyber Division.[242]

Each FBI field office has at least one InfraGard chapter and includes participants from critical infrastructure sectors.[243] InfraGard provides subject-matter experts from those sectors with opportunities to exchange information with each other and with the US intelligence community. The Cyber Division's National Industry Partnership Unit uses the InfraGard network to facilitate the transfer of information between the public and private sectors.[244]

The InfraGard program has elicited praise from certain quarters. In 2011 it was highlighted to Congress as "a prime example of the success of public private partnerships."[245] The 9/11 Review Commission, in 2015, assessed that InfraGard had achieved "good traction with industry" and that it was sufficiently independent of the FBI to be seen as "an honest broker."[246]

However, InfraGard has not completely avoided criticism. The 9/11 Review Commission believed that InfraGard's mission was unnecessarily limited. According to the commission, InfraGard was viewed as cyber-focused although it could have a broader portfolio.[247] (Even before 9/11, InfraGard had attempted to expand its bailiwick—apparently without success judging by the commission's report—by addressing not only cyber threats but physical threats to critical infrastructure as well.)[248] Furthermore, the 9/11 Review Commission expressed surprise at discovering that the Cyber Division, rather than the director of private sector engagement, was responsible for the InfraGard program.[249]

The Commission's questions are valid. However, an even more fundamental one is: Why is InfraGard still associated with the FBI at all? Shortly after the creation of the DHS, the FBI attempted to offload the program onto the new department. Certainly the DHS—if it ever manages to function cohesively—would be the practical home for information-sharing about the cyber threats to critical infrastructure.

Furthermore, the DHS already fields a work force that can help mitigate the threats encountered by InfraGard members. A program to create the position of protective security adviser began in 2004 as a DHS pilot project. These advisers function as the DHS's field liaisons and coordinators for critical infrastructure protection.[250] As the DHS described in 2012, these advisers—who were largely responsible for the Office of Infrastructure Protection's field activities—function as on-site critical infrastructure and vulnerability assessment specialists.[251] (In case the DHS's shell game has confused anyone, the Office of Infrastructure Protection was part of the NPPD.)[252] In 2015, the DHS advised Congress that it had been offering cybersecurity training to the security advisers. The security adviser program was significant in the formation of the DHS's cybersecurity adviser position. The DHS modeled this position after the protective security adviser position. Similar to the latter, which provided a DHS presence in the field, cybersecurity advisers were the NPPD's deployed cyber work force. They have multiple functions including assisting with the adoption of best practices, information-sharing, and incident response. The risk management function includes conducting cyber risk assessments.[253]

Going beyond human-to-human liaison to address cyber vulnerabilities, the DHS has attempted to automate how it shares information about vulnerabilities with the private sector by developing its Automated Indicator Sharing (AIS) program.[254] Established in 2016, the AIS connects participating entities, including those in the private sector, to a CISA-managed system that facilitates the bidirectional sharing of cyber threat indicators. A server at each participant's location allows participants to share threat indicators that they have observed on their networks with CISA, which will in turn share these indicators with all AIS participants.[255] CISA cyber analysts review the indicators and defensive measures submitted through AIS; remove any personally identifiable information and other sensitive information; and disseminate the information—in machine-readable form—to AIS participants.[256] Conceptually, AIS, according to the DHS, is part of an effort to create an "ecosystem" in which indicators of cyber compromises will be shared in real time and incur costs on adversaries who will only be able to use an exploit once before information-sharing shuts it down.[257]

Although AIS appears to be a step in the right direction in terms of information-sharing, its implementation has suffered from many of the same problems that have plagued previous programs, such as the NIPC's efforts. Some of the information that CISA pushes to participants does not address participants' needs. According to a 2020 DHS inspector general report, the information lacked sufficient details—including context and data on which users could take action—to fully mitigate threats. Furthermore, CISA appears to suffer the same pox of limited personnel resources that previous information-sharing efforts, which are supposed to add value, have suffered. In 2017 and 2018, CISA had no dedicated staff to manage the AIS capability and perform outreach.[258]

Furthermore, the DHS has struggled to get private sector buy-in. Cybersecurity advisers—consistent with the DHS's role as the US government's primary portal for automated cyber threat information-sharing—encourage companies to sign up for AIS.[259] However, as of 2018, only six nonfederal entities were actually sharing threat information through AIS with CISA.[260] A 2020 report by the DHS inspector general painted a similarly anemic picture of participation.[261] Without an inflow of data, CISA cannot—even if it was better able to disseminate information—facilitate the emergence of a free-flowing ecosystem necessary to thwart threats in real time.

Breaking and Entering by the DHS

Since its early days, the DHS has sought to enhance its analysis and ultimately its ability to protect critical infrastructure through the use of "red team" techniques. In 2004, the DHS explained that the Information Analysis and Infrastructure Protection Directorate had established both physical and cyber target risk analysis teams (i.e., red teams). These red teams emulated terrorist doctrine, mindsets, and priorities, along with nonconventional strategy to test other measures of Information Analysis and Infrastructure Protection Directorate component to protect critical infrastructure.[262] The NPPD continued this practice. In 2016, the assistant secretary for the Office of Cybersecurity and Communications in the NPPD advised Congress that cybersecurity advisers could link infrastructure owners and operators with NCCIC-based assessment teams, which would test the critical infrastructure by trying to hack it. The assistant secretary was quick to emphasize that the DHS "[did] this only at the invitation of the company."[263] Under CISA, the NCCIC continued its red-teaming. According to 2019 congressional testimony, the NCCIC red-teamed operational technology included industrial control systems that operated critical infrastructure.[264]

Notes

1. US Senate, *Threats to the Homeland,* before the Committee on Homeland Security and Governmental Affairs, 113th Congress (Washington, DC, 2013), https://www.govinfo.gov/content/pkg/CHRG-113shrg86635/pdf./CHRG-113shrg86635.pdf.

2. National Counterintelligence Executive, *Foreign Spies Stealing US Economic Secrets in Cyberspace: Report to Congress on Foreign Economic Collection and Industrial Espionage, 2009–2011* (Washington, DC, 2011), https://www.dni.gov/files/documents/Newsroom/Reports%20and%20Pubs/20111103_report_fecie.pdf.

3. Department of Justice, "U.S. Charges Five Chinese Military Hackers for Cyber Espionage Against U.S. Corporations and a Labor Organization for Commercial Advantage," May 19, 2014, https://www.justice.gov/opa/pr/us-charges-five-chinese-military-hackers-cyber-espionage-against-us-corporations-and-labor.

4. Department of Justice, "Chinese Telecommunications Device Manufacturer and Its US Affiliate Indicted for Theft of Trade Secrets, Wire Fraud, and Obstruction of Justice," January 28, 2019, https://www.justice.gov/opa/pr/chinese-telecommunications-device-manufacturer-and-its-us-affiliate-indicted-theft-trade.

5. James J. Wirtz, "The Cyber Pearl Harbor Redux: Helpful Analogy or Cyber Hype?" *Intelligence and National Security* 33, no. 5 (2018), https://www.tandfonline.com/doi/abs/10.1080/02684527.2018.1460087?journalCode=fint20.

6. Danny Yadron, "Iranian Hackers Infiltrated New York Dam in 2013," *Wall Street Journal,* December 20, 2015, https://www.wsj.com/articles/iranian-hackers-infiltrated-new-york-dam-in-2013-1450662559.

7. Department of Justice, "Update to Sony Investigation," December 19, 2014, https://www.justice.gov/opa/pr/update-sony-investigation.

8. US Senate, *Current and Projected National Security Threats to the United States,* before the Select Committee on Intelligence, 113th Congress (Washington,

DC, 2013), https://www.govinfo.gov/content/pkg/CHRG-113shrg82721/pdf./CHRG-113shrg82721.pdf.

9. Government Accountability Office, *Critical Infrastructure Protection: Department of Homeland Security Faces Challenges in Fulfilling Cybersecurity Responsibilities* (Washington, DC, 2005), https://www.gao.gov/new.items/d05434.pdf.

10. Center for Strategic and International Studies, *Significant Cyber Incidents Since 2006* (Washington, DC, undated), https://csis-website-prod.s3.amazonaws.com/s3fs-public/201020_Significant_Cyber_Events_List.pdf.

11. Ibid.

12. Department of Justice, "NOAA National Weather Service Employee Indicted for Allegedly Downloading Restricted Government Files," October 20, 2014, https://www.justice.gov/usao-sdoh/pr/noaa-national-weather-service-employee-indicted-allegedly-downloading-restricted.

13. Nicole Perlroth, "Accused of Spying for China, Until She Wasn't," *New York Times,* May 9, 2015, https://www.nytimes.com/2015/05/10/business/accused-of-spying-for-china-until-she-wasnt.html.

14. US Senate, *Departments of State, Justice, the Judiciary, and Related Agencies Appropriations, 1962,* before a subcommittee of the Committee on Appropriations, 87th Congress (Washington, DC, 1961).

15. The reporter, Ali Watkins, who authored the *Politico* story, was playing her own spy game—a three-year affair—with James Wolfe, a senior aide to the Senate Select Committee on Intelligence (Michael M. Grynbaum, Scott Shane, and Emily Flitter, "How an Affair Between a Reporter and a Security Aide Has Rattled Washington Media," *New York Times,* June 24, 2018, https://www.nytimes.com/2018/06/24/business/media/james-wolfe-ali-watkins-leaks-reporter.html). Wolfe later admitted to discussing unclassified but nonpublic information and lying to the FBI (Josh Gerstein and Matthew Choi, "Ex-Senate Aide Gets 2 Months in Prison for Lying to FBI," *Politico,* December 20, 2018, https://www.politico.com/story/2018/12/20/james-wolfe-sentencing-senate-intelligence-committee-leaking-1071960.

16. Ali Watkins, "Russia Escalates Spy Games After Years of US Neglect," *Politico,* June 1, 2017, https://www.politico.com/story/2017/06/01/russia-spies-espionage-trump-239003.

17. US Senate, *Meeting the Espionage Challenge: A Review of United States Counterintelligence and Security Programs,* report of the Select Committee on Intelligence (Washington, DC, 1986), https://www.cia.gov/library/readingroom/docs/CIA-RDP90-00530R000300620021-3.pdf.

18. "The Soviets and U.S. High Technology," January 5, 1982, https://www.cia.gov/library/readingroom/docs/CIA-RDP88-01070R000100030002-0.pdf.

19. William Overend, "FBI Also a Resident of S.F. Neighborhood: Soviet Consulate—Cow Hollow Intrigue," *Los Angeles Times,* July 28, 1985, https://www.latimes.com/archives/la-xpm-1985-07-28-mn-5383-story.html.

20. National Security Agency, *American Cryptology During the Cold War, 1945–1989,* book 3, *Retrenchment and Reform, 1972–1980* (Washington, DC, 1998), https://www.nsa.gov/Portals/70/documents/news-features/declassified-documents/cryptologic-histories/cold_war_iii.pdf.

21. US Senate, *Meeting the Espionage Challenge.*

22. General Accounting Office, *Critical Infrastructure Protection: Challenges for Selected Agencies and Industry Sectors* (Washington, DC, 2003), https://www.gao.gov/assets/240/237449.pdf.

23. US Senate, *Current and Projected National Security Threats to the United States,* before the Select Committee on Intelligence, 106th Congress (Washington, DC, 2000), https://www.govinfo.gov/content/pkg/CHRG-106shrg65329/pdf./CHRG-106shrg65329.pdf.

24. US Senate, *Current and Projected National Security Threats to the United States,* before the Select Committee on Intelligence, 108th Congress (Washington, DC, 2003), https://www.govinfo.gov/content/pkg/CHRG-108shrg89797/pdf./CHRG-108shrg89797.pdf.

25. US Senate, *Current and Projected National Security Threats to the United States,* before the Select Committee on Intelligence, 109th Congress (Washington, DC, 2005), https://www.govinfo.gov/content/pkg/CHRG-109shrg22379/pdf./CHRG-109shrg22379.pdf.

26. Christopher Bing and Joel Schectman, "Project Raven: Inside the UAE's Secret Hacking Team of American Mercenaries," Reuters, January 30, 2019, https://www.reuters.com/investigates/special-report/usa-spying-raven.

27. US Senate, *Departments of State, Justice, and Commerce, the Judiciary, and Related Agencies Appropriations for Fiscal Year 1975,* pt. 1, before a subcommittee of the Committee on Appropriations, 93rd Congress (Washington, DC, 1974).

28. US Congress, *Departments of State, Justice, and Commerce, the Judiciary, and Related Agencies Appropriations for 1976,* pt. 2, before the Committee on Appropriations, House of Representatives, 94th Congress (Washington, DC, 1975).

29. US Congress, *Departments of Commerce, Justice and State, the Judiciary, and Related Agencies Appropriations for 1984,* pt. 6, before a subcommittee of the Committee on Appropriations, House of Representatives, 98th Congress (Washington, DC, 1983).

30. US Congress, *Departments of Commerce, Justice, and State, the Judiciary, and Related Agencies Appropriations for 1985,* pt. 8, before a subcommittee of the Committee on Appropriations, House of Representatives, 98th Congress (Washington, DC, 1984).

31. US Senate, *Critical Infrastructure Protection: Toward a New Policy Directive,* before the Committee of the Judiciary, 105th Congress (Washington, DC, 1998).

32. US Congress, *Departments of Commerce, Justice, and State, the Judiciary, and Related Agencies Appropriations for 1993,* pt. 2B, before a subcommittee of the Committee on Appropriations, House of Representatives, 102nd Congress (Washington, DC, 1992).

33. US Congress, *Department of Commerce, Justice, and State, the Judiciary, and Related Agencies Appropriations for 1997,* pt. 2, before a subcommittee of the Committee on Appropriations, House of Representatives, 104th Congress (Washington, DC, 1996).

34. US Congress, *Departments of Commerce, Justice, and State, the Judiciary, and Related Agencies Appropriations for 2000,* pt. 6, before a subcommittee of the Committee on Appropriations, House of Representatives, 106th Congress (Washington, DC, 1999).

35. Executive Order 13010, "Critical Infrastructure Protection," July 15, 1996, https://www.hsdl.org/?view&did=1613.

36. Ibid.

37. Ibid.

38. US Congress, *Departments of Commerce, Justice, and State, the Judiciary, and Related Agencies Appropriations for 1999,* pt. 6, before a subcommittee of the Committee on Appropriations, House of Representatives, 105th Congress (Washington, DC, 1998).

39. US Congress, *Departments of Commerce, Justice, and State, the Judiciary, and Related Agencies Appropriations for 1998,* before a subcommittee of the Committee on Appropriations, House of Representatives, 105th Congress (Washington, DC, 1997).

40. US Congress, *The Activities of the Federal Bureau of Investigation,* pt. 2, before the Subcommittee on Crime of the Committee on the Judiciary, House of Representatives, 105th Congress (Washington, DC, 1997).

41. US Congress, *Departments of Commerce, Justice, and State, the Judiciary, and Related Agencies Appropriations for 1998.*

42. "The National Information Infrastructure: Agenda for Action," September 15, 1993, https://clintonwhitehouse6.archives.gov/1993/09/1993-09-15-the-national-information-infrastructure-agenda-for-action.html.

43. US Congress, *Departments of Commerce, Justice, and State, the Judiciary, and Related Agencies Appropriations for 1999,* pt. 6.

44. US Congress, *Departments of Commerce, Justice, and State, the Judiciary, and Related Agencies Appropriations for 1998.*

45. US Senate, *Current and Projected National Security Threats to the United States,* before the Select Committee on Intelligence, 105th Congress (Washington, DC, 1998).

46. US Congress, *The Activities of the Federal Bureau of Investigation,* pt. 2.

47. US Congress, *Departments of Commerce, Justice, and State, the Judiciary, and Related Agencies Appropriations for 1999,* pt. 6.

48. Ibid.; Presidential Decision Directive / NSC-63, May 22, 1998, https://fas.org/irp/offdocs/pdd/pdd-63.htm.

49. US Senate, *Cyber Attack: Improving Prevention and Prosecution,* before the Committee on the Judiciary, 106th Congress (Washington, DC, 2000).

50. US Congress, *Departments of Commerce, Justice, and State, the Judiciary, and Related Agencies Appropriations for 1999,* pt. 6; Presidential Decision Directive / NSC-63.

51. US Senate, *Cyber Attack.*

52. Presidential Decision Directive / NSC-63.

53. US Congress, *Departments of Commerce, Justice, and State, the Judiciary, and Related Agencies Appropriations for 2001,* pt. 6, before the Committee on Appropriations, House of Representatives, 106th Congress (Washington, DC, 2000).

54. For an NIPC organization chart, see General Accounting Office, *Critical Infrastructure Protection: Significant Challenges in Developing National Capabilities* (Washington, DC, 2001), p. 32, https://www.gao.gov/new.items/d01323.pdf.

55. US Senate, *Department of Defense Authorization for Appropriations for Fiscal Year 2000 and the Future Years Defense Program,* pt. 5, before the Committee on Armed Services, 106th Congress (Washington, DC, 1999).

56. The cutesy name sent an unfortunately defeatist message—after all, a resistance is usually fought against an adversary that already has an upper hand. Was the NIPC suggesting that the US government was at a disadvantage vis-à-vis cyber threats?

57. General Accounting Office, *Critical Infrastructure Protection: Significant Challenges in Developing National Capabilities* (Washington, DC, 2001), https://www.gao.gov/assets/160/157052.pdf.

58. US Senate, *Department of Defense Authorization for Appropriations for Fiscal Year 2000 and the Future Years Defense Program,* pt. 5.

59. US Congress, *Departments of Commerce, Justice, and State, the Judiciary, and Related Agencies Appropriations for 2001,* pt. 2.

60. US Senate, *Improving Our Ability to Fight Cybercrime: Oversight of the National Infrastructure Protection Center,* before the Committee on the Judiciary, 107th Congress (Washington, DC, 2001).

61. US Senate, *Securing Our Infrastructure: Private/Public Information Sharing,* before the Committee on Governmental Affairs, 107th Congress (Washington, DC, 2002).

62. US Senate, *Improving Our Ability to Fight Cybercrime.*

63. US Senate, *Securing Our Infrastructure.*

64. US Senate, *Improving Our Ability to Fight Cybercrime.*

65. US Senate, *Department of Defense Authorization for Appropriations for Fiscal Year 2000 and the Future Years Defense Program,* pt. 5.

66. General Accounting Office, *Critical Infrastructure Protection: Significant Challenges in Developing National Capabilities.*

67. Ibid.

68. US Senate, *Improving Our Ability to Fight Cybercrime.*

69. Ted Bridis, "FBI Unit Fails to React on Time to Electronic Threats, Report Says," *Wall Street Journal,* May 22, 2001.

70. General Accounting Office, *Critical Infrastructure Protection: Significant Challenges in Developing National Capabilities.*

71. US Senate, *Critical Infrastructure Protection: Who's in Charge?* before the Committee on Governmental Affairs, 107th Congress (Washington, DC, 2001).

72. US Senate, *Department of Defense Authorization for Appropriations for Fiscal Year 2000 and the Future Years Defense Program,* pt. 5; General Accounting Office, *Critical Infrastructure Protection: Significant Challenges in Developing National Capabilities.*

73. US Senate, *Department of Defense Authorization for Appropriations for Fiscal Year 2000 and the Future Years Defense Program,* pt. 5.

74. General Accounting Office, *Critical Infrastructure Protection: Significant Challenges in Developing National Capabilities.*

75. US Senate, *Department of Defense Authorization for Appropriations for Fiscal Year 2000 and the Future Years Defense Program,* pt. 5.

76. General Accounting Office, *Critical Infrastructure Protection: Significant Challenges in Developing National Capabilities.*

77. US Senate, *Department of Defense Authorization for Appropriations for Fiscal Year 2000 and the Future Years Defense Program,* pt. 5.

78. US Senate, *Improving Our Ability to Fight Cybercrime.*

79. US Senate, *Critical Information Infrastructure Protection: The Threat Is Real,* before the Committee on the Judiciary, 106th Congress (Washington, DC, 1999).

80. Ibid.

81. General Accounting Office, *Critical Infrastructure Protection: Significant Challenges in Developing National Capabilities.*

82. US Senate, *Critical Infrastructure Protection: Toward a New Policy Directive.*

83. General Accounting Office, *Critical Infrastructure Protection: Significant Challenges in Developing National Capabilities.*

84. US Congress, *Departments of Commerce, Justice, and State, the Judiciary, and Related Agencies Appropriations for 2000,* pt. 6.

85. General Accounting Office, *Critical Infrastructure Protection: Significant Challenges in Developing National Capabilities.*

86. Ibid.

87. Ibid.

88. US Senate, *Department of Defense Authorization for Appropriations for Fiscal Year 2000 and the Future Years Defense Program,* pt. 5.

89. General Accounting Office, *Critical Infrastructure Protection: Significant Challenges in Developing National Capabilities.*

90. US Senate, *Critical Infrastructure Protection: Who's in Charge?*

91. Presidential Decision Directive / NSC-63.

92. General Accounting Office, *Critical Infrastructure Protection: Significant Challenges in Developing National Capabilities.*

93. Bridis, "FBI Unit Fails to React on Time to Electronic Threats."

94. US Senate, *Cyber Attack.*

95. General Accounting Office, *Critical Infrastructure Protection: Significant Challenges in Developing National Capabilities.*

96. US Senate, *Securing our Infrastructure.*

97. US Senate, *Department of Defense Authorization for Appropriations for Fiscal Year 2000 and the Future Years Defense Program,* pt. 5.

98. US Senate, *Critical Infrastructure Protection.*

99. US Senate, *Department of Defense Authorization for Appropriations for Fiscal Year 2000 and the Future Years Defense Program,* pt. 5.

100. General Accounting Office, *Critical Infrastructure Protection: Significant Challenges in Developing National Capabilities.*

101. Charles Piller, "Federal Agency Created to Combat the Rise in Cyber-Crime Is Viewed with Distrust by Firms It Is Supposed to Protect," *Los Angeles Times,* March 5, 2000, https://www.latimes.com/archives/la-xpm-2000-mar-05-mn-5606-story.html.

102. US Senate, *Critical Infrastructure Protection: Toward a New Policy Directive.*

103. Piller, "Federal Agency Created to Combat the Rise in Cyber-Crime."

104. US Senate, *Cyber Attacks: The National Protection Plan and Its Privacy Implications,* before the Committee on the Judiciary, 106th Congress (Washington, DC, 2000), https://www.govinfo.gov/content/pkg/CHRG-106shrg68776/pdf./CHRG-106shrg68776.pdf.

105. US Senate, *Critical Infrastructure Protection: Toward a New Policy Directive.*

106. Advisory Panel to Assess Domestic Response Capabilities for Terrorism Involving Weapons of Mass Destruction, *Second Annual Report* (Washington, DC, 2000), https://www.rand.org/content/dam/rand/www/external/nsrd/terrpanel/terror2.pdf.

107. US Senate, *Critical Infrastructure Protection.*

108. Ibid.

109. Advisory Panel to Assess Domestic Response Capabilities for Terrorism Involving Weapons of Mass Destruction, *Second Annual Report.*

110. "FBI Is Considering a Plan to Terminate Cyber-Security Unit," *Wall Street Journal,* March 21, 2002.

111. Larry Mefford, assistant director, Cyber Division, Federal Bureau of Investigation, testimony before the House Committee on Government Reform, Subcommittee on National Security, Veterans Affairs, and International Relations, June 11, 2002, https://archives.fbi.gov/archives/news/testimony/nipcs-role-in-the-new-department-of-homeland-security.

112. "Panel Chairman Seeks Study of FBI Reorganization; Rep Wolf Cites Concerns on Impact of Shifting Agents to Counterterrorism Effort," *Washington Post,* June 5, 2002; US Senate, *Counterterrorism,* before the Committee on the Judiciary, 107th Congress (Washington, DC, 2002).

113. National Security Agency, *American Cryptology During the Cold War, 1945–1989,* book 1, *The Struggle for Centralization, 1945–1960* (Ft. Meade, MD, 1995).

114. "Tempest: A Signal Problem" (undated), https://www.nsa.gov/Portals/70/documents/news-features/declassified-documents/cryptologic-spectrum/tempest.pdf.

115. "Static Magic or the Wonderful World of Tempest," *Cryptolog,* November 1983, https://www.nsa.gov/Portals/70/documents/news-features/declassified-documents/cryptologs/cryptolog_84.pdf.

116. National Security Agency, *American Cryptology During the Cold War, 1945–1989,* https://www.nsa.gov/Portals/70/documents/news-features/declassified-documents/cryptologic-histories/cold_war_ii.pdf.

117. National Security Agency, "TEMPEST," https://apps.nsa.gov/iaarchive/programs/iad-initiatives/tempest.cfm.

118. US Senate, *Meeting the Espionage Challenge.*

119. General Accounting Office, *Economic Espionage: The Threat to US Industry* (Washington, DC, 1992), https://www.gao.gov/assets/110/104477.pdf.

120. US Congress, *Going Dark: Lawful Electronic Surveillance in the Face of New Technologies,* before the Committee on the Judiciary, House of Representatives, 112th Congress (Washington, DC, 2011), https://www.govinfo.gov/content/pkg/CHRG-112hhrg64581/pdf./CHRG-112hhrg64581.pdf.

121. Jack Corrigan, "NSA Cyber Chief Wants to Share Digital Threats Early and Often," September 5, 2019, https://www.nextgov.com/cybersecurity/2019/09/nsa-cyber-chief-wants-share-digital-threats-early-and-often/159673.

122. National Security Agency, "Strengthening the Front Line: NSA Launches New Cybersecurity Directorate" (October 1, 2019), https://www.nsa.gov/Press-Room/News-Highlights/Article/Article/1973871/strengthening-the-front-line-nsa-launches-new-cybersecurity-directorate/.

123. Ibid.

124. USA Patriot Act, Public Law 107-56, October 26, 2001, https://www.congress.gov/107/plaws/publ56/PLAW-107publ56.pdf.

125. Charles Clancy and Emily Frye, "Is It Time to Designate Social Media as 'Critical Infrastructure'?" *The Hill,* July 27, 2020, https://thehill.com/opinion/cybersecurity/509154-is-it-time-to-designate-social-media-as-critical-infrastructure?sf125852197=1.

126. Homeland Security Act of 2002, Public Law 107-296, November 25, 2002, https://www.dhs.gov/sites/default/files/publications/hr_5005_enr.pdf.

127. Ibid.

128. Homeland Security Act of 2002; US Senate, *Critical Infrastructure Protection: Who's in Charge?*

129. Presidential Decision Directive / NSC-63.

130. "Protecting America's Critical Infrastructures: PDD 63," February 8, 1999, https://www.hsdl.org/?view&did=3544; General Accounting Office, *Critical Infrastructure Protection: Significant Challenges in Developing National Capabilities;* General Accounting Office, *Critical Infrastructure Protection: Significant Homeland Security Challenges Need to Be Addressed* (Washington, DC, 2002), https://www.gao.gov/assets/110/109467.pdf.

131. Homeland Security Act of 2002.

132. Daniel Morgan, *Research and Development in the Department of Homeland Security* (Washington, DC: Congressional Research Service, 2003).

133. US Congress, *Department of Homeland Security Appropriations for 2017,* pt. 1C, before the Committee on Appropriations, House of Representatives, 114th Congress (Washington, DC, 2016).

134. US Congress, *Department of Homeland Security Appropriations for 2005,* pt. 1, before a subcommittee of the Committee on Appropriations, House of Representatives, 108th Congress (Washington, DC, 2004).

135. Presidential Decision Directive / NSC-63.

136. US Congress, *Department of Homeland Security Appropriations for 2005,* pt 1.

137. Government Accountability Office, *Critical Infrastructure Protection: Department of Homeland Security Faces Challenges,* p. 24; US Congress, *Cyber Security: U.S. Vulnerability and Preparedness,* before the Committee on Science, 109th Congress (Washington, DC, 2005), https://www.govinfo.gov/content/pkg/CHRG-109hhrg23332/pdf./CHRG-109hhrg23332.pdf.

138. Government Accountability Office, *Critical Infrastructure Protection: Department of Homeland Security Faces Challenges.*

139. Ibid.

140. US Congress, *Cyber Security: U.S. Vulnerability and Preparedness.*

141. US Senate, *Department of Homeland Security's Budget Submission for Fiscal Year 2005,* before the Committee on Governmental Affairs, 108th Congress (Washington, DC, 2004).

142. US Senate, *Securing our Infrastructure.*

143. US Congress, *Cyber Security: U.S. Vulnerability and Preparedness.*

144. US Congress, *Cybersecurity: Protecting America's Critical Infrastructure, Economy, and Consumers,* before the Committee on Energy and Commerce, House of Representatives, 109th Congress (Washington, DC, 2006), https://www.govinfo.gov/content/pkg/CHRG-109hhrg31464/pdf./CHRG-109hhrg31464.pdf; US Congress, *Cyber Security: U.S. Vulnerability and Preparedness;* Democratic Staff of the Committee on Homeland Security, *The State of Homeland Security, 2006* (Washington, DC); Government Accountability Office, *Critical Infrastructure Protection: Department of Homeland Security Faces Challenges.*

145. US Congress, *Cyber Security: U.S. Vulnerability and Preparedness.*

146. Government Accountability Office, *Critical Infrastructure Protection: Department of Homeland Security Faces Challenges.*

147. US Congress, *Cyber Security: U.S. Vulnerability and Preparedness.*

148. Government Accountability Office, *Critical Infrastructure Protection. Key Private and Public Cyber Expectations Need to Be Consistently Addressed* (Washington, DC, 2010), https://www.gao.gov/assets/310/307222.pdf.

149. US Congress, *Department of Homeland Security, Appropriations for 2012,* pt. 4, before a subcommittee of the Committee on Appropriations, House of Representatives, 112th Congress (Washington, DC, 2012).

150. US Congress, *Department of Homeland Security's Information Analysis and Infrastructure Protection Budget Proposal for Fiscal Year 2005,* before the Select Committee on Homeland Security, 108th Congress (Washington, DC, 2004).

151. US Senate, *The Homeland Security Department's Budget Submission for Fiscal Year 2011,* before the Committee on Homeland Security and Governmental Affairs, 111th Congress (Washington, DC, 2010).

152. US Senate, *Homeland Security Department's Budget Submission for Fiscal Year 2012,* before the Committee on Homeland Security and Governmental Affairs, 112th Congress (Washington, DC, 2011).

153. US Senate, *Ten Years after 9/11, 2011,* before the Committee on Homeland Security and Governmental Affairs, 112th Congress (Washington, DC, 2012).

154. Department of Homeland Security, *Preventing and Defending Against Cyber Attacks* (Washington, DC, 2011), https://www.dhs.gov/xlibrary/assets/preventing-and-defending-against-cyber-attacks.pdf.

155. US Congress, *Cybersecurity Recommendations for the Next Administration,* before the Committee on Homeland Security, House of Representatives, 110th Congress (Washington, DC, 2008), https://www.govinfo.gov/content/pkg/CHRG-110hhrg48089/pdf./CHRG-110hhrg48089.pdf.

156. National Security Presidential Directive 54, "Cybersecurity Policy," January 8, 2008, https://fas.org/irp/offdocs/nspd/nspd-54.pdf.

157. Steve Barrett, "National Communications System Joins Homeland Security Department," March 10, 2003, https://archive.defense.gov/news/newsarticle.aspx?id=29323.

158. US Congress, *Department of Homeland Security Appropriations for 2017,* pt. 1C.

159. Department of Homeland Security, *DHS Can Enhance Efforts to Protect Commercial Facilities from Terrorism and Physical Threats* (Washington, DC, 2020), https://www.oig.dhs.gov/sites/default/files/assets/2020-06/OIG-20-37-Jun20.pdf.

160. Christopher Krebs, May 1, 2019, https://docs.house.gov/meetings/AP/AP15/20190501/109345/HHRG-116-AP15-Wstate-KrebsC-20190501.pdf.

161. Christopher Krebs, director, Cybersecurity and Infrastructure Security Agency, testimony for the hearing "CISA Fiscal Year 2021 President's Budget," before the House Committee on Homeland Security, March 11, 2020, https://homeland.house.gov/imo/media/doc/Testimony%20-%20Krebs.pdf.

162. Jory Heckman, "Launch of DHS Cyber Agency 'More of a Groundbreaking Than a Ribbon Cutting,'" *Federal News Radio,* November 16, 2018, https://federalnewsnetwork.com/cybersecurity/2018/11/launch-of-dhs-cyber-agency-more-of-a-groundbreaking-than-a-ribbon-cutting.

163. US Congress, *Examining the Mission, Structure, and Reorganization Effort of the National Protection and Programs Directorate,* before the Committee on Homeland Security, House of Representatives, 114th Congress (Washington, DC, 2015).

164. Cyberspace Solarium Commission, 2020, https://drive.google.com/file/d/1ryMCIL_dZ30QyjFqFkkf10MxIXJGT4yv/view.

165. John Kelly, "Why St. Elizabeth? More Musing on Washington's Hospital for the Mentally Ill," *Washington Post,* July 6, 2019, https://www.washingtonpost.com/local/why-st-elizabeth-more-musing-on-washingtons-hospital-for-the-mentally-ill/2019/07/05/f18c38be-9f78-11e9-b27f-ed2942f73d70_story.html.

166. US Congress, *Department of Homeland Security Appropriations for 2017,* pt. 1C.

167. US Senate, *Fifteen Years After 9/11: Threats to the Homeland,* before the Committee on Homeland Security and Governmental Affairs, 114th Congress (Washington, DC, 2016), https://www.govinfo.gov/content/pkg/CHRG-114shrg25160/pdf./CHRG-114shrg25160.pdf.

168. Center for Strategic and International Studies, *Significant Cyber Events Since 2006.*

169. US Congress, *Worldwide Threats to the Homeland: ISIS and the New Wave of Terror,* before the committee on Homeland Security, House of Representatives, 114th Congress (Washington, DC, 2016), https://www.govinfo.gov/content/pkg/CHRG-114hhrg25265/pdf./CHRG-114hhrg25265.pdf.

170. Tom Coburn, "A Review of the Department of Homeland Security's Missions and Performance," January 2015, https://www.hsgac.senate.gov/imo/media/doc/Senator%20Coburn%20DHS%20Report%20FINAL.pdf.

171. Christopher Krebs, March 13, 2019, https://docs.house.gov/meetings/AP/AP15/20190313/109080/HHRG-116-AP15-Wstate-KrebsC-20190313.pdf.

172. US Congress, *A DHS Intelligence Enterprise: Still Just a Vision or Reality?* before the Committee on Homeland Security, House of Representatives, 111th Congress (Washington, DC, 2010).

173. Department of Homeland Security, *Survey of the Information Analysis and Infrastructure Protection Directorate* (Washington, DC, 2004), https://www.oig.dhs.gov/sites/default/files/assets/Mgmt/OIG_SurveyIAIP_0204.pdf.

174. US Senate, *Department of Homeland Security Status Report: Assessing Challenges and Measuring Progress,* before the Committee on Homeland Security and Governmental Affairs, 110th Congress (Washington, DC, 2007); Government Accountability Office, *DHS Intelligence Analysis: Additional Actions Needed to Address Analytic Priorities and Workforce Challenges* (Washington, DC, 2014), https://www.gao.gov/assets/670/663794.pdf.

175. Department of Homeland Security, *Efforts to Identify Critical Infrastructure Assets and Systems* (Washington, DC, 2009), https://www.oig.dhs.gov/sites/default/files/assets/Mgmt/OIG_09-86_Jun09.pdf.

176. US Congress, *The Progress of the DHS Chief Intelligence Officer,* before the Subcommittee on Intelligence, Information Sharing, and Terrorism Risk Assessment of the Committee on Homeland Security, House of Representatives, 109th Congress (Washington, DC, 2006).

177. Department of Homeland Security, *Quadrennial Homeland Security Review 2014* (Washington, DC), https://www.dhs.gov/sites/default/files/publications/2014-qhsr-final-508.pdf.

178. US Congress, *Is the Office of Intelligence and Analysis Adequately Connected to the Broader Homeland Communities?* before the Subcommittee on Intelligence, Information Sharing, and Terrorism Risk Assessment of the Committee on Homeland Security, House of Representatives, 111th Congress (Washington, DC, 2010).

179. US Senate, *Cybersecurity, Terrorism, and Beyond: Addressing Evolving Threats to the Homeland,* before the Committee on Homeland Security and Governmental Affairs, 113th Congress (Washington, DC, 2014), https://www.govinfo.gov/content/pkg/CHRG-113shrg92903/pdf./CHRG-113shrg92903.pdf.

180. Mark A. Randol, *The Department of Homeland Security Intelligence Enterprise: Operational Overview and Oversight Challenges for Congress* (Washington, DC: Congressional Research Service, 2010), https://fas.org/sgp/crs/homesec/R40602.pdf.

181. US Senate, *Department of Homeland Security Appropriations for Fiscal Year 2007,* pt. 2.

182. US Congress, *The Progress of the DHS Chief Intelligence Officer.*

183. Randol, *The Department of Homeland Security Intelligence Enterprise.*

184. "About the Homeland Infrastructure Threat and Risk Analysis Center," https://www.cisa.gov/node/18.

185. Randol, *The Department of Homeland Security Intelligence Enterprise.*

186. US Congress, *Cyber Security: U.S. Vulnerability and Preparedness.*

187. Randol, *The Department of Homeland Security Intelligence Enterprise.*

188. US Congress, *Is the Office of Intelligence and Analysis Adequately Connected?*

189. US Congress, *A DHS Intelligence Enterprise.*

190. US Congress, *Is the Office of Intelligence and Analysis Adequately Connected?*

191. US Congress, *A DHS Intelligence Enterprise.*

192. US Congress, *Cybersecurity Recommendations for the Next Administration.*

193. Presidential Decision Directive / NSC-63.

194. General Accounting Office, *Critical Infrastructure: Significant Challenges in Developing National Capabilities.*

195. US Senate, *Critical Infrastructure Protection: Toward a New Policy Directive.*

196. US Senate, *Securing Our Infrastructure.*

197. US Senate, *Improving Our Ability to Fight Cybercrime.*

198. Mefford, testimony before the House Committee on Government Reform, Subcommittee on National Security, Veterans Affairs, and International Relations, June 11, 2002.

199. General Accounting Office, *Critical Infrastructure Protection: Significant Challenges in Developing National Capabilities.*

200. Presidential Decision Directive 63, "Critical Infrastructure Protection," May 22, 1998, https://fas.org/irp/offdocs/pdd/pdd-63.htm.

201. Homeland Security Presidential Directive 7, "Critical Infrastructure Identification, Prioritization, and Protection," June 27, 2008, https://www.cisa.gov/homeland-security-presidential-directive-7.

202. Department of Homeland Security, "Critical Infrastructure Sectors," March 24, 2020, https://www.cisa.gov/critical-infrastructure-sectors.

203. Congressional Research Service, *Critical Infrastructure: Emerging Trends and Policy Considerations for Congress* (Washington, DC, 2020).

204. NIPP 2013, "Partnering for Critical Infrastructure Security and Resilience" (Washington, DC: Department of Homeland Security, 2013), https://www.cisa.gov/sites/default/files/publications/national-infrastructure-protection-plan-2013-508.pdf.

205. National Council of ISACs, https://www.nationalisacs.org.

206. Department of Homeland Security, *Preventing and Defending Against Cyber Attacks* (Washington, DC, 2011), https://www.dhs.gov/xlibrary/assets/preventing-and-defending-against-cyber-attacks.pdf.

207. Robert Kolasky for a hearing on *Public-Private Initiatives to Secure the Supply Chain,* before the Committee on Homeland Security, House of Representatives (Washington, DC, October 16, 2019), https://homeland.house.gov/imo/media/doc/Testimony-Kolasky1.pdf.

208. Christopher Krebs for a hearing on *Defending Our Democracy: Building Partnerships to Protect America's Elections,* before the Committee on Homeland Security, House of Representatives (Washington, DC, February 13, 2019), https://homeland.house.gov/imo/media/doc/Testimony-Krebs.pdf.

209. Christopher Krebs, May 1, 2019.

210. US Congress, *Cybersecurity Recommendations for the Next Administration.*

211. Government Accountability Office, *Critical Infrastructure Protection: Key Private and Public Cyber Expectations.*

212. White House, "Executive Order—Improving Critical Infrastructure Cybersecurity," February 12, 2013, https://obamawhitehouse.archives.gov/the-press-office/2013/02/12/executive-order-improving-critical-infrastructure-cybersecurity; White House, "Cybersecurity—Executive Order 13636," https://obamawhitehouse.archives.gov/issues/foreign-policy/cybersecurity/eo-13636.

213. Ibid.

214. Government Accountability Office, *Critical Infrastructure Protection: Key Private and Public Cyber Expectations.*

215. US Congress, *A DHS Intelligence Enterprise.*

216. Government Accountability Office, *DHS Intelligence Analysis.*

217. Ibid.

218. Ibid.

219. US Congress, *Cyber Security: U.S. Vulnerability and Preparedness.*

220. US Senate, *Department of Homeland Security's Budget Submission for Fiscal Year 2005.*

221. US Congress, *Cyber Security: U.S. Vulnerability and Preparedness.*

222. US Congress, *A DHS Intelligence Enterprise.*

223. Government Accountability Office, *DHS Intelligence Analysis.*

224. Government Accountability Office, *Information Sharing: DHS Is Assessing Fusion Center Capabilities and Results, but Needs to More Accurately Account for Federal Funding Provided to Centers* (Washington, DC, 2014), https://www.gao.gov/assets/670/666760.pdf; Government Accountability Office, *Information Sharing: Federal Agencies Are Helping Fusion Centers Build and Protect Privacy, but Could Better Manage Results* (Washington, DC, 2010), https://www.gao.gov/assets/320/310268.pdf; Department of Homeland Security, *Implementing 9/11 Commission Recommendations: Progress Report 2011* (Washington, DC), https://www.dhs.gov/xlibrary/assets/implementing-9-11-commission-report-progress-2011.pdf.

225. Ben Bain, "DHS, Industry to Try Fusion Centers for Classified Data Swap," *Federal Computer Week,* March 16, 2010, https://fcw.com/articles/2010/03/16/web-cyber-threat-fusion-center.aspx.

226. US Senate, *The Homeland Security Department's Budget Submission for Fiscal Year 2012.*

227. Department of Security, *Preventing and Defending Against Cyber Attacks* (Washington, DC, 2011), https://www.dhs.gov/xlibrary/assets/preventing-and-defending-against-cyber-attacks.pdf.
228. Bain, "DHS, Industry to Try Fusion Centers."
229. US Congress, *Departments of Commerce, Justice, and State, the Judiciary, and Related Agencies Appropriations for 1999,* pt. 6; US Senate, *Department of Defense Authorization for Appropriations for Fiscal Year 2000 and the Future Years Defense Program,* pt. 5.
230. US Senate, *Critical Infrastructure Protection: Toward a New Policy Directive.*
231. US Senate, *Cyber Attack: Improving Prevention and Prosecution.*
232. US Senate, *Critical Infrastructure Protection: Toward a New Policy Directive.*
233. US Senate, *Securing Our Infrastructure.*
234. US Senate, *Critical Information Infrastructure Protection: The Threat Is Real.*
235. Ibid.
236. White House, "National Plan for Information Systems Protection: Version 1.0" (Washington, DC, 2000), https://fas.org/irp/offdocs/pdd/CIP-plan.pdf.
237. US Senate, *Current and Projected National Security Threats to the United States,* 105th Congress.
238. Ronald L. Dick, director, National Infrastructure Protection Center, Federal Bureau of Investigation, "Information Technology," testimony before the House Committee on Government Reform, Subcommittee on Government Efficiency, Financial Management, and Intergovernmental Relations, September 26, 2001.
239. US Senate, *Improving Our Ability to Fight Cybercrime.*
240. Mefford, testimony before the House Committee on Government Reform, June 11, 2002.
241. US Senate, *Counterterrorism.*
242. 9/11 Review Commission, *The FBI: Protecting the Homeland in the 21st Century* (Washington, DC, 2015), https://www.fbi.gov/file-repository/final-9-11-review-commission-report-unclassified.pdf/view.
243. Darren E. Tromblay and Robert Spelbrink, *Securing US Innovation* (Lanham: Rowman and Littlefield, 2016).
244. Department of Justice, *Audit of the Federal Bureau of Investigation's Implementation of Its Next Generation Cyber Initiative* (Washington, DC, 2015).
245. US Senate, *Cyber Security: Responding to the Threat of Cyber Crime and Terrorism,* before the Committee on the Judiciary, 112th Congress (Washington, DC, 2011).
246. 9/11 Review Commission, *The FBI.*
247. Ibid.
248. US Senate, *Department of Defense Authorization for Appropriations for Fiscal Year 2000 and the Future Years Defense Program,* pt. 5.
249. 9/11 Review Commission, *The FBI.*
250. Department of Homeland Security, *Protective Security Advisor Program Efforts to Build Effective Critical Infrastructure Partnerships: Oil and Gas Subsector* (Washington, DC, 2010), https://www.hsdl.org/?view&did=13056.
251. US Congress, *Department of Homeland Security, Appropriations for 2012,* pt. 4, before a subcommittee of the Committee on Appropriations, House of Representatives, 112th Congress (Washington, DC, 2012).
252. Department of Homeland Security, "Protective Security Advisor Program," https://www.cisa.gov/sites/default/files/publications/PSA-Program-Fact-Sheet-05-15-508.pdf.
253. US Congress, *DHS's Vulnerability Assessments in Protecting Our Nation's Critical Infrastructure,* before the Committee on Homeland Security, House of Rep-

resentatives, 114th Congress (Washington, DC, 2016), https://www.govinfo.gov/content/pkg/CHRG-114hhrg25264/html./CHRG-114hhrg25264.htm.

254. Department of Homeland Security, *DHS Made Limited Progress to Improve Information Sharing Under the Cybersecurity Act in Calendar Years 2017 and 2018* (Washington, DC, 2020), https://www.oig.dhs.gov/sites/default/files/assets/2020-09/OIG-20-74-Sep20.pdf.

255. Department of Homeland Security, "Automated Indicator Sharing," https://www.cisa.gov/automated-indicator-sharing-ais.

256. Department of Homeland Security, *DHS Made Limited Progress;* US Congress, *DHS's Vulnerability Assessments.*

257. Department of Homeland Security, "Automated Indicator Sharing."

258. Department of Homeland Security, *DHS Made Limited Progress.*

259. US Congress, *DHS's Vulnerability Assessments.*

260. Joseph Marks, "Only 6 Non-Federal Groups Share Cyber Threat Info with Homeland Security," June 27, 2018, https://www.nextgov.com/cybersecurity/2018/06/only-6-non-federal-groups-share-cyber-threat-info-homeland-security/149343.

261. Department of Homeland Security, *DHS Made Limited Progress.*

262. US Congress, *Department of Homeland Security Appropriations for 2005,* pt. 1.

263. US Congress, *DHS's Vulnerability Assessments.*

264. Christopher Krebs, March 13, 2019.

6

Addressing Global Necessities and Domestic Shortcomings

TWO CHANGING APERTURES HAVE SIGNIFICANT IMPLICATIONS FOR the ability of the United States to secure the private sector. The first is the expanding aperture of globalization. Innovation and business practices span the globe and interface with countries—such as China—that have long track records of coercion and theft against US industry. In order to protect US interests, Washington must make counterproliferation, counterintelligence, and counterterrorism—in defense of the private sector—an internationally oriented enterprise. Historically, the United States, notably through the Federal Bureau of Investigation (FBI), has endeavored to secure private sector assets abroad since at least the 1940s.

At the same time that the United States is expanding its aperture to address an increasingly complex international business environment, it is encountering a closing aperture—the work force necessary to secure the private sector—at home. Certainly the lure of private industry is not new. Then–FBI director J. Edgar Hoover lamented the Bureau's loss of personnel as early as the 1940s. However, two factors—the increasingly stifling culture of government coupled with competition for specialized expertise—have exacerbated the inability to attract and retain talent.

Counterproliferation

The US government, primarily through the FBI, has combated proliferation since at least World War II. In 1940, the FBI established its Special Intelligence Service (SIS), which had responsibility for intelligence collection throughout most of the Western Hemisphere.[1] The Bureau's collection, via the SIS, was not simply in furtherance of its parochial interests; it agreed to provide a number of other US government agencies with information responsive to intelligence requirements. Of specific interest—in regard to counterproliferation—was the agreement to provide the Office of the Coordinator of Inter-American Affairs with information about individuals dealing commercially with sympathizers or subjects of Axis nations. Additionally, the FBI agreed to provide information to the Board of Economic Warfare concerning strategic materials.[2] Throughout Latin America, the Bureau's counterproliferation mission focused on preventing raw commodities from falling into Axis hands. Much of the FBI's counterproliferation work centered around platinum, diamonds, and rubber.[3]

Understanding the Target

The FBI's SIS was responsible for disrupting Axis acquisition of strategic materials in a variety of ways, but first it had to understand the threat. Platinum, which was used in the manufacture of electrical equipment and in the production of nitric acid (which is used in the manufacture of explosives such as nitroglycerin), was an Axis target for acquisition and thus the FBI provided urgent attention to developing an awareness about what the threat to it looked like throughout Latin America.[4] Various US government entities viewed the smuggling of platinum out of Colombia and into Axis hands as "one of the principal forms of subversive activity," according to the FBI.[5] Colombia was reportedly a significant source of the platinum—which made its way through Ecuador and Argentina—that illicitly arrived in Europe.[6] An official in the US Office of Economic Warfare had assessed that the FBI representatives in Colombia knew about all of the smugglers in Colombia and had developed complete files on them. In early 1944, the US ambassador to Colombia made a request of the FBI to dispatch four agents to investigate platinum smuggling in that country.[7] A special squad of agents arrived in May 1944 and worked with local officials to identify many of the important smugglers.[8]

Rubber was another commodity that Axis operatives made multiple attempts to acquire and thus was another item of concern to the FBI. During World War II, Bolivia had agreed to sell to the United States all of Bolivia's locally produced rubber, with the exception of a small quantity for commercial use in that country. The urgent need for rubber, according to the FBI, made controlling the smuggling of it from Bolivia "vitally necessary."[9] Throughout the war, the offices of the legal attachés (the FBI's representative at embassies) in La Paz, Bolivia, and Buenos Aires, Argentina (Bolivian rubber made its way into Argentina) helped the Bolivian government to detect and recover contraband rubber shipments. (One example that illustrates the extent of smuggling occurred in April 1945, when Bolivian customs authority seized 876 pounds of rubber from trunks that belonged to a Bolivian senator who was en route to Argentina on a diplomatic mission.)[10]

With apologies to Jule Styne and Leo Robin, diamonds may have been a Nazi's best friend, but the FBI was determined to break up that friendship. According to a history of the SIS operations that the Bureau prepared for internal use, the "prevention of industrial diamonds from reaching Germany was one of the major objectives in economic warfare and was also one of the chief concerns of SIS representatives in Venezuela."[11] The FBI's internal SIS history described Venezuela as one of the world's largest producers of both industrial and other forms of diamonds. A large black market for these was oriented around Caracas and was the source of many of the diamonds that were smuggled to Germany.[12]

Finally, the FBI took steps to secure tin ore. Bolivia was of particular importance due to the presence of this resource in that country. Access to Bolivian tin ore became especially valuable following Japan's seizure of Malayan tin deposits. The FBI's role in Bolivia was twofold: not only did it need to protect tin ore resources from Axis agents, but it also had to closely follow Bolivia's turbulent internal politics in order to assess the extent to which Germany and Argentina were attempting to gain control over the entire country by fomenting revolution.[13]

A commodity-oriented approach to counterproliferation meant that FBI agents, at times, had to locate not simply the end-user of materials but rather the source of materials. For instance, an agent traveling in Colombia discovered that indigenous peoples were panning a local river and selling gold and platinum to Germans who had begun to appear in the region.[14]

Interdiction

The FBI, lacking law enforcement powers in foreign countries, sometimes relied on extrajudicial measures to prevent the Axis from acquiring essential commodities. For instance, when a German spy ring attempted to smuggle insulin to Europe, SIS representatives swapped the insulin for talcum powder.[15] The SIS also apparently considered confiscation a measure of success. For instance, an SIS representative in Ecuador managed to divert 14 kilograms of platinum that had been smuggled in from Colombia.[16] One SIS agent recalled that the platinum interdicted in Ecuador over the course of several years had been stored in fruit jars stashed inside the US embassy's safe.[17]

Blacklisting

The SIS also disrupted the activities of proliferators by making it more difficult for them to operate. On July 17, 1941, Franklin D. Roosevelt created a Proclaimed List of Certain Blocked Nationals (also known as the Black List), which consisted of records created by the State Department's Division of World Trade Intelligence and collected by the Department of Commerce.[18] This list documented companies and individuals located in Central and South America whose accounts were frozen during World War II under the provisions of the Trading with the Enemy Act.[19] SIS intelligence collection contributed to this list. For instance, according to the FBI's internal history of the SIS, its representatives in Argentina were "credited with obtaining and submitting considerable information about commercial firms which subsequently appeared on [the Black List]."[20] Similarly, the work of the legal attaché in Santiago, Chile, directly resulted in the Black List gaining several new entries. SIS-facilitated additions to the Black List included "influential and wealthy individuals of Nazi sentiment who were acting as Nazi supply agents, assisting the Nazi cause in the storage of materials and equipment throughout [Chile]."[21]

Arrest

In certain situations, the FBI, by sharing information with its foreign counterparts (often after the Bureau had collected sufficient intelligence), effected the arrest of proliferators. An FBI memorandum stressed that the Bureau's representatives were not in Latin America simply to "report the rumors and information given to them by professional informants and other sources, such as other intelligence operating in Latin America."[22] Instead the FBI "actually conduct[ed] investigations with regard to the information and reports received. . . . Not only [was the Bureau] locating and definitely identifying [enemy operatives] but by ordinary investigative means [the FBI was] obtaining specific, positive evidence which [was] being turned over to the local governments for use in arresting or interning the agents in question."[23]

The FBI focused significant effort on obtaining local law enforcement action to stop platinum smuggling. For instance, the squad of agents that the FBI had dispatched to Colombia in May 1944 were sufficiently successful in their intelligence-gathering that, by October 1944, the US ambassador was able to present the Colombian government with enough information to warrant the designation of four individuals for deportation, the declaration of four additional aliens as persona non grata (the polite way of saying that an individual has been booted out of a country), and the denaturalization of a naturalized Colombian resident.[24]

However, local law enforcement, while removing specific individuals from the process of proliferation, did not always bring about the desired end result. One agent who was assigned to Guayaquil, Ecuador, recalled the porousness of the local law enforcement. According to the agent, the country's national police would search out platinum smugglers and seize the platinum. Despite the legal attaché's augmentation of the national police officers' salaries, several weeks after the police seized platinum from one smuggler, the police would arrest another smuggler who was carrying the same platinum. In other words, seized platinum, despite the ostensible cooperation of Ecuadorian officers, was reentering the black market.[25] Arresting smugglers meant little if the commodity remained available.

Policies

Finally, SIS counterproliferation activities helped countries—through the promulgation of new policies—to tighten the overall control over vulnerabilities that facilitated smuggling. According to one Bureau document, SIS agents, working in cooperation with the US embassy, advised "local government officials in devising and preparing completely new laws and regulations designed to protect allied war interests in connection with the smuggling of . . . strategic materials."[26] For instance, following the Colombian government's SIS-facilitated crackdown on platinum smuggling in late 1944, the government promulgated a decree that established rigid control over platinum production and trafficking.[27]

The FBI ceded its SIS operation to the Central Intelligence Group (CIG), which would become the Central Intelligence Agency (CIA), in 1947. However, it has continued to engage, internationally, on counterintelligence and counterproliferation issues. For instance, at the behest of the State Department as well as other government and private sector organizations, the FBI provided Awareness of National Security Incidents and Response program presentations to audiences abroad.[28] As of 1998, the Bureau had presented its economic espionage program at lectures in ten countries—including Australia, Ireland, New Zealand, Panama, South Korea, and the United Kingdom—and reached an audience of more than 80,000 people.[29]

Critical Infrastructure Protection

Although the FBI ended its Plant Survey Program on the domestic front relatively soon after the United States entered World War II, it deployed these capabilities throughout Latin America, where they complemented the SIS's intelligence-gathering by hardening important targets against Axis activities. According to the Bureau's internal history of the SIS, it became responsible for "plant surveys with respect to plants and other facilities vital to the production, manufacture, and transportation of strategic materials" vital to the war effort. These responsibilities included "furnishing technical advisers and instructors to the local governments, industries, and officials with regard to plant surveys and plant protection matters."[30]

The FBI initiated its first Latin American plant survey in late 1941. An undercover SIS operative, employing the guise of a security consultant, proceeded to assess facilities. Companies surveyed adopted a number of the agent's recommendations.[31] The Bureau initiated its official plant surveys in July 1942, at the request of the US ambassador to Bolivia, who asked the FBI to assess the industrial installations in that country. Agents in Bolivia made surveys of two large tin-mining interests and provided recommendations to the facilities about protective measures that could be implemented.[32] Then, in January 1943—at the request of the US State Department—the FBI established an extensive Plant Survey Program throughout Latin America. Specifically, the State Department was interested in having the FBI assign "security officers" capable of providing instructions to authorities and industrialists about the protection of facilities against sabotage. The Bureau dispatched eighteen agents who were specially qualified in this work.

Unlike other aspects of the SIS, which with the exception of the legal attachés operated under nonofficial cover, the plant survey "security officers," also referred to as "security liaison men," implemented a program that filled overt educational and advisory functions. The eighteen agents—and five stenographers—surveyed a total of 104 companies, installations, ports, and organizations with a total of 150 branch facilities.[33] This was probably a more difficult assignment than might be imagined, since, according to an FBI memorandum, the eighteen agents were "completely without prior training in foreign work and had no knowledge whatsoever with regard to the Spanish or Portuguese language."[34]

Expeditions to remote locations could cause unexpected complications. In August 1943, the Bureau's Executive Conference considered the circumstances of two agents in Peru who had lost $102.50 worth of property after upsetting a canoe that was "the only means of public transportation" to a plant they were supposed to survey. The Executive Conference refused to reimburse the agents for the lost property "because of the feeling that such losses were a natural risk in connection with any agent's work."[35] (However, the Executive Conference did agree to reimburse the cost of a camera, belonging to a clerk in the office of the legal attaché, that the agents had borrowed—in lieu of the Bureau's bulkier Speed Graphic camera—to reduce the weight of the canoe.)[36]

The State Department—as indicated by its role in organizing the FBI's Plant Survey Program during World War II—has engaged in efforts to protect the US private sector abroad. In 1985, it established the Overseas Security Advisory Council (OSAC), a public sector–private sector partnership under the auspices of the Department's Bureau of Diplomatic Security.[37] Its creation was the formalization of ad hoc exchanges that were occurring between regional security officers at embassies and elements from the US private sector that had concerns about their people and facilities.[38] In 1986, the Omnibus Diplomatic Security and Antiterrorism Act designated the State Department's assistant secretary for diplomatic security as responsible for "liaison with American overseas private sector security interests."[39] Representatives from the US Agency for International Development and the Departments of Commerce, State, and Treasury compose OSAC. Additionally, OSAC has technical advisers from the FBI, the National Security Agency, the National Counterintelligence Executive, and the US Secret Service.[40]

Because OSAC is a public-private partnership, the exchange of information is one of its primary functions. On the government side, Diplomatic Security's Research and Information Support Center is OSAC's operational component. The center is devoted exclusively to US private sector entities that have interests overseas and employs multiple regional security experts. Additionally, OSAC serves as a clearinghouse for information from US businesses and organizations that may be of value to the security of other

US organizations' personnel and facilities.[41] OSAC also exchanges information with the Shield Unit of the New York Police Department, the Australian Security Intelligence Organization, and the United Nations Department of Safety and Security.[42]

Four committees compose OSAC: the Transnational Crime Committee, which provides information and case studies on transnational criminals and organizations; the Protection of Information and Technology Committee, which deals with intellectual property issues; the Security Awareness and Education Committe; and the Country Council Support Committee, which promotes communications between OSAC and country councils, which represent OSAC abroad. Each country council is a forum for the US private sector and embassy representatives to address security issues of mutual concern and, when necessary, facilitate a unified approach to the host government.[43] As of 2007, OSAC had country councils active in more than a hundred countries worldwide.[44]

Cyber

The US government has pursued international engagement on issues of cybersecurity as technology has made borders increasingly irrelevant to foreign intelligence, terrorist, and criminal threats. One aspect of this is establishing norms that facilitate cooperative foreign governments' ability to disrupt threat actors. The FBI's National Infrastructure Protection Center (NIPC) took steps vis-à-vis international partners including outreach—through conferences and cyber-crime training classes—to raise awareness about cyber threats; provide advice on how to handle the threats; and furnish guidance on how to deal with the threats via substantive legislation.[45] (The last item is an interesting corollary to the work that the FBI, premised on its SIS operations, did in advising governments on how they could take policy action to create an inhospitable environment for Axis operatives.)

Additionally, the US government has engaged in exchanges of personnel in order to facilitate disruption of threats. Several FBI entities handling cyber issues have established positions to facilitate cooperation with other countries. For instance, as of 2001, the NIPC, in addition to interagency personnel, also included representatives from the United Kingdom, Canada, and Australia.[46] The FBI, in 2007, established its National Cyber Investigative Joint Task Force (NCIJTF).[47] This is an interagency information-sharing body that fulfills functions similar to the NIPC's interagency aspects. Like the NIPC, the NCIJTF includes representatives of foreign governments; in 2013, Australia and the United Kingdom established representation at the task force, and in 2014, Canada became a member.[48]

The US government has also bridged the gap through international intelligence collaboration. The NIPC established information-sharing connectivity with foreign watch and warning centers, including those in the United Kingdom, Canada, Australia, New Zealand, and Sweden.[49] More recently, in 2014, the Bureau advised Congress that it was providing support to and working with newly established Interpol and Europol cyber-crime centers.[50] The Department of Homeland Security (DHS) has been a participant in the International Watch and Warning Network, an international body established in 2004 that fosters international collaboration on addressing cyber threats, attacks, and vulnerabilities among countries in the Americas, Europe, and the Asia Pacific region. Government participants represent critical information infrastructure protection organizations; computer security incident response teams; and law enforcement agencies that handle cyber crimes.[51]

Finally, the US government—especially the FBI—places personnel abroad in order to share information and engage in collaborative relationships. The FBI's legal attachés—the Bureau's representatives abroad—are often the first officials whom foreign law enforcement contacts when an event occurs that requires US assistance.[52] In acknowledgment of the increasing cyber aspect involved with the work of legal attachés, the FBI, in 2015, established three permanent cyber assistant legal attachés, in London, Ottawa, and Canberra. These attachés are actually embedded with foreign host nations' law enforcement and intelligence agencies.[53] Even prior to the establishment of the cyber assistant legal attachés, the FBI had embedded personnel with police departments in Romania, Estonia, Ukraine, and the Netherlands in order to identify emerging threats.[54] Then–FBI director James Comey explained to Congress, in 2016, that these personnel filled an essentially "old-fashioned" function of sitting with police in order to "get the evidence to make the case" and then arrange for the FBI's foreign counterparts to make arrests.[55] This is the high-tech analogue to the work of disrupting threats that the SIS accomplished through its furnishing of threat information to the governments of the countries in which it was working.

The Need for Expertise

Expansion of the risk to which the US private sector is exposed beyond US borders demands the development of a national security work force that possesses acumen in a variety of fields ranging from international relations (to assess the objectives of foreign state and nonstate actors that may prompt them to target US industry); to cyber knowledge (in order to assess vulnerabilities inherent to new technology employed by the private sector); to business (in order to assess how practices such as international joint ventures can put assets at risk to foreign collection or manipulation). Beyond these distinct silos of knowledge, a first-rate work force capable of countering threats and protecting US resources must have the difficult-to-measure but essential qualities of creativity and innovation.

Increasingly, the US government has experienced difficulty with attracting the right people for the right jobs. This problem starts at the hiring stage—government cannot provide the same incentives that the private sector can offer. However, the problem does not end with getting people in the door. The government must compete with industry—as well as other agencies—to keep talent in positions that are necessary for securing the private sector. Part of retention is ensuring that agencies' cultures encourage the development of expertise. Unfortunately, some agencies have been bureaucratically sluggish—to their detriment—in this area. Finally, part of culture is ensuring that the human capital that an agency has is used to maximum efficacy.

Necessity of Expertise

The US national security enterprise needs access to the best and the brightest in order to ensure that it can engage with—and assess the implications of developments within—the private sector. This is particularly true when discussing bleeding-edge technology sectors. For instance, the *Quadrennial Homeland Security Review* noted, in 2010, that there was a need, across the government, for a work force that had sufficient "capacity and expertise" to manage emerging cybersecurity risks.[56] Similarly, the Department of Commerce's Bureau of Industry and Security noted, in its fiscal year 2021 report, that it needed engineers capable of identifying and making recommendations about export controls on new technologies.[57]

The challenge is not simply understanding technology but also assessing what adversaries of the United States will do with it. As then–FBI director Christopher Wray explained to Congress in 2020, the problem was not simply "low-IQ bad guys."[58] The Bureau of Industry and Security explained that it too was up against adversaries employing "evolving and increasingly sophisticated tactics" in order to acquire US technology.[59]

Hiring

Agencies responsible for addressing the implications of technology that the private sector has released on the world are at multiple disadvantages when seeking to hire sufficiently talented personnel. Some of this is by necessity—such as security requirements—but other disadvantages are self-inflicted—such as the inability to offer competitive incentives.

The security clearance/expertise tradeoff continues to be a problem for government's acquisition of personnel. Former FBI official Clint Watts advised Congress that the United States took a narrow view of potential hires, by excessively focusing on security clearances while allowing for rudimentary training, a combination that screened out numerous top-talent individuals.[60] The Government Accountability Office has similarly acknowledged the impact of the security clearance process on attracting high-caliber talent, stating in a 2014 report that it could take anywhere from several months to a year for the Department of Homeland Security's Office of Intelligence and Analysis (OIA) to bring a new hire on board once the OIA had extended a hiring offer. This, according to the Government Accountability Office, drove many applicants to seek employment elsewhere.[61]

Even when an agency can hurry a candidate through the door, there is often insufficient vetting for knowledge. According to a 2015 report by the Partnership for Public Service, agencies lacked effective tools with which to screen applicants for those who are "truly qualified and warrant more extensive examination."[62] A former acting undersecretary of defense for policy noted that the government faced a deficit of people who were talented at working in the information environment and that the ones who may have at one point been capable were not always able to keep up with the technological advances.[63]

Certain agencies have unilaterally lowered their standards for analytic personnel and have no one but themselves to blame for the results. For instance, an agency with the profile of the FBI should be able to take its pick of applicants, yet in 2004 it advised Congress that it had provided field offices with 5,061 "minimally qualified" candidates from which the field offices could fill positions, as well as another 1,210 minimally qualified candidates from which FBI headquarter divisions could select personnel. In 2004, the FBI advised Congress that it had decided to waive its requirement that hires possess a college education.[64] (This set the Bureau on a worse footing than it had been on more than twenty years prior. In 1980, the FBI had proudly told Congress that all of the analysts whom it had hired for the Terrorist Research and Analytical Center were college graduates.)[65] It is hard to believe that the Bureau did not have its choice of applicants. Yet in 2005 it advised Congress that less than half of its new analytic hires had advanced degrees.[66]

The Bureau further shortchanged its prospects of building a consistently high-caliber work force by pursuing a scattershot hiring effort that did not necessarily target the desired personnel. For instance, in 2005 the FBI advised Congress that its recruiting campaign included television and radio spots, as well as billboards. This hardly set a high bar for the people whose attention this publicity would attract. Furthermore, it may have even scared off people who might otherwise have joined the Bureau but rejected it based on the premise that it was pandering to the lowest common denominator. The lowest-common-denominator aspect was especially apparent in the FBI's pride in having run an advertisement, targeting potential analysts, during the 2005 Super Bowl.[67] Nothing indicates analytic aptitude like a passive affinity for football.

Agencies also stumble in their efforts to attract the right people by not having a clear idea of those agencies' needs. In 2014, Congress enacted the Cybersecurity Workforce Assessment Act, which required the secretary of the DHS to assess the readiness of the DHS's cybersecurity work force to meet its mission and to develop a comprehensive work force strategy. Although the act directed the DHS to submit a cybersecurity work force assessment annually, the department's chief human capital officer routinely submitted the report almost a year after each one was due.[68]

Organizational vacuums also create environments that stymie planning for and hiring of expertise. For instance, the DHS's Office of Intelligence and Analysis has—from its outset—been plagued by a confused mission.[69] The FBI has had a long-term difficulty in this area and it is not certain at all that it has cleared the hurdles that it set for itself. In 1990, for instance, it viewed analysts in field offices as a means by which agents would be "relieved of many of their non-investigative functions."[70] This appeared to be a continuing theme. In 1994, the Bureau advised Congress that "support personnel in FBI field offices and Headquarters have recently undertaken additional tasks to alleviate some non-investigative type tasks from Agents."[71] This line of thinking seemed to set the FBI up to hire and retain individuals based not on what they were but rather on what they were not—a rather low bar for admittance.

In combination with the dual difficulties of long hiring processes and ineffective screening of applicants for talent, the US national security enterprise is further impeded by private industry's luring of potential hires even before they reach an agency's doorstep. According to a 2015 Department of Justice inspector general report, the FBI encountered challenges in hiring computer scientists to fill advanced technical positions due to competition from the private sector that could offer higher compensation to individuals with requisite skills.[72] The problem of insufficient compensation for needed skills is by no means new. For instance, in 1947 the FBI, desperate for Russian-language experience, indicated that the limited number of potential hires coupled with a low entrance salary presented a difficulty.

Retention

Even when an agency has successfully hired the personnel it needs, it must contend with a new problem: the retention of talent. Self-congratulatory intelligence services have relied on their reputation and mission as sufficient for attracting talent. A 2000 CIA report, for instance, noted that "even in an era of stiff competition from the private sector, interest in employment with the Agency remains high, allowing us to be quite selective in those we bring aboard."[73] The FBI has exhibited similar hubris. Speaking to Congress in 2015, then-director James Comey claimed that "if you're interested in dough, you don't want to work in the FBI," since one did not "come to [the Bureau] to get rich" and the work's "value proposition" was "totally different."[74] Furthermore, Comey was convinced that once people came to the FBI, they almost never left.[75]

However, Comey's glibness was not consistent with the historical record. As early as 1941, then–FBI director J. Edgar Hoover bemoaned the resignation of two agents-in-charge for higher salaries in private corporations engaged in national defense contract work.[76] The end of World War II and the postwar economic boom caused the Bureau to again worry about the problem of private sector enticements. According to a late-1945 memo, several special-agents-in-charge and numerous "experienced" agents had resigned in the previous weeks.[77] The memo indicated concern that "business prospects for the next several years should be extremely good and there will no doubt be numerous attractive jobs open for which many of the Bureau personnel in the upper brackets will be qualified to fill and [the Bureau] may reasonably expect to continue losing some of [its] personnel."[78] Approximately sixty years later nothing had changed. A 2007 congressional hearing highlighted that agents with FBI experience could often obtain lucrative private sector jobs and had little incentive to remain with the federal government.[79]

The long-term competition of human capital is also indicated by Hoover's letters to various Bureau officials who declined offers from the private sector. In 1943, after learning that FBI official D. M. Ladd had turned down a position with a New York chemical firm, Hoover sent Ladd a note acknowledging that Ladd had "contributed much to the work of the Bureau and [the FBI needed his] services."[80] (Ladd, in 1939—while special-agent-in-charge of the Chicago field office—had also received an offer from Montgomery and Ward to become the head of a plant protection staff.)[81] In 1947, Hoover commended another Bureau official for turning down the offer of a position in Minneapolis that would have paid him "a greater salary than that which [he] received from the Bureau."[82] According to Hoover it was "indeed good that [the agent's] interest in the work of the Bureau led to [the agent's] declination of this offer."[83]

From Cover Company to Career

During World War II, the FBI assigned a number of its agents to a variety of bona fide companies as cover for the agents' work throughout the Western Hemisphere as part of the Bureau's Special Intelligence Service. According to the Bureau's internal history of the Special Intelligence Service, the agents often worked enthusiastically for the cover companies and were considered "vital assets" to these firms. Companies indicated a desire to permanently employ certain agents—due to those agents' efficacy—under any conditions that the FBI imposed.[84]

Federal agencies not only have to address competition from private industry but also often compete with each other for personnel. This, in turn, may deprive private sector–facing organizations such as the FBI of expertise. The problem is not remotely new. In 1946, as the FBI was planning to turn its Special Intelligence Service operation over to the Central Intelligence Group (the short-lived predecessor to the CIA), Hoover noted that he was "entirely justified in taking a very firm position that [he would] not agree to the transferring of Bureau personnel" to the CIG, since this would cause a serious curtailment of the FBI's domestic work.[85] Although CIG agreed that it would not extend offers of employment to any current FBI personnel, it would "employ any capable and desirable individual who has terminated his FBI connections."[86]

Hoover was skeptical about the CIG/CIA assurances. In 1946, he wrote to the attorney general that despite statements to the contrary, the CIG would "definitely endeavor to proselytize Bureau personnel presently serving upon [Special Intelligence Service] assignments."[87] These personnel were "men of outstanding qualifications with exceptional backgrounds" who had "performed very meritorious service," and thus it was "logical to assume" that the CIG would attempt to employ their services.[88]

Hoover's concerns proved to be justified. A 1952 FBI memo assessed that the CIA had "actively endeavored to proselytize" Bureau personnel.[89] According to the memo, certain agency officials would go so far as to ask former FBI personnel who had taken positions with the CIA to contact present and former Bureau employees and "entice them" with offers of higher pay at the Agency.[90] According to Hoover, these overtures were "in violation of all agreements and decency."[91] Hoover was further annoyed to learn that certain Bureau employees had been "shopping around" the CIA for positions.[92]

Additionally, the FBI seemed suspicious and envious of the CIA's ability to hire personnel. A 1951 Bureau memo noted that the Agency was "very liberal" with their high-grade positions.[93] Furthermore, the FBI expressed disbelief that the CIA was willing to hire a clerk-typist as a GS-4 and a stenographer as a GS-5. An FBI official seemed to be aghast at the proposal, writing in a 1953 memo that "there is no clerk-typist position in the Bureau as high as a grade GS-4 and [the FBI had] no starting stenographer as high as grade GS-5."[94] "Obviously," the Bureau official noted, the "CIA [was] either assigning these people to jobs which are over-classified or else the titles of the jobs [were] not fully descriptive of their duties."[95] The memo recommended that the Bureau engage in some intelligence collection against its bureaucratic rival by suggesting the agent responsible for liaison with the CIA "endeavor to obtain job descriptions or in lieu thereof detailed information concerning the duties and responsibilities of these two positions."[96] It appears that the collection did occur, as a subsequent memo set forth "results of discreet inquiries and observations made by the Liaison Agent."[97]

More than half a century later, it was amusing when the FBI, perhaps in jest or perhaps a little more serious than not, hinted at a similar competition with the Department of Homeland Security. During a 2014 congressional hearing on worldwide threats to the homeland, Comey hinted that he did not want the DHS to know his secrets, since the FBI was "competing for the same talent" but conceded that the DHS secretary had figured out Comey's secret: it was "much cooler to work for the FBI."[98]

Culture

An agency's ability—or inability—to establish a culture that values expertise plays a role in determining whether experts remain with that agency. Presented with other options, especially when they are more lucrative ones, individuals may feel less of a need to remain in a position that is no longer fulfilling. The CIA, for instance, has experienced attrition due to employees' desire to make better use of their skills and knowledge.[99] A related problem is the lack of opportunities to develop expertise. This has had a negative impact on the maturation of the DHS's Office of Intelligence and Analysis. Historically—which does not cover a significant time frame, for a department that came into existence only in 2002 and a component that has existed for an even shorter period of time—the OIA lacked a commitment to investing in its work force. According to the Government Accountability Office, managers' decisions to send employees to training and provide rotational opportunities were inconsistent.[100]

It is not surprising that this lack of consistent dedication toward the careers of OIA employees has created a sense of dissatisfaction. According to the 2012 Intelligence Community Climate Survey, only 36 percent of OIA employees would recommend their organization as a good place to work. Even within the DHS, the OIA had one of the lowest morale scores among the department's components (and this is the department that includes such elements as the Transportation Security Administration; Customs and Border Protection; and the Federal Protective Service). Such dissatisfaction leads to a "grass is greener" attitude, and analysts who left OIA gave the impression that they were leaving it for other intelligence community components that they perceived to be higher-profile.[101] The same motivation may also be driving losses from OIA to private industry. In 2014, for instance, then–DHS secretary Jeh Johnson advised Congress that he had recently lost a "very, very valued member of his cybersecurity team to Citigroup."[102]

Departures from the OIA are troubling because the unique mission of the office has been undercut by the trend. The OIA has an important responsibility as the primary element within the US intelligence community analyzing threats to domestic critical infrastructure. For instance, the OIA was responsible for supporting cybersecurity customers including the DHS's National Cybersecurity and Communications Integration Center (NCCIC), US Computer Emergency Readiness Team (US-CERT), and

Industrial Control Systems CERT. Furthermore, the OIA has access to unique data sets—such as the information collected by the CERTs—which it has at least aspired to using for predictive analysis.[103] However, as of 2016 little actual "analysis" was occurring. According to a US congressional report, the DHS's intelligence enterprise frequently resorted to repackaging the work of other agencies. The report found that the OIA frequently copied analytical products from intelligence community agencies and simply republished them in the days following their release.[104] These factors, by 2014, created the impression that the OIA was not as prestigious as other intelligence community elements and negatively impacted the OIA's ability to recruit top talent.[105] The OIA therefore appears to be in an analytical death spiral: people who can get out leave for other agencies or private industry and analysis decreases in the value it provides customers, and the lack of efficacy harms the OIA's reputation, dissuading talent from considering it as a meaningful career.

Several trends—that are even more historically entrenched—within the FBI put that Bureau at risk of losing talent. As of the 2019 FBI climate survey, employees rated the Bureau at 4.28 out of 5.0 as a good place to work. This is certainly better than the far-less-than half score that the DHS's Office of Intelligence and Analysis garnered from its employees. Further attesting to employees' dedication, FBI employees gave a response of 4.56 out of 5.0 regarding their pride in working for the organization. Although the personnel of the FBI have a positive view of their agency, the dedication to fostering expertise, in both the Bureau's intelligence disciplines of collection and analysis, has had a more uneven trajectory. For instance, Bureau employees, on the 2019 climate survey, scored the FBI at only 3.98 out of 5.0 on the subject of their participation in development opportunities during the preceding year, and employees gave only a 3.89 out of 5.0 on the topic of whether development was valued.[106] In other words, Bureau employees are a dedicated lot who do not necessarily receive dedication to their development in return.

The FBI's commitment to the education and continued integration of its personnel into the academic/professional fields that would keep them at the forefront as experts is a lengthy and varied story. While it is not a disaster, it does merit watching and warrants steps to mitigate movement in the wrong direction.

Hoover definitely saw the value of ensuring that the FBI's personnel had academic aptitude. On the hiring line he demanded initial expertise, stating in 1937 that agents must be "graduate lawyers" or "expert accountants" or possess "extensive investigative experience."[107] By 1950 he had pared this list to the fields of law and accounting as acceptable precursors to the agent position.[108] Hoover also supported the continuing education of Bureau personnel. As early as 1936, he highlighted the fact that the FBI had "a faculty of experts, 60 leading men in their fields in various universities, who assist[ed] us in training our special agents."[109] Starting in 1972, the year Hoover died, the FBI's training division maintained a formal affiliation with the University of Virginia, which accredited the division's courses, certified its instructors, and provided instructional resources.[110] The commitment to academic excellence appears to have precipitously eroded. In 2004 the FBI advised Congress that it had provided personnel with funds to attend for-profit, only-slightly-above-degree-mill institutions such as American Intercontinental University, Capella University, DeVry University, National University, and Strayer University.[111] Certainly the Bureau's personnel should have higher academic aspirations, considering the availability of online courses from far more prestigious institutions.

The FBI, under Hoover, demonstrated a dedication to ensuring that the Bureau's personnel were recognized as expert figures in the academic world. For instance, in 1961 an FBI memorandum noted that the FBI could make a positive contribution to academic programs in Eastern European and Russian history by providing speakers who could furnish accurate and detailed information about communism.[112] In 1967, another FBI memo noted that four FBI officials—William Sullivan, C. D. Brennan, Arbor Gray, and Fern Stukenbroeker—were making speeches, regarding communism, on college and university campuses. The memo acknowledged that these officials' work in this matter had been "outstanding" and "strongly recommended that [these efforts] be continued and even expanded."[113]

Professional engagement has been a stronger suit for the Bureau. For instance, in 1943, after the Bureau had bowed out of conducting plant surveys, an FBI special-agent-in-charge received an invitation to deliver an address on the FBI's relation to plant protection at a meeting of plant protection officials. The majority of the Bureau's Executive Conference believed that the FBI should be represented (a view that received Director Hoover's standard, terse "O.K.").[114] Furthermore, Bureau personnel have been longtime participants in the American Society for Industrial Security (ASIS). Fern Stukenbroeker, an FBI agent, spoke on multiple occasions at this organization's functions on topics including espionage.[115] Additionally, in 1966, Stukenbroeker had a role, meriting a commendation, in placing a story by Hoover, titled "The Modern Day Soviet Spy—A Profile," in the ASIS publication *Industrial Security*.[116] Several decades later, once the Bureau had established its Awareness of National Security Issues and Response program, it indicated that each of the coordinators for this program were members of ASIS, which facilitated public-private communication and cooperation.[117]

Use of Expert Resources

Consistent with its inconsistency, where human capital is concerned, the FBI has sent mixed messages over the decades about how it uses expertise.

This is reflected in the FBI's inability to provide incentives to put the right people where they can employ their talents. For instance, in 1977 the FBI advised Congress that the field of counterintelligence was "a highly specialized field requiring a careful selection of high-caliber personnel, followed by intensive training and substantial practical experience."[118] Less than a decade later it pointed to New York as an area with a large hostile intelligence presence but admitted that it was difficult to maintain counterintelligence specialists at the New York field office due to a "problem in providing sufficient financial incentives to its agents who must work in New York City."[119] Although the Bureau recognized a need, it could not provide incentives that would help to meet it.

The impression that the FBI does not consistently value the development of expertise in its agents is indicated by the way in which it treats that group of personnel interchangeably. This can be placed at Hoover's feet. Early on, Hoover, in 1932, justified transfers of the Bureau's personnel based on the "fluctuating character of crime conditions."[120] Such a philosophy left little room for specialization. The outlook of interchangeability has pervaded the Bureau's approach to how it assigns its personnel. For instance—in early 1992—the Bureau reassigned 300 agents from the counterintelligence program (the field that it had previously said was in need of specific personnel) to the Violent Crimes and Major Offenders Program.[121] Then, in 2003, the FBI shifted 400 agents from counternarcotics work to counterterrorism.[122]

Its ethos of interchangeability has caused problems for the FBI, especially in the area of technological expertise. For instance, according to a 2011 Government Accountability Office assessment, when investigative personnel who had been working on cyber investigations rotated to a new office, they were not necessarily assigned to cyber investigations in the new location. Not only did this allow expertise to languish, but according to the Government Accountability Office, the agents who rotated in to replace the agents who rotated out might lack the necessary expertise to do the work. The churn of experts is particularly troubling given the lack of qualified resources. According to a 2011 report by the US Department of Justice, 36 percent of the FBI's field agents felt that they lacked networking and counterintelligence expertise necessary for pursuing national security–related cyber investigations.[123]

The FBI's apparent inability to consistently leverage its expertise runs counter to the narrative it has directed at Congress since at least 9/11. Then–FBI director Robert Mueller III claimed that the FBI needed to recruit and hire people who were capable of addressing cyber crime.[124] In 2014, then-director James Comey said (at least partially in jest) that the FBI might have to loosen its no-tolerance policy for marijuana smokers in order to bring employees on board capable of dealing with cybersecurity challenges.[125] Then-director Chris Wray echoed Mueller's and Comey's concerns in 2020 when he advised Congress that the FBI needed more computer scientists and data analysts.[126] However, if the Bureau is not going to consistently use its personnel in the fields for which it hired and trained them, there is little reason to recruit them.

Interchangeability is one of two competing narratives within the FBI. An emphasis on expertise is also discernible if one parses the Bureau's history. In 1946, for instance, the Bureau assessed Fern Stukenbroeker, an agent, as "one of the best-read agents in the office," with specialties in history and political science.[127] This made him "well equipped for his work in connection with Communist matters to which he [was] assigned."[128] In 1950, Hoover seemed to suggest the need for specialization when he stated that the "counter-espionage assignments of the FBI require an objective different from the handling of criminal cases," since "the identification of a wrongdoer [was] only the first step," as opposed to a criminal case, in which the "identification and arrest of the wrongdoer [were] the ultimate objectives."[129] Hoover reemphasized the need for a unique perspective in counterintelligence when he stated, in 1958, that its "effectiveness [could not] necessarily be measured in terms of convictions secured or sentences imposed."[130]

The post-9/11 FBI attempted to address the need for specialization. In March 2004, the FBI established four career paths—intelligence, counterterrorism/counterintelligence, cyber, and criminal investigative—for its agents. One of the reasons for these career paths was to provide agents with "an opportunity to develop specialized skills, experience, and aptitudes."[131] Furthermore, the Bureau indicated that it would take into account agents' background and experience in order to determine the agents' career path.[132] In 2006, the FBI stated explicitly that the career path program was a direct response to the recommendations of the 9/11 Commission and that the Bureau developed this career path to establish a work force that was "recruited, trained, rewarded, and retained to ensure the development of an institutional culture imbued with a deep expertise in intelligence and national security."[133]

Analytic expertise within the FBI has evolved in a similarly conflicted manner. As with the question of expertise within the ranks of intelligence collectors during Hoover's tenure, the Bureau, during the Hoover era, sent mixed messages about the role of analysis and analysts. Hoover, in 1945, concurred with a proposal to create a unit within the FBI that would prepare monographs on various facets of the Bureau's work. According to the proposal, a large part of the effectiveness of the Bureau's intelligence coverage would be lost if the results from intelligence and investigative operations were not "written up in interesting form."[134] In 1956, Hoover advised Congress that "constant research [in the field of domestic security] has been conducted to put to effective use the vast reservoir of valuable information concerning the internal security of this country which has been built up in our files."[135] On the other hand, Hoover's tolerance for "interesting" write-ups seemed to be, at times, limited. For instance, in 1950 he advised Congress that "the Bureau's responsibility is that of being solely a fact-finding agency. We merely gather the information. We get the facts both favorable and unfavorable and put them into report form."[136]

The Bureau's establishment of a dedicated analytic work force also indicated ambivalence. In 1951 the FBI established a "research analyst" position, which was clerical in nature.[137] However, the idea was informed not by a desire for expertise but rather a desire for economy. The Bureau embarked on this project thinking that "experienced clerical employees could prepare competent memoranda at a cost less than that occasioned by the use of [agent supervisors]."[138] Five years later, the FBI reassessed this decision, stating that "there is no doubt an experienced Special Agent Supervisor can prepare a better memorandum than a clerical Research Analyst."[139] This mentality prevailed through much of the 1970s. As a 1977 FBI report explained: "Service and support personnel are utilized to collate information both in the field and at Headquarters but their activity is limited to support and not operational input. It is the responsibility of the case Agent or the program coordinator to insure necessary correlation of all investigative and intelligence data pertinent to ongoing or contemplated investigations."[140] In other words, nonagent personnel were fine for scutwork, but analysis was too important to be handled by anyone but FBI agents.

However, by the late 1970s, the FBI had returned to the idea that had informed its 1951 decision: efficiency (but not necessarily expertise). By the end of the decade, the Bureau had established additional clerical programs meant to release agent manpower that was being squandered on conducting clerical functions.[141] The FBI created a number of pseudo-analytic positions, starting in the late 1970s, with the expectation that analysis could be automated. In 1979, the Bureau told Congress that information analysts would staff the FBI's new Organized Crime Information System.[142] Analysts were responsible for placing intelligence into the system, to make it available nationwide, and prepared data to give agents "an improved focused insight into their investigations."[143] However, the Organized Crime Information System did not require subject-matter expertise and was, instead, a system that emphasized link analysis and relationships between groups.[144] (Such tasks are data manipulation rather than data analysis; one can draw charts all day, but they mean nothing if they do not explain anything.) Similarly, the Intelligence Information System analyst existed to handle functions pertaining to the Bureau's Intelligence Information System counterintelligence application, which the Bureau deployed in 1980.[145] These functions provided a starting point for—rather than an outcome of—analysis.

By the end of the twentieth century, the FBI's analytical work force was in desperate need of reform. During the development of the Bureau's 1998–2003 strategic plan, nearly every FBI program indicated that the Bureau's analytic functions were deficient.[146] In 1998, an internal Bureau document assessed that two-thirds of the FBI's analysts were unqualified for their positions.[147] This reflected a long-standing promotion policy predilection that was finally being recognized as detrimental to the organization's well-being. The previously mentioned 1956 memo gave an early hint of the trouble that would follow when it stated that the establishment of research analysts "would give promotional opportunities for older, experienced clerical employees."[148] The research analyst experiment had expired but the reasoning laid bare by this memo had not. The 9/11 Commission encountered a number of situations in which "poorly qualified administrative personnel were promoted to analyst position, in part as a reward for good performance in other positions."[149]

During the years leading up to 9/11, the FBI scrambled to reinvigorate its analytical work force with an infusion of qualified personnel. In a 2000 statement to Congress, the Bureau acknowledged that it needed to "create a professional intelligence cadre of experienced and trained analysts."[150] Accordingly, it planned to "overhaul the existing analytical infrastructure based on standardized core skills, competencies, and personnel attributes."[151] As part of this new look, the FBI's Administrative Services Division—in conjunction with the Criminal Investigative, National Security, and Counterterrorism Divisions—planned to develop "fair, valid, and legally defensible selection and promotion criteria."[152] Creation of the new criteria seemed to be a direct rebuke to the previous, capricious promotion of clerical personnel as a reward, rather than as a recognition of capability.

Even after the FBI recognized the problem, it struggled to address the prevalence of overpromoted, undereducated personnel in its ranks. The Bureau's budget justification for fiscal year 2001 very clearly acknowledged the need to "reexamine the existing [analytic] cadre to assure they have the skills required for the new challenges."[153] In order to cull the intellectual dead weight represented by uncredentialed analysts, the FBI, in 2003, made an attempt to "improve the quality of its analytical corps by requiring a college degree for the intelligence analyst position . . . [and] also mandat[ed] that analysts reapply for their current jobs."[154] Unfortunately, this was canceled before the process was completed.[155] The 9/11 Commission staff painted a grim picture, stating that "there appear[ed] to be no process for evaluating and reassigning unqualified analysts."[156]

The FBI found itself in the unenviable position of attempting to establish a cadre of qualified, competent analysts on a largely rotten foundation. It had been down this road already. In 1984, then-director William Webster claimed that the FBI was "lightyears" ahead in terms of "education, training, and experience" compared to where it had been in the late 1970s.[157] For instance, as of 1983, the FBI was staffing its Terrorist Research and Analytical Center at headquarters with analysts who had graduate degrees, experience in analysis, and geopolitical backgrounds.[158] However, historically, according to the 9/11 Commission, when the FBI had hired or promoted people with the appropriate analytical skills, the lack of professional opportunities often caused them to leave the Bureau.[159]

Despite implementing reforms, the FBI was doomed to revisit the clash of qualified and unqualified analysts. As of 2004, the FBI continued to grapple with a problem posed by "legacy" analysts—the ones who had been promoted into analyst intelligence positions though lacking a bachelor's degree.[160] The FBI continued to allow internal candidates—including those lacking a bachelor's degree (consistent with its problematic decision to scuttle that requirement)—to apply for intelligence analyst positions.[161] The outcome was a repeat of the pre-9/11 dynamic. A 2005 report by the Department of Justice's inspector general assessed that the higher an analyst's level of education, the more likely the analyst was to leave the FBI.[162] This is problematic given the deep background knowledge and sophisticated research and critical-thinking skills that would seem to be prerequisites for an analytic position. Furthermore, even after the inspector general made this assessment in 2005, the FBI apparently continued to struggle with getting candidates with advanced degrees into the organization. (While advanced degrees are not the end-all, be-all, they are a useful shorthand for specific areas of expertise.) According to a 2007 follow-up report, the inspector general noted that according to the available data, fewer than 50 percent of the FBI's new analytic hires had advanced degrees.[163] These data suggest that the Bureau continued to undervalue academic credentials—a view consistent with its pre-9/11 concept of "analysts."

Notes

1. Darren E. Tromblay, *The FBI Abroad* (Boulder: Lynne Rienner, 2020).
2. Federal Bureau of Investigation, *History of the S.I.S. Division,* vol. 1 (undated).
3. Federal Bureau of Investigation, *Annual Report: Special Intelligence Service* (Washington, DC, 1944–1945).
4. Ibid.
5. Federal Bureau of Investigation, *History of the S.I.S. Division,* vol. 2, *Accomplishment Argentina-Japan* (undated).
6. Federal Bureau of Investigation, *History of the S.I.S. Division,* vol. 2; Federal Bureau of Investigation, *Annual Report: Special Intelligence Service.*
7. Federal Bureau of Investigation, *History of the S.I.S. Division,* vol. 2.
8. Federal Bureau of Investigation, *History of the S.I.S. Division,* vol. 1.
9. Federal Bureau of Investigation, *History of the S.I.S. Division,* vol. 2.
10. Ibid.

11. Federal Bureau of Investigation, *History of the S.I.S. Division,* vol. 3, *Accomplishment Mexico-Venezuela* (undated).
12. Ibid.
13. Federal Bureau of Investigation, *History of the S.I.S. Division,* vol. 2.
14. Society of Former Special Agents of the FBI, "Interview of Chester J. Peterson (1940–1947)," March 24, 2005.
15. Federal Bureau of Investigation, *History of the S.I.S. Division,* vol. 2.
16. Federal Bureau of Investigation, *Annual Report: Special Intelligence Service.*
17. Society of Former Special Agents of the FBI, "Interview of Wallace F. Estill (1941–1974)," August 3, 2004.
18. Department of State, memorandum to Diplomatic and Consular Office in the American Republics, September 20, 1941, https://history.state.gov/historicaldocuments/frus1941v06/d317; National Archives, "Records of the Office of the Director, Office of Trade Promotion, Bureau of Foreign Commerce and Its Predecessors," https://www.archives.gov/files/records-mgmt/rcs/schedules/departments/department-of-commerce/rg-0151/n1-151-88-008_sf115.pdf.
19. National Archives, "Records."
20. Federal Bureau of Investigation, *History of the S.I.S. Division,* vol. 2.
21. Ibid.
22. Federal Bureau of Investigation, C. H. Carson, memorandum to D. M. Ladd, October 9, 1942 (Washington, DC, National Archives and Records Administration).
23. Ibid.
24. Federal Bureau of Investigation, *History of the S.I.S. Division,* vol. 2.
25. Society of Former Special Agents of the FBI, "Interview of Wallace F. Estill."
26. Federal Bureau of Investigation, V. P. Keay, memorandum to D. M. Ladd, "World-Wide Intelligence Coverage," September 1, 1948.
27. Federal Bureau of Investigation, *History of the S.I.S. Division,* vol. 2.
28. Michael J. Waguespack, testimony before the House Committee on Government Reform, April 3, 2001, https://archives.fbi.gov/archives/news/testimony/fbis-ansir-program.
29. US Senate, *Current and Projected National Security Threats to the United States,* before the Select Committee on Intelligence, 105th Congress (Washington, DC, 1998).
30. Federal Bureau of Investigation, *History of the S.I.S. Division,* vol. 1.
31. Ibid.
32. Federal Bureau of Investigation, *History of the S.I.S. Division,* vol. 2.
33. Federal Bureau of Investigation, *History of the S.I.S. Division,* vol. 1.
34. Federal Bureau of Investigation, Keay, memorandum to Ladd, "World-Wide Intelligence Coverage."
35. Federal Bureau of Investigation, memorandum to the director, August 20, 1943.
36. Ibid.
37. Henry H. Willis, Genevieve Lester, and Gregory F. Treverton, "Information Sharing for Infrastructure Risk Management: Barriers and Solutions," *Intelligence and National Security,* July 15, 2009, https://doi.org/10.1080/02684520903036925.
38. US Congress, *Protecting American Interests Abroad: U.S. Citizens, Businesses, and Nongovernmental Organizations,* before the Committee on Government Reform, House of Representatives, 107th Congress (Washington, DC, 2001), https://www.govinfo.gov/content/pkg/CHRG-107hhrg75955/pdf./CHRG-107hhrg75955.pdf.
39. Ibid.
40. Ibid.

41. Ibid.

42. US Congress, *Homeland Security Beyond Our Borders: Examining the Status of Counterterrorism Coordination Overseas,* before the Committee on Homeland Security, House of Representatives, 110th Congress (Washington, DC, 2007), https://www.govinfo.gov/content/pkg/CHRG-110hhrg48970/pdf./CHRG-110hhrg48970.pdf.

43. US Congress, *Protecting American Interests Abroad.*

44. US Congress, *Homeland Security Beyond Our Borders.*

45. US Senate, *Securing Our Infrastructure: Private/Public Information Sharing,* before the Committee on Governmental Affairs, 107th Congress (Washington, DC, 2002); General Accounting Office, *Critical Infrastructure Protection: Significant Challenges in Developing National Capabilities* (Washington, DC, 2001), https://www.gao.gov/assets/160/157052.pdf; US Senate, *Cyber Attack: Improving Prevention and Prosecution,* before the Committee on the Judiciary, 106th Congress (Washington, DC, 2000).

46. US Congress, *Departments of Commerce, Justice, and State, the Judiciary, and Related Agencies Appropriations for 2001,* pt. 6, before a subcommittee of the Committee on Appropriations, House of Representatives, 106th Congress (Washington, DC, 2000).

47. US Congress, *Commerce, Justice, Science, and Related Agencies Appropriations for 2009,* pt. 1, before a subcommittee of the Committee on Appropriations, House of Representatives, 103rd Congress (Washington, DC, 2009).

48. US Senate, *Cybersecurity, Terrorism, and Beyond: Addressing Evolving Threats to the Homeland,* before the Committee on Homeland Security and Governmental Affairs, 113th Congress (Washington, DC, 2014), https://www.govinfo.gov/content/pkg/CHRG-113shrg92903/pdf./CHRG-113shrg92903.pdf.

49. Ronald L. Dick, director, National Infrastructure Protection Center, Federal Bureau of Investigation, "Information Technology," testimony before the House Committee on Government Reform, Subcommittee on Government Efficiency, Financial Management, and Intergovernmental Relations, September 26, 2001.

50. US Senate, *Cybersecurity, Terrorism, and Beyond.*

51. US Congress, *Cybersecurity: Protecting America's Critical Infrastructure, Economy, and Consumers,* before the Committee on Energy and Commerce, House of Representatives, 109th Congress (Washington, DC, 2006), https://www.govinfo.gov/content/pkg/CHRG-109hhrg31464/pdf./CHRG-109hhrg31464.pdf.

52. US Senate, *Securing Our Infrastructure.*

53. White House, "Fact Sheet: Administration Cybersecurity Efforts 2015," July 9, 2015, https://obamawhitehouse.archives.gov/the-press-office/2015/07/09/fact-sheet-administration-cybersecurity-efforts-2015.

54. US Senate, *Threats to the Homeland,* before the Committee on Homeland Security and Governmental Affairs, 113th Congress (Washington, DC, 2013).

55. US Congress, *Commerce, Justice, Science, and Related Agencies Appropriations for 2017,* before a subcommittee of the Committee on Appropriations, House of Representatives, 104th Congress (Washington, DC, 2016).

56. Department of Homeland Security, *Quadrennial Homeland Security Review 2010* (Washington, DC), https://www.dhs.gov/sites/default/files/publications/2010-qhsr-report.pdf.

57. Department of Commerce, Bureau of Industry and Security, "Fiscal Year 2021: Congressional Budget Submission," https://www.commerce.gov/sites/default/files/2020-02/fy2021_bis_congressional_budget_justification.pdf.

58. US Congress, *Global Terrorism: Threats to the Homeland,* pt. 2, before the Committee on Homeland Security, House of Representatives, 116th Congress

(Washington, DC, 2020), https://www.govinfo.gov/content/pkg/CHRG-116hhrg40463/pdf./CHRG-116hhrg40463.pdf.

59. Department of Commerce, Bureau of Industry and Security, "Fiscal Year 2021."

60. US Senate, *Cyber Enabled Information Operations,* before the Committee on Armed Services, 115th Congress (Washington, DC, 2017).

61. Government Accountability Office, *DHS Intelligence Analysis: Additional Actions Needed to Address Analytic Priorities and Workforce Challenges* (Washington, DC, 2014), https://www.gao.gov/assets/670/663794.pdf.

62. Partnership for Public Service, *Cyber In-Security* (Washington, DC, 2015), https://ourpublicservice.org/wp-content/uploads/2015/04/5a6ae63596cc99f7039b9e409c70891a-1429280031.pdf.

63. US Senate, *Cyber Enabled Information Operations.*

64. US Congress, *Departments of Commerce, Justice, and State, the Judiciary, and Related Agencies Appropriations for 2005,* pt. 10, before a committee of the Committee on Appropriations, House of Representatives, 108th Congress (Washington, DC, 2004).

65. US Senate, *FBI Oversight and Budget Authorization for Fiscal Year 1986,* before the Subcommittee on Security and Terrorism of the Committee on the Judiciary, 99th Congress (Washington, DC, 1985).

66. US Senate, *Federal Bureau of Investigation Oversight,* before the Committee on the Judiciary, 109th Congress (Washington, DC, 2005).

67. Ibid.

68. Department of Homeland Security, *DHS Needs to Improve Cybersecurity Workforce Planning* (Washington, DC, 2019), https://www.oig.dhs.gov/sites/default/files/assets/2019-09/OIG-19-62-Sep19.pdf.

69. US Congress, *A DHS Intelligence Enterprise: Still Just a Vision or Reality?* before the Subcommittee on Intelligence, Information Sharing, and Terrorism Risk Assessment of the Committee on Homeland Security, House of Representatives, 111th Congress (Washington, DC, 2010).

70. US Congress, *Departments of Commerce, Justice, and State, the Judiciary, and Related Agencies Appropriations for 1991,* before a subcommittee of the Committee on Appropriations, House of Representatives, 101st Congress (Washington, DC, 1990).

71. US Congress, *Departments of Commerce, Justice, and State, the Judiciary, and Related Agencies Appropriations for 1995,* pt. 2A, before a subcommittee of the Committee on Appropriations, House of Representatives, 103rd Congress (Washington, DC, 1994).

72. Department of Justice, *Audit of the Federal Bureau of Investigation's Implementation of Its Next Generation Cyber Initiative* (Washington, DC, 2015), https://oig.justice.gov/reports/2015/a1529.pdf.#page=1.

73. Central Intelligence Agency, *Semiannual Report to the Director of Central Intelligence* (Langley, July–December 2000), https://www.cia.gov/library/readingroom/docs/DOC_0001311476.pdf.

74. US Senate, *Counterterrorism, Counterintelligence, and the Challenges of "Going Dark,"* before the Select Committee on Intelligence, 114th Congress (Washington, DC, 2015), https://www.govinfo.gov/content/pkg/CHRG-114shrg27896/pdf./CHRG-114shrg27896.pdf.

75. Ibid.

76. US Congress, *Department of Justice Appropriation Bill for 1942,* before a subcommittee of the Committee on Appropriations, House of Representatives, 77th Congress (Washington, DC, 1941).

77. Federal Bureau of Investigation, G. A. Nease, memorandum to the director, "Personnel Turnover," December 28, 1945 (Nease personnel file FOIA release),

https://ia801708.us.archive.org/4/items/foia_Nease_Gordon_A._-2/Nease_Gordon_A._-2.pdf.

78. Ibid.

79. US Congress, *Science, the Departments of State, Justice, and Commerce, and Related Agencies Appropriations for 2007,* pt. 10, before the Committee on Appropriations, House of Representatives, 109th Congress (Washington, DC, 2006).

80. Federal Bureau of Investigation, J. Edgar Hoover, memorandum to D. M. Ladd, November 16, 1943 (D. M. Ladd personnel file FOIA release), https://ia801802.us.archive.org/7/items/foia_Ladd_D._Milton-3/Ladd_D._Milton-3.pdf.

81. Federal Bureau of Investigation, Clyde Tolson, memorandum to the director, August 14, 1939 (D. M. Ladd personnel file FOIA release), https://ia801802.us.archive.org/0/items/foia_Ladd_D._Milton-2/Ladd_D._Milton-2.pdf.

82. Federal Bureau of Investigation, J. Edgar Hoover, memorandum to Wick, January 10, 1947 (Wick personnel file FOIA release).

83. Ibid.

84. Federal Bureau of Investigation, *History of the S.I.S. Division,* vol. 1.

85. Federal Bureau of Investigation, J. Edgar Hoover, memorandum to William D. Pawley, September 19, 1946.

86. Donald Galloway, memorandum to E. A. Tamm, "South and Central American Turnover F.B.I. to C.I.G.," August 7, 1946.

87. Federal Bureau of Investigation, director, memorandum to the attorney general, August 9, 1946.

88. Ibid.

89. Federal Bureau of Investigation, J. P. Mohr, memorandum to Clyde Tolson, "CIA Proselyting of Bureau Personnel," September 30, 1952.

90. Ibid.

91. Ibid.

92. Federal Bureau of Investigation, J. Edgar Hoover, memorandum to Clyde Tolson, February 6, 1951 (Belmont personnel file FOIA release).

93. Federal Bureau of Investigation, V. P. Keay, memorandum to A. H. Belmont, Central Intelligence Agency, July 16, 1951.

94. Federal Bureau of Investigation, H. L. Edwards, memorandum to Glavin, December 7, 1953.

95. Ibid.

96. Ibid.

97. Federal Bureau of Investigation, V. P. Keay, memorandum to A. H. Belmont, "Recruitment by CIA," January 8, 1954.

98. US Congress, *Worldwide Threats to the Homeland,* before the Committee on Homeland Security, House of Representatives, 113th Congress (Washington, DC, 2014), https://www.govinfo.gov/content/pkg/CHRG-113hhrg93367/pdf./CHRG-113hhrg93367.pdf.

99. Central Intelligence Agency, *Report of Follow-Up Inspection: Retention in the Agency* (Langley, 2010), https://assets.documentcloud.org/documents/741885/cia-employee-retention-report.pdf.

100. Government Accountability Office, *DHS Intelligence Analysis.*

101. Ibid.

102. US Congress, *Worldwide Threats to the Homeland.*

103. US Congress, *A DHS Intelligence Enterprise.*

104. US Congress, *Reviewing the Department of Homeland Security's Intelligence Enterprise,* House Homeland Security Committee majority staff report (Washington, DC, 2016).

105. Government Accountability Office, *DHS Intelligence Analysis.*

106. Federal Bureau of Investigation, *2019 Climate Survey* (Washington, DC), https://assets.documentcloud.org/documents/6279656/Climate-Survey-2019.pdf.

107. US Congress, *Department of Justice Appropriation Bill for 1938,* before a subcommittee of the Committee on Appropriations, House of Representatives, 75th Congress (Washington, DC, 1937).

108. US Senate, *Departments of State, Justice, Commerce, and the Judiciary Appropriations for 1951,* pt. 1, before the Committee on Appropriations, 81st Congress (Washington, DC, 1950).

109. US Congress, *Department of Justice Appropriation Bill for 1937,* before the House Committee on Appropriations, 74th Congress (Washington, DC, 1936).

110. US Congress, *FBI Oversight and Authorization Request for Fiscal Year 1989,* before the Subcommittee on Civil and Constitutional Rights of the Committee on the Judiciary, House of Representatives, 100th Congress (Washington, DC, 1988).

111. US Congress, *Departments of Commerce, Justice, and State, the Judiciary, and Related Agencies Appropriations for 2005,* pt. 6, before a subcommittee of the Committee on Appropriations, House of Representatives, 108th Congress (2004).

112. Federal Bureau of Investigation, M. A. Jones, memorandum to DeLoach, "Bureau Speeches on Communism," May 22, 1961, https://ia801805.us.archive.org/6/items/foia_Stukenbroeker_Fern_C.-2/Stukenbroeker_Fern_C.-2.pdf.

113. Federal Bureau of Investigation, M. A. Jones to Wick, "Speeches on Communism," March 21, 1967, https://ia801706.us.archive.org/20/items/foia_Stukenbroeker_Fern_C.-5/Stukenbroeker_Fern_C.-5.pdf.

114. Federal Bureau of Investigation, Clyde Tolson to the director, October 15, 1943, https://ia600402.us.archive.org/25/items/FBIExecutivesConference/FBI%20Executives%20Conference%20Sections%2020%2C%2021%2C%20and%2026%2C%201943-1945.pdf.

115. Letter from ASIS to JEH, 22 October 1965, https://ia801803.us.archive.org/29/items/foia_Stukenbroeker_Fern_C.-4/Stukenbroeker_Fern_C.-4.pdf; Federal Bureau of Investigation, SAC Memphis, memorandum to the director, February 18, 1971, https://ia801706.us.archive.org/20/items/foia_Stukenbroeker_Fern_C.-5/Stukenbroeker_Fern_C.-5.pdf.

116. Federal Bureau of Investigation, M. A. Jones to Wick and Fern C. Stukenbroeker, "Commendation Matter," August 30, 1966, https://ia801706.us.archive.org/20/items/foia_Stukenbroeker_Fern_C.-5/Stukenbroeker_Fern_C.-5.pdf.

117. Federation of American Scientists, *Awareness of National Security Issues and Response* (1998), https://fas.org/irp/ops/ci/ansir.htm.

118. US Senate, *Departments of State, Justice, and Commerce, the Judiciary, and Related Agencies Appropriations for Fiscal Year 1978,* pt. 1, before a subcommittee of the Committee on Appropriations, 95th Congress (Washington, DC, 1977).

119. US Senate, *Meeting the Espionage Challenge: A Review of United States Counterintelligence and Security Programs,* report of the Select Committee on Intelligence (Washington, DC, 1986), https://www.cia.gov/library/readingroom/docs/CIA-RDP90-00530R000300620021-3.pdf.

120. US Congress, *Department of Justice Appropriation Bill for 1933,* before a subcommittee of the House Committee on Appropriations, 72nd Congress (Washington, DC, 1932).

121. US Congress, *FBI Oversight and Authorization, Fiscal Year 1993,* before the Subcommittee on Civil and Constitutional Rights of the Committee on the Judiciary, House of Representatives, 102nd Congress (Washington, DC, 1992).

122. US Congress, *Departments of Commerce, Justice, and State, the Judiciary, and Related Agencies Appropriations for 2004,* pt. 10, before a subcommittee of the Committee on Appropriations, House of Representatives, 108th Congress (Washington, DC, 2003).

123. Department of Justice, *The Federal Bureau of Investigation's Ability to Address the National Security Cyber Intrusion Threat* (Washington, DC, 2011).

124. US Senate, *Current and Projected National Security Threats to the United States,* before the Select Committee on Intelligence, 112th Congress (Washington, DC, 2012), https://www.govinfo.gov/content/pkg/CHRG-112shrg74790/pdf./CHRG-112shrg74790.pdf.

125. Charles Levinson, "Comey: FBI 'Grappling with Hiring Policy Concerning Marijuana,'" *Wall Street Journal,* May 20, 2014, https://www.wsj.com/articles/BL-LB-48089.

126. US Congress, *Global Terrorism,* pt. 2.

127. Federal Bureau of Investigation, Fern C. Stukenbroeker, special agent, special efficiency rating, "Being Submitted Pursuant to SAC Letter Number 135, Series 1946, Dated December 17, 1946," https://ia801801.us.archive.org/33/items/foia_Stukenbroeker_Fern_C.-1/Stukenbroeker_Fern_C.-1.pdf.

128. Federal Bureau of Investigation, "Inspection Report," November 22, 1946, (Stukenbroeker personnel file), https://ia801801.us.archive.org/33/items/foia_Stukenbroeker_Fern_C.-1/Stukenbroeker_Fern_C.-1.pdf.

129. US Senate, *Departments of State, Justice, Commerce, and the Judiciary Appropriations for 1951,* pt. 1, before a subcommittee of the Committee on Appropriations, 81st Congress (Washington, DC, 1950).

130. US Congress, *Departments of State and Justice, the Judiciary, and Related Agencies Appropriations for 1959,* before a subcommittee of the Committee on Appropriations, House of Representatives, 85th Congress (Washington, DC, 1958).

131. US Senate, *Current and Projected National Security Threats to the United States,* before the Select Committee on Intelligence, 109th Congress (Washington, DC, 2005), https://www.govinfo.gov/content/pkg/CHRG-109shrg22379/pdf./CHRG-109shrg22379.pdf.

132. US Senate, *Federal Bureau of Investigation Oversight,* before the Committee on the Judiciary, 109th Congress (Washington, DC, 2005).

133. US Senate, *FBI Oversight,* before the Committee on the Judiciary, 109th Congress (Washington, DC, 2006).

134. Federal Bureau of Investigation, Executive Conference, memorandum to the director, "SIS Survey," January 16, 1945.

135. US Congress, *Departments of State and Justice, the Judiciary, and Related Agencies Appropriations for 1957,* before a subcommittee of the Committee on Appropriations, House of Representatives, 84th Congress (Washington, DC, 1956).

136. US Senate, *Departments of State, Justice, Commerce, and the Judiciary Appropriations for 1951,* pt. 1.

137. Federal Bureau of Investigation, A. H. Belmont, memorandum to L. V. Boardman, "Special Memoranda Unit, Liaison Section," September 1956 (Belmont personnel file FOIA release).

138. Ibid.

139. Ibid.

140. Federal Bureau of Investigation, *Assumption of Federal Drug Enforcement: A Feasibility Study* (Washington, DC, 1977).

141. US Senate, *Departments of State, Justice, and Commerce, the Judiciary, and Related Agencies Appropriations for Fiscal Year 1978,* pt. 6.

142. US Senate, *Departments of State, Justice, and Commerce, the Judiciary, and Related Agencies Appropriations for Fiscal Year 1980,* before a subcommittee of the Committee on Appropriations, 96th Congress (Washington, DC, 1979).

143. US Congress, *Departments of Commerce, Justice, and State, the Judiciary, and Related Agencies Appropriations for 1992,* pt. 2, before a subcommittee of the

Committee on Appropriations, House of Representatives, 102nd Congress (Washington, DC, 1991).

144. US Congress, *Departments of Commerce, Justice, and State, the Judiciary, and Related Agencies Appropriations for 1990,* pt. 2, before a subcommittee of the Committee on Appropriations, House of Representatives, 101st Congress (Washington, DC, 1989).

145. US Senate, *FBI Budget and Oversight for Fiscal Year 1987,* before the Committee on the Judiciary, 99th Congress (Washington, DC, 1986); William H. Webster, memorandum to the director of central intelligence, "Report for the Administration," December 1, 1980, https://www.cia.gov/library/readingroom/docs/CIA-RDP05T00644R000601780002-1.pdf.

146. US Congress, *Departments of Commerce, Justice, and State, the Judiciary, and Related Agencies Appropriations for 2000,* pt. 6, before a subcommittee of the Committee on Appropriations, House of Representatives, 106th Congress (1999).

147. US Senate, *Federal Bureau of Investigation Oversight.*

148. Federal Bureau of Investigation, Belmont to Boardman, "Special Memoranda Unit, Liaison Section."

149. 9/11 Commission, "Law Enforcement, Counterterrorism, and Intelligence Collection in the United States Prior to 9/11," Staff Statement no. 9, undated, http://govinfo.library.unt.edu/911/staff_statements/staff_statement_9.pdf.

150. US Congress, *Departments of Commerce, Justice, and State, the Judiciary, and Related Agencies Appropriations for 2001,* pt. 2.

151. Ibid.

152. Ibid.

153. Ibid.

154. Department of Justice, *The Federal Bureau of Investigation's Efforts to Hire, Train, and Retain Intelligence Analysts* (Washington, DC, 2005).

155. Ibid.

156. 9/11 Commission, "Reforming Law Enforcement, Counterterrorism, and Intelligence Collection in the United States," Staff Statement no. 12, undated, https://govinfo.library.unt.edu/911/staff_statements/staff_statement_12.pdf.

157. US Senate, *FBI Oversight and Budget Authorization,* before the Subcommittee on Security and Terrorism of the Committee on the Judiciary, 98th Congress (Washington, DC, 1984).

158. Ibid.

159. 9/11 Commission, "Law Enforcement, Counterterrorism, and Intelligence Collection in the United States Prior to 9/11."

160. US Congress, *Departments of Commerce, Justice, and State, the Judiciary, and Related Agencies Appropriations for 2005,* pt. 10.

161. US Senate, *FBI Oversight.*

162. Department of Justice, *The Federal Bureau of Investigation's Efforts to Hire, Train, and Retain Intelligence Analysts.*

163. Department of Justice, *Follow-Up Audit of the Federal Bureau of Investigation's Efforts to Hire, Train, and Retain Intelligence Analysts* (Washington, DC, 2007), https://oig.justice.gov/reports/FBI/a0730/findings.htm.

7

Reassessing the Public-Private National Security Relationship

PRIVATE INDUSTRY HAS AN INCREASINGLY UNILATERAL IMPACT ON national security. In order to address new challenges, government must engage corporations to understand and keep pace with the implications of new technologies and ways of doing business. However, because the private sector is less reliant on government patronage, and may actually find its credibility with customers enhanced by eschewing cooperation with Washington, the government must find new ways to elicit cooperation rather than demanding assistance from industry.

Reaching a Security Consensus

In order to effectively work with the private sector, the government must concede that private sector entities are independent actors that will not always be in lockstep with Washington. The gap between how each sector thinks about its obligations (assuming industry thinks about them at all) to national security has widened. Therefore, rather than demanding specific behaviors, the government should engage industry in a discussion of what security (without nationalism) looks like. Companies pay lip service to human rights, civil liberties, sustainability, and the like, and many of these values align with what the United States seeks to promote internationally. Framing national security concerns in these terms will help industry to understand the implications of its decisions.

Discussion of national security with the private sector should use the established concepts of corporate social responsibility and, more specifically, the "triple bottom line" as frameworks for deliberation. Although a US-based company might be leery about being seen as a geopolitical proxy for Washington, there are values—consistent with the proclaimed interests of the United States—that are beneficial to companies' brands. Perhaps the most prominent example of companies seeking to claim the moral high ground is Google's dictum of "don't be evil."[1] The most cynical might see such stances as a marketing ploy, but holding companies to their stated values is a way to curb practices that fuel the world's dangerous regimes and undermine national security.

Corporate social responsibility is an issue with which companies continue to grapple. The concept of establishing a positive relationship between corporate financial performance and corporate social performance has been in circulation since the 1960s.[2] However, well-managed companies have been less interested in integrating corporate social responsibility with their business strategies and have instead developed their corporate social responsibility programs to conform with the companies' existing purpose and values. (In other words, corporate social responsibility sometimes may be more veneer than virtue.) What companies define as corporate social responsibility has, consequently, covered a wide range of issues.[3] This means the concept remains malleable and open to continued interpretation, which leaves the discussion open to how national security might be incorporated into corporate social responsibility. Linking national security with corporate social responsibility is not entirely new. Discussion, though, has been limited in scope and primarily focused on the private sector's obligation to protect the American public rather than advancing US interests globally.[4]

Triple bottom line accounting—which examines the private sector's impact on both "people" and "planet," in addition to "profits"—provides a framework for discussing corporate social responsibility. The focus of triple bottom line accounting is on values—rather than the national interest—which means that companies can promote widespread, transnational norms without being perceived as shills for Washington, a perception that might exclude them, whether through official prohibitions or public distaste, from markets. Many of these norms are consistent with the values that the United States seeks to export (e.g., democracy, freedom of expression).

The first aspect of the triple bottom line is profits.[5] While the stereotype of the private sector–national security nexus is the defense-industrial complex, the globalized nature of business means that fueling conflict has deleterious impacts on much of the private sector in the form of disrupted supply chains and the like. Furthermore, making concessions to repressive regimes has the potential to limit market share. For instance, allowing countries such as China to dominate 5G technology and influence global standards deprives the US private sector of opportunities for expansion and also strengthens repressive regimes. Everyone, except the bad guys, loses.

The second aspect of triple bottom line accounting is people. This refers to both the company's work force as well the wider community where the company does business.[6] Again, the globalized nature of production and markets aligns an emphasis on people with national security. The most obvious aspect of this is in the area of production. Companies that defy the people component of triple bottom line accounting undercut national security in a variety of ways. For instance, access to inexpensive labor in countries such as China often means turning a blind eye in order to remain on good terms with repressive regimes, which in turn are adversarial to US interests.

Furthermore, seeking markets in countries with authoritarian regimes—again, China comes to mind—means capitulating to censorship requirements. Multiple US firms have placed market share above people in ways that enhance the power of US competitors and adversaries. Google's flirtation with Beijing, in the form of Project Dragonfly (a modified search engine), is just one example of a company knuckling under to geopolitics, over its concern for the rights of a broader audience.[7] Similarly, Facebook, which wants to be the internet's "town square," has helped to suppress dissent in Vietnam.[8] On the domestic front, it is important for the private sector to ensure that its actions are consistent with the values that the United States, or at least a significant portion of the American people, hope to credibly export (e.g., racial equity).

Finally, the third layer of the triple bottom line is the planet. Anyone who views climate change as irrelevant to national security should take note that the US Department of Defense (DoD) thinks differently. According to a 2015 DoD report, climate change "increases the risk of instability and conflict overseas and has implications for DoD in areas of operations; personnel installations; and the stability, development, and human security of other nations."[9]

These factors provide a starting point for a discourse among government, industry, academia, and other nonprofit entities that examine transnational threats. This dialogue should focus on identifying the values that are widely held across responsible governments (and that would help to bring about responsible governments). The 2010 *Quadrennial Homeland Security Review* hinted at the need for this, highlighting an inspiration to develop a "common security mindset" emphasizing the shared responsibility across society—presumably including private industry—for homeland security.[10] Once issues of common concern are identified, both industry and the US government need to improve their collaboration to ensure that necessary information is traveling in both directions.

Investors are a crucial constituency in the process of holding the private sector accountable to the values to which it has publicly committed (and for pressing the government to share information that will help the private sector to uphold security-related values). The investment community has already started to take some of these issues into account when assessing a company's risk. According to a Government Accountability Office study, twelve of fourteen institutional investors indicated that they sought information about companies' environmental and social aspects as factors that could impact risk and financial performance. Institutions have developed around this process of assessment. For instance, in 2011 the Sustainability Accounting Standards Board was formed and has since created a voluntary reporting framework in consultation with companies, investors, and subject-matter experts.[11] Furthermore, organizations that focus on corporate social responsibility within specific industry sectors have emerged over the past two decades. In 2008, US information and communications technology companies, in collaboration with nongovernmental organizations, investors, and universities, formed the Global Network Initiative to promoted best practices related to the conduct of the US private sector in countries with records of poor internet freedom.[12]

Certain areas of the private sector have already started to incorporate aspects of security into their public image. For instance, according to the Government Accountability Office, companies in the internet and banking industries have disclosed information about their data security practices.[13] Such practices align with multiple aspects of triple bottom line accounting: loss of information (and loss of consumer confidence) certainly has ramifications for profits; and such losses also have the potential to undermine security of a population, in a variety of ways, including helping authoritarian governments to more effectively repress their populations if that government acquires specific types of information. The US government should be using programs such as the Federal Bureau of Investigation's Counterintelligence Strategic Partnership Program and the Department of Homeland Security's Project Shield America to engage the investment community and provide an understanding of security so that investors can use the information in judging companies' commitment to values (and, implicitly, their public appeal).

Historically, the US has pursued global private sector engagement security through incremental, tactical activities. The Federal Bureau of Investigation (FBI) helped to create standards through its plant survey activities in Latin America and has sought to create similar norms in the modern field of cybersecurity. The State Department has—through the Overseas Security Advisory Council—attempted to create and support the understanding of security among multinational industries. Ultimately, the US government should scale up its international efforts to encourage the development of norms among US industry that will support human security. It should encourage linking corporate social responsibility to the internationally recognized concept of human security, which the United Nations characterizes as focusing on "widespread and cross-cutting challenges to the survival, livelihood and dignity of the people."[14] In other words, the US government should treat corporate social responsibility as a tool with which it can advance policy objectives through business.

The Role of Private Industry in Advancing Security

If the government is going to keep pace with the private sector, it clearly needs to leverage external expertise in the thick of industry's developments, which can keep government intelligence and analytical operations up to speed with current realities beyond federal agencies. However, there has been a reticence from industry to consistently engage in the sharing of information. For instance, in 2001, the head of the FBI's National Infrastructure Protection Center noted that it was difficult to get private sector experts "at the table" to share information about vulnerabilities.[15]

As an act of good corporate citizenry as well as self-interest, industry should be willing to share information with the government. Two main categories of data are of interest. The first is information about threats encountered by the private sector. Since industry is increasingly in the direct line of fire from foreign intelligence services seeking to illicitly acquire sensitive information and from nonstate actors bent on attacking symbols of US power, it will likely be positioned to provide early warning of threat actors' behavior. Second, the private sector should work with the government to assess, through exercises such as red-teaming, unidentified consequences that new technologies and business practices might introduce for national security.

Government expects greater participation from industry. According to the 2010 *Quadrennial Homeland Security Review,* "government must work creatively and collaboratively with the private sector to identify solutions that take into account both public and private interests."[16] Furthermore, the Department of Homeland Security (DHS) conceded that the private sector "must be fully empowered to see and solve ever larger parts of the problem set."[17] Industry has indicated its willingness to collaborate and its frustration at government's inability to do so effectively. For instance, according to the Government Accountability Office, industry wanted more information about the technology that specific threat actors intended to use against the private sector rather than broad descriptions and alerts.[18] However, for industry to obtain relevant information, it must meet government halfway in order to help decisionmakers who have limited subject-matter expertise about new technologies (or the analytic resources to assess the convergence of established ones) to understand vulnerabilities that threat actors might target.

Lack of Government Support

US government agencies have a hurdle of perception to overcome. The problem is not simply that they do not effectively share information; it is that a perception of ineffective cooperation now exists. Perhaps the most egregious, yet unsurprising, lack of consideration comes from the Department of Homeland Security—truly problematic given its primary role in the protection of critical infrastructure. As of 2014, representatives from nine private critical infrastructure sectors and subsectors believed that the DHS's Office of Intelligence and Analysis (OIA) often failed to generate products that were useful or relevant. Speaking to the failure of the government to field a work force that can keep pace with developments in technology and business practices, the critical infrastructure representatives assessed that the OIA did not understand industries' intelligence needs.[19] This is troubling considering the DHS's consistent lip service to working with industry. In 2006, for instance, the DHS advised Congress that "the government must deliver real value to our private sector partners."[20] To paraphrase a line from the movie *Cool Hand Luke,* what the DHS and the private sector clearly have is a failure to communicate. The FBI is not untainted. A CEO of the National Cyber Forensics and Training Alliance once complained that the Bureau would accept unclassified information from the private sector and then classify it, which then prevented the Bureau from sharing it with other entities in the private sector.[21]

Structuring Engagement

Once government and the private sector are working from the same sheet of music, the government should be willing to provide information, in an easily accessible way, that will facilitate responsible decisionmaking. A primary problem with information-sharing is the fragmentation of functions across multiple government agencies. Counterintelligence, counterproliferation, and counterterrorism are directed at ensuring that the United States maintains information and physical (e.g., critical infrastructure) advantages that will support US elements of national power. Disparate agencies (e.g., the FBI's Office of Private Sector, the DHS's Project Shield America outreach) handle dialogue with industry about these topics. Each additional conduit through which information must flow either to or from government is one more potential point of failure. Furthermore, information sharing and analysis centers (ISACs) are not structured for broad information-sharing. Rather, they focus inward on sector-specific problems and interface with a specific agency that may, or may not, have access to, or the ability to effectively communicate, information that may have implications for sectors beyond the ISAC for which they serve as a liaison.

No single government agency has a sufficiently broad bailiwick to address the breadth of security issues that the private sector faces. Furthermore, information that the private sector can furnish about its experiences dealing with threat actors can serve an interagency constituency. Therefore, the US government should establish an interagency hub for security outreach. This entity should be a public-private partnership, along the lines of the National Endowment for Democracy. By incorporating private sector entities as stakeholders, this new entity would remain responsive to industry needs and concerns, drawing on US government agencies in response to these needs and concerns, rather than leaving government agencies to determine what works best for a milieu with which it has limited experience.

This new partnership would not engage in clandestine or law enforcement activities (both of which might deter private sector participation), but would instead function as an honest broker between industry and government. One of its primary functions would be to translate sensitive concerns identified by the US intelligence community and other collectors into publicly distributable products and assistance. It should align these outreach efforts with industry sectors, identifying the implications of broad national security concerns for distinct subsets of customers. To provide assistance, the partnership should assume responsibility for and then deconflict and streamline existing outreach programs (and should receive the personnel and other resources associated with those programs—for instance, taking direct responsibility for initiatives such as the FBI's InfraGard and the DHS's cybersecurity advisers).[22]

Industry has already indicated its desire for consolidation of information-sharing functions. According to a 2014 General Accounting Office report, private sector officials stated their preference for a single or centralized government source in order to avoid confusion about authoritative sources.[23] By following what the market suggests, the US government may find that the private sector is more cooperative and collaborative in the ensurance of national security.

The Need to Bridge the Government–Private Sector Knowledge Gap

One of the ways in which the breakdown of understanding between the private sector and the public sector can be resolved is to facilitate greater cross-pollination between the sectors. Multiple agencies have pointed to the possibility of using public-private exchange programs. For instance, the National Geospatial Intelligence Agency (NGA) allows its staff to work temporarily for private firms while drawing a government salary. According to then–NGA director Robert Cardillo, employees returning are "invaluable" since they bring skills to the NGA. Furthermore, according to Cardillo, firms provided reciprocal assistance to the NGA by quietly lending their information technology experts to the agency.[24] This practice is consistent with the NGA's broader relationship with industry. For instance, the agency acquires commercial imagery and other remote sensing data and analysis to supplement the data it receives from the National Reconnaissance Office.[25]

Other agencies have indicated their willingness to engage in similar initiatives. For instance, the FBI, in 2005, stated that it was working on a fellows program to exchange staff with the private sector.[26] In 2014, the DHS's OIA indicated that—although it was not considering long-term deployments to the private sector—it was contemplating the concept of shorter-term deployment to interested companies.[27] Private sector officials in the critical infrastructure sector believed that detailing OIA analysts to companies would better position the OIA to understand their industries and provide effective intelligence analysis.[28] The Office of the Director of National Intelligence has also explored similar exchanges. The Public-Private Talent Exchange involves assignments from the intelligence community to the private sector.[29]

Whether these programs can create an overarching culture of excellence remains to be seen. The intelligence community must get its own house in order if it is going to effectively send its best and brightest to represent it in the private sector (and bring them back again). According to then–director of national intelligence John Negroponte, the intelligence community had established an *Analytic Resources Catalog* providing a detailed inventory of analysts throughout the intelligence community, tied to their expertise and experience. Negroponte believed that this would enable and encourage informal sharing of information and knowledge.[30]

However, the logical next step, formalization of these relationships as well as enhancements of expertise based on experience, has been stymied by bureaucratic impediments. Even though, according to the Government Accountability Office, chief human capital officers in various agencies concurred that lateral mobility opportunities such as rotations, details, and opportunities to gain experience in other sectors can help employees gain new skills more cost-effectively than training, few employees moved horizontally, because managers were reluctant to lose employees.[31] This parochial view, which effectively allows immediate managers to undercut the good of the larger intelligence enterprise, would likely impact rotations to and from the private sector.

Agencies—as opposed to individual managers—have appeared sluggish in developing and implementing processes to effect the rotation of talent in and out of agencies. For instance, although the FBI claimed in 2005 that it was establishing a fellows program, a 2015 Department of Justice inspector general report indicated that the FBI was still considering a working group's recommended measures, including "supporting and encouraging mobility of personnel between the public and private sectors to bring knowledgeable and seasoned professionals back in to the FBI."[32]

To encourage the development of talent, the US government should remove as many distortions as possible, in order to allow talent to circulate and grow. Employees should be permitted to apply for, and accept, rotational assignments based on their merits, rather than on the acquiescence of their management. Instead, managers should be required to make a business case for preventing an employee from taking a rotational assignment (a business case that includes guarantees of steps that an agency will take to ensure that the employee is not denied opportunities for professional development). Agencies should create a culture that generally discourages the obstruction of career development in service of short-term resource retention.

An underlying premise for this new approach is that the labor market, if allowed to ebb and flow throughout the government, will not only reward talent but also clearly identify poor management. If employees are consistently making efforts to remove themselves from a specific chain of command, human resources professionals within agencies should ask why. A similar indicator of managerial trouble is a manager's inability to attract talent. The outcome would be either that the manager changes their behavior in order to create a more positive work environment that will attract employees to apply for positions, or that an agency carefully scrutinizes clearly apparent dysfunction and focuses its efforts either at remediating a manager's performance or taking more severe steps such as demotion or termination of employment.

Both the FBI and the DHS have a long way to go in improving their employees' satisfaction with the agencies' work environment. In an analysis of the Office of Personnel Management's Federal Employees Viewpoints Survey, the Partnership for Public Service determined that the FBI ranked at 242 out of 420 as a desirable place to work. The DHS's Office of Intelligence and Analysis ranked at 406 out of 420. The partnership specifically used the following survey statements: "I recommend my organization as a good place to work"; "Considering everything, how satisfied are you with your job?"; and "Considering everything, how satisfied are you with your organization?"—as the basis for the results.[33] The middling-to-poor performance of the domestic intelligence entities responsible for securing the private sector certainly reflects conditions that leadership/management is responsible for creating and perpetuating.

The government can also improve its human capital by facilitating the rotation of talent into and out of government jobs (as opposed to temporary tours of duty). Expert employees may depart government service due to their perception that they cannot make sufficient career advancement in their government billets. For instance, a 2010 Central Intelligence Agency (CIA) inspector general report found that a need to make better use of skills and knowledge was among the top ten reasons for individuals to leave the CIA.[34] Employees who leave government for positions that add to their expertise in certain sectors should be courted back into government service. However, this seems to be a novel concept in agencies across the government. For instance, in 2015, while testifying to Congress, then–FBI director James Comey mulled: "Should there be a model where [employees] come, then they go and do something in the private sector, then come back? That's something we haven't done before, but that may be a model I want to look at."[35]

If the government is unable to facilitate the rotation of personnel into and out of the private sector, it should lean heavily on the concept of a national intelligence reserve corps. The Intelligence Reform and Terrorism Prevention Act of 2004, which established the position of the director of national intelligence, included a provision to establish an intelligence reserve corps similar to the one maintained by US military services. Through a 2006 policy memorandum, the Office of the Director of National Intelligence established this corps, which would serve during a period of emergency, for the intelligence community. The 2006 memorandum delegated the authority to make national intelligence reserve corps appointments to heads of individual intelligence community agencies.[36] The memo directed the agencies to provide all professional, technical, managerial, and executive employees who separate from agencies' rolls with an opportunity to place their name on a roster of national intelligence reserve corps volunteers. Counterterrorism expert Daniel Byman has suggested expanding this concept. In addition to individuals who separated from intelligence community agencies, Byman also suggests that the national intelligence reserve corps draw on individuals from relevant private sector entities who would receive training on the agency to which they were assigned.[37] The statute should be amended to make the national intelligence reserve corps a continuous operation—Byman suggests several days of service in the government per month—rather than one that operates only in time of an emergency.

Filling the Void

Industry has consistently demonstrated its willingness to resort to nongovernment security providers in the absence of effective government programs. For instance, in the late nineteenth and early twentieth centuries, industries such as General Motors enlisted the services of the Pinkerton Company for security. During the Cold War, the American Security Council collected information from former FBI agents and assembled "subversive" files, which it provided to subscribers.[38] Private sector security firms have proliferated around cyber vulnerabilities. These firms offer the same expertise as government by hiring savvy, former government officials. For instance, Shawn Henry, an FBI executive decamped from the Bureau to join CrowdStrike.[39] Furthermore, these private firms seem to offer customers information that the US government appears to treat as authoritative. Joel Brenner, who served as head of Office of the National Counterintelligence Executive, noted that there was a "seismic shift" toward increasing reliance on the private sector in the intelligence world.[40] As an example, in a 2011 report, the office specifically cited information from the cybersecurity firm Mandiant.[41]

Such efforts have, on occasion, baffled the government by offering services for which customers paid, that the government believed it could handle with similar efficacy. For instance, according to a 1939 FBI memorandum, the Cleveland and Safety Council's Industrial Safety Committee, under the direction of none other than Eliot Ness (of *The Untouchables* fame), took upon itself the investigation of suspicious employees and other individuals who were engaged in sabotage and espionage activities. According to the Bureau, the only function of the Industrial Safety Committee was "to receive and correlate reports as to sabotage and espionage. This is strictly the function of the Federal Bureau of Investigation and is handled by the Bureau without cost to the industrial plants, where the Industrial Safety Committee [was] charging these plants a large amount of money for the same service."[42] More recently, the DHS noted that an industrial facility had paid $5,000 to contract for a site assessment that a protective security adviser could have performed free of charge.[43]

Furthermore, private sector security firms are not immune from the sort of attacks that they are hired to protect their clients from. In December 2020, FireEye, a cybersecurity outfit, announced that it had been the target of a successful hack. According to the firm, the "highly sophisticated" perpetrators were likely state-sponsored.[44] Although FireEye did not name the state sponsor, media accounts pointed to Russia.[45] The perpetrators succeeded in pilfering red-team assessment tools that FireEye used to test customers' security.[46]

Closing Thoughts

The US government has a very real reason to be concerned about national security threats to the private sector. Increasingly, industry is a driver of innovation that impacts national security and yet does not always do so in collaboration with the government. This is not undesirable. In fact, given the challenges in the field of government human capital, it is probably a good thing for the sake of technology. However, industry, for its own sake (e.g., to protect proprietary information), should understand that there is a business case to be made for working with the government to protect the private sector from a multitude of threats and to assist the government in understanding those threats.

Government, in turn, needs to readjust how it operates in order to make it a partner capable of adding value in a relationship with the private sector. First, it must reform its human capital management, especially in the analytic field, in order to engage in cutting-edge assessment methodologies such as red-teaming. This cannot be effectively accomplished if the wrong people are in the wrong places (or in any place, given some of the deficiencies that bodies responsible for government oversight have identified) and do not have the latitude to engage in out-of-the-box thinking. As the 9/11 Commission pointed out, "failures of imagination" contributed to the inability of the United States to anticipate the September 11, 2001, attacks. Second, institutions must be in place to facilitate the most effective use of human capital. These include the facilitation of effective management within agencies and the creation of robust mechanisms to close the gap between the government and private industry through exchanges of talent.

Not implementing these remedies will lead to intelligence failures. Industry's advances will become Pandora's boxes rather than boons for humanity. Government will waste resources in trying to independently keep pace with or, in some situations, cleaning up after disasters created by technology. On industry's side of the equation, failing to work collaboratively with government makes its aspirations toward corporate social responsibility read as disingenuous. Furthermore, from a practical, market-based viewpoint, it puts investors at risk, since loss of knowledge through economic espionage and loss of access to markets through disrupted supply chains or terrorism against facilities means reduced revenue.

Notes

1. Kate Conger, "Google Removes 'Don't Be Evil' Clause from Its Code of Conduct," May 18, 2018, https://gizmodo.com/google-removes-nearly-all-mentions-of-dont-be-evil-from-1826153393.

2. Matteo Torello, "The Business Case for Corporate Social Responsibility," Harvard Law School Forum on Corporate Governance, June 26, 2011, https://corpgov.law.harvard.edu/2011/06/26/the-business-case-for-corporate-social-responsibility.

3. K. Katsuri Rangan, Lisa Chase, and Sohel Karim, "The Truth About CSR," *Harvard Business Review,* January–February 2016, https://hbr.org/2015/01/the-truth-about-csr.

4. Darren E. Tromblay, "Tech Pressure on Privacy: National Security Requires a Fuller View of Corporate Social Responsibility," July 19, 2018, https://www.justsecurity.org/59546/tech-pressure-privacy-national-security-requires-fuller-view-corporate-social-responsibility.

5. University of Wisconsin, "The Triple Bottom Line," https://sustain.wisconsin.edu/sustainability/triple-bottom-line.

6. Ibid.

7. Christopher Mims, "Google Outgrows Its Youthful Ideals," *Wall Street Journal,* August 17, 2018, https://www.wsj.com/articles/googles-risky-pragmatism-1534427069.

8. David S. Cloud and Shashank Bengali, "Facebook Touts Free Speech; In Vietnam, It's Aiding in Censorship," *Los Angeles Times,* October 22, 2020, https://www.latimes.com/world-nation/story/2020-10-22/facebook-censorship-suppress-dissent-vietnam.

9. Department of Defense, *National Security Implications of Climate-Related Risks and a Changing Climate* (Washington, DC, 2015), https://archive.defense.gov/pubs/150724-congressional-report-on-national-implications-of-climate-change.pdf.

10. Department of Homeland Security, *Quadrennial Homeland Security Review 2010* (Washington, DC), https://www.dhs.gov/sites/default/files/publications/2010-qhsr-report.pdf.

11. Government Accountability Office, *Public Companies: Disclosure of Environmental, Social, and Governance Factors and Options to Enhance Them* (Washington, DC, 2020), https://www.gao.gov/assets/710/707949.pdf.

12. Congressional Research Service, *Internet Freedom in China: US Government Activity, Private Sector Initiatives, and Issues of Congressional Interest* (Washington, DC, 2018).

13. Government Accountability Office, *Public Companies.*

14. United Nations, "What Is Human Security?" https://www.un.org/humansecurity/what-is-human-security.

15. US Senate, *Critical Infrastructure Protection: Who's in Charge?* before the Committee on Governmental Affairs, 107th Congress (Washington, DC, 2001).

16. Department of Homeland Security, *Quadrennial Homeland Security Review 2010* (Washington, DC), https://www.dhs.gov/sites/default/files/publications/2010-qhsr-report.pdf.

17. Ibid.

18. Government Accountability Office, *Critical Infrastructure Protection: Key Private and Public Cyber Expectations Need to Be Consistently Addressed* (Washington, DC, 2010), https://www.gao.gov/assets/310/307222.pdf.

19. Government Accountability Office, *DHS Intelligence Analysis: Additional Actions Needed to Address Analytic Priorities and Workforce Challenges* (Washington, DC, 2014).

20. US Congress, *Cybersecurity: Protecting America's Critical Infrastructure, Economy, and Consumers,* before the Committee on Energy and Commerce, House of Representatives, 109th Congress (Washington, DC, 2006), https://www.govinfo.gov/content/pkg/CHRG-109hhrg31464/pdf./CHRG-109hhrg31464.pdf.

21. Department of Justice, *Audit of the Federal Bureau of Investigation's Implementation of Its Next Generation Cyber Initiative* (Washington, DC, 2015).

22. Darren E. Tromblay, *Protecting Partners or Preserving Fiefdoms?* (Washington, DC: Information Technology and Innovation Foundation, 2017), http://www2.itif.org/2017-counterintelligence-outreach-industry.pdf.

23. Government Accountability Office, *Critical Infrastructure Protection.*

24. "Spooks for Hire: America's Intelligence Agencies Find Creative Ways to Compete for Talent," *The Economist,* March 3, 2018.

25. Congressional Research Service, *Commercial Space: Federal Regulation, Oversight, and Utilization* (Washington, DC, 2018).

26. Federal Bureau of Investigation, "Statement of the FBI Regarding the Office of the Inspector General's Report *The Federal Bureau of Investigation's Efforts to Hire, Train, and Retain Intelligence Analysts,*" May 4, 2005, https://archives.fbi.gov/archives/news/pressrel/press-releases/statement-of-the-fbi-regarding-the-office-of-the-inspector-generals-report-the-federal-bureau-of-investigations-efforts-to-hire-train-and-retain-intelligence-analysts.

27. Government Accountability Office, *DHS Intelligence Analysis.*

28. Ibid.

29. Office of the Director of National Intelligence, "ODNI Private Sector Programs," https://www.dni.gov/index.php/who-we-are/organizations/national-security-partnerships/ps-engagement/odni-private-sector-programs.

30. US Senate, *Current and Future Worldwide Threats to the National Security of the United States,* before the Committee on Armed Services, 109th Congress (Washington, DC, 2006), https://www.govinfo.gov/content/pkg/CHRG-109shrg32745/pdf./CHRG-109shrg32745.pdf.

31. Government Accountability Office, *Improving Federal Recruiting and Hiring Efforts* (Washington, DC, 2019), https://www.gao.gov/assets/710/700657.pdf.

32. Department of Justice, *Audit of the Federal Bureau of Investigation's Implementation of Its Next Generation Cyber Initiative.*

33. Partnership for Public Service, *Best Places to Work: Agency Rankings* (Washington, DC, 2020), https://bestplacestowork.org/rankings/overall/sub.

34. Central Intelligence Agency, *Report of Follow-Up Inspection: Retention in the Agency 2010,* https://assets.documentcloud.org/documents/741885/cia-employee-retention-report.pdf.

35. US Senate, *Counterterrorism, Counterintelligence, and the Challenges of "Going Dark,"* before the Select Committee on Intelligence, 114th Congress (Washington, DC, 2015), https://www.govinfo.gov/content/pkg/CHRG-114shrg27896/pdf./CHRG-114shrg27896.pdf.

36. Office of the Director of National Intelligence, "National Intelligence Reserve Corps," 2006, https://fas.org/irp/dni/icpm/2006-600-1.pdf.

37. Daniel Byman, *An Intelligence Reserve Corps to Counter Terrorist Use of the Internet* (Stanford: Hoover Institution, 2018), https://www.hoover.org/sites/default/files/research/docs/byman_webreadypdf.pdf.

38. Joan M. Jensen, *Army Surveillance in America, 1775–1980* (New Haven: Yale University Press, 1991).

39. Darren E. Tromblay and Robert Spelbrink, *Securing US Innovation* (Lanham: Rowman and Littlefield, 2016).

40. Ariana Eunjung Cha, "Even Spies Embrace China's Free Market," *Washington Post,* February 15, 2008.

41. Office of the National Counterintelligence Executive, *Foreign Spies Stealing US Economic Secrets in Cyberspace: Report to Congress on Foreign Economic Collection and Industrial Espionage, 2009–2011* (Washington, DC, 2011).

42. Federal Bureau of Investigation, memorandum to E. A. Tamm, "Industrial Safety Committee of Cleveland Safety Council," November 18, 1939, https://vault.fbi.gov/Eliot%20Ness/Eliot%20Ness%20Part%203%20of%205.

43. Department of Homeland Security, *DHS Can Enhance Efforts to Protect Commercial Facilities from Terrorism and Physical Threats* (Washington, DC, 2020), https://www.oig.dhs.gov/sites/default/files/assets/2020-06/OIG-20-37-Jun20.pdf.

44. "FireEye Shares Details of Recent Cyber Attack, Actions to Protect Community," December 8, 2020, https://www.fireeye.com/blog/products-and-services/2020/12/fireeye-shares-details-of-recent-cyber-attack-actions-to-protect-community.html.

45. David E. Sanger, "Russian Hackers Broke Into Federal Agencies, U.S. Officials Suspect," *New York Times,* December 13, 2020, https://www.nytimes.com/2020/12/13/us/politics/russian-hackers-us-government-treasury-commerce.html.?action=click&module=Top%20Stories&pgtype=Homepage; David E. Sanger and Nicole Perlroth, "FireEye, a Top Cybersecurity Firm, Says It Was Hacked by a Nation-State," *New York Times,* December 8, 2020, https://www.nytimes.com/2020/12/08/technology/fireeye-hacked-russians.html.

46. "FireEye Shares Details of Recent Cyber Attack."

Appendix

Appendix: Key US Government Entities Engaged in Securing the Private Sector

Department-Level	Agency-Level	Components
Office of the Director of National Intelligence	Cyber Threat Intelligence Integration Center National Counter-intelligence and Security Center (formerly the National Counter-intelligence Executive and the National Counter-intelligence Center) National Counter-proliferation Center National Counter-terrorism Center	
US Department of Commerce	Bureau of Industry and Security	
US Department of Defense	Defense Investigative Service	

continues

Appendix: Continued

Department-Level	Agency-Level	Components	
US Department of Justice	Federal Bureau of Investigation	Counterintelligence Division Weapons of Mass Destruction Directorate Directorate of Intelligence	Counterproliferation Center
		Cyber Division	
US Department of Homeland Security	Customs and Border Protection		
	Cybersecurity and Infrastructure Security Agency (formerly the National Protection and Programs Directorate and the Information Analysis and Infrastructure Protection Directorate)	National Cybersecurity and Communications Integration Center	
		National Risk Management Center	
	Immigration and Customs Enforcement	Homeland Security Investigations	Counterproliferation Investigative Centers Export Enforcement Coordination Center
	Office of Intelligence and Analysis (formerly the Information Analysis component of the Information Analysis and Infrastructure Protection Directorate)		
US Department of Treasury	Committee on Foreign Investment in the United States		

Acronyms

ACLU	American Civil Liberties Union
AEC	Atomic Energy Commission
AI	artificial intelligence
AIS	Automated Indicator Sharing program
ANSIR	Awareness of National Security Incidents and Response program
ASIS	American Society for Industrial Security
AUSA	Association of the United States Army
BIS	Bureau of Industry and Security (Department of Commerce)
BRAG	Bioterrorism Risk Assessment Group
CAIEP	China Association for the International Exchange of Personnel
CFATS	Chemical Facility Anti-Terrorism Standards
CFIUS	Committee on Foreign Investment in the United States
CIA	Central Intelligence Agency
CIG	Central Intelligence Group
CISA	Cybersecurity and Infrastructure Security Agency (DHS)
CISPP	Counterintelligence Strategic Partnerships Program
CITAC	Computer Investigations and Infrastructure Threat Assessment Center (FBI)
DCI	director of central intelligence
DECA	Development of Counterintelligence Awareness program
DHS	Department of Homeland Security
DIS	Defense Investigative Service
DNI	director of national intelligence
DoD	Department of Defense
DoE	Department of Energy

DSS	Defense Security Service
FBI	Federal Bureau of Investigation
GAO	General Accounting Office
GEOINT	geospatial intelligence
HITRAC	Homeland Infrastructure Threat and Risk Analysis Center (DHS)
HUMINT	human intelligence
ICE	Immigration and Customs Enforcement
IMINT	imagery intelligence
IPTF	Infrastructure Protection Task Force (FBI)
ISAC	information sharing and analysis center
NACIC	National Counterintelligence Center
NCCIC	National Cybersecurity and Communications Integration Center
NCIJTF	National Cyber Investigative Joint Task Force (FBI)
NCIX	National Counterintelligence Executive
NCSC	National Counterintelligence and Security Center
NCSD	National Cyber Security Division (DHS)
NGA	National Geospatial Intelligence Agency
NICC	National Infrastructure Coordinating Center (DHS)
NIPC	National Infrastructure Protection Center (FBI)
NIPCI	National Infrastructure Protection and Computer Intrusion (FBI)
NPIC	National Photographic Interpretation Center
NPPD	National Protection and Programs Directorate (DHS)
NRP	National Reconnaissance Program
NSA	National Security Agency
NSTL	National Security Threat List
ODNI	Office of the Director of National Intelligence
OIA	Office of Intelligence and Analysis (DHS)
ONCIX	Office of the National Counterintelligence Executive
OPEC	Organization of the Petroleum Exporting Countries
OSAC	Overseas Security Advisory Council
OSINT	open-source intelligence
PIS	Polish Intelligence Service
POLAMCO	Polish American Machinery Corporation
SAFEA	State Administration of Foreign Experts Affairs (China)
SIGINT	signals intelligence
SIS	Special Intelligence Service (FBI)
UAV	unmanned aerial vehicle
UNICEF	United Nations International Children's Fund
US-CERT	US Computer Emergency Readiness Team
WMD	weapon of mass destruction

Bibliography

Advisory Panel to Assess Domestic Response Capabilities for Terrorism Involving Weapons of Mass Destruction. *Second Annual Report* (Santa Monica, CA: 2000) https://www.rand.org/content/dam/rand/www/external/nsrd/terrpanel/terror2.pdf.

———. *Fourth Annual Report to the President and the Congress.* 2002. https://www.rand.org/content/dam/rand/www/external/nsrd/terrpanel/terror4.pdf.

Ahmed, Murad. "Palantir Goes from CIA-Funded Start Up to Big Business." *Financial Times,* June 24, 2015. https://www.ft.com/content/926af768-1a4c-11e5-a130-2e7db721f996.

Alba, Davey. "A Google VP Told the US Senate the Company Has 'Terminated' the Chinese Search App, Dragonfly." *BuzzFeed,* July 16, 2019.

Andrew, Christopher, and Vasili Mitrokhin. *The Sword and the Shield: The Mitrokhin Archive and the Secret History of the KGB* (New York: Basic, 1999).

Andrews, Edmund L. "U.S. Plans to Push Giving F.B.I. Access in Computer Codes." *New York Times,* February 5, 1994.

Ante, Spencer, and William Mauldin. "IBM, Lenovo Deal Likely to Spark Security Review." *Wall Street Journal,* January 24, 2014.

Bain, Ben. "DHS, Industry to Try Fusion Centers for Classified Data Swap." *Federal Computer Week,* March 16, 2010. https://fcw.com/articles/2010/03/16/web-cyber-threat-fusion-center.aspx.

Barbash, Fred, and Ellen Nakashima. "Chinese Hackers May Have Breached the Federal Government's Personnel Office, US Officials Say." *Washington Post,* July 10, 2014. https://www.washingtonpost.com/news/morning-mix/wp/2014/07/09/report-chinese-hacked-into-the-federal-governments-personnel-office.

Barrett, David M. *The CIA and Congress* (Lawrence: University Press of Kansas, 2005).

Barry, Ellen, and Gina Kolata. "China's Lavish Funds Lured U.S. Scientists; What Did It Get in Return?" *New York Times,* February 6, 2020. https://www.nytimes.com/2020/02/06/us/chinas-lavish-funds-lured-us-scientists-what-did-it-get-in-return.html.

Batvinis, Raymond. *The Origins of FBI Counterintelligence* (Lawrence: University Press of Kansas, 2007).

Best, Richard A., and Jennifer K. Elsea. *Satellite Surveillance: Domestic Issues* (Washington, DC, 2011). https://fas.org/sgp/crs/intel/RL34421.pdf.

Bing, Christopher, and Joel Schectman. "Project Raven: Inside the UAE's Secret Hacking Team of American Mercenaries." Reuters, January 30, 2019. https://www.reuters.com/investigates/special-report/usa-spying-raven.

Bjelopera, Jerome P. *Homeland Security Investigations: A Directorate Within U.S. Immigration and Customs Enforcement—In Brief* (Washington, DC: Congressional Research Service, 2015).

Boffey, Phillip M. "Assessing Technology Leaks." *New York Times,* January 2, 1985. https://www.nytimes.com/1985/01/02/business/assessing-technology-leaks.html.

Bogdanich, Walt, and Michael Forsythe. "How McKinsey Has Helped Raise the Stature of Authoritarian Governments." *New York Times,* December 15, 2018. https://www.nytimes.com/2018/12/15/world/asia/mckinsey-china-russia.html.

Brenner, Susan W., and Anthony C. Crescenze. "State-Sponsored Crime: The Futility of the Economic Espionage Act." *Houston Journal of International Law* no. 28 (2006): 389–465.

Bridis, Ted. "FBI Unit Fails to React on Time to Electronic Threats, Report Says." *Wall Street Journal,* May 22, 2001.

Brown, Michael, and Pawneet Singh. *China's Technology Transfer Strategy* (Washington, DC: Defense Innovation Unit, 2018). https://admin.govexec.com/media/diux_chinatechnologytransferstudy_jan_2018_(1).pdf.

Bucci, Steven, James Carafano, and Jessica Zuckerman. *60 Terrorist Plots Since 9/11: Continued Lessons in Domestic Counterterrorism* (Washington, DC: Heritage Foundation, 2013). https://www.heritage.org/terrorism/report/60-terrorist-plots-911-continued-lessons-domestic-counterterrorism.

Buckley, Chris, and Keith Bradsner. "China's Communists to Private Business: You Heed Us, We'll Help You." *New York Times,* September 17, 2020. https://www.nytimes.com/2020/09/17/business/china-communist-private-business.html.

Bureau of Industry and Security. *Fiscal Year 2019 President's Submission.* http://www.osec.doc.gov/bmi/budget/FY19CBJ/BIS_FY19_President%27s_Budget_FINAL.pdf.

———, Office of Export Enforcement. *A Brief History of United States Export Controls.* N.d. https://www.governmentattic.org/34docs/BIShistoryABHOUSEC_undated.pdf.

Byman, Daniel. *An Intelligence Reserve Corps to Counter Terrorist Use of the Internet* (Stanford: Hoover Institution, 2018). https://www.hoover.org/sites/default/files/research/docs/byman_webreadypdf.pdf.

Center for Strategic and International Studies. *Significant Cyber Incidents Since 2006* (Washington, DC, undated). https://csis-website-prod.s3.amazonaws.com/s3fs-public/201106_Significant_Cyber_Events_List.pdf.

Central Intelligence Agency. *National Intelligence Estimate.* No. 4-68 (June 18, 1968).

———. *National Intelligence Estimate: Soviet Capabilities for Clandestine Attack Against the US with Weapons of Mass Destruction and the Vulnerability of the US to Such Attack, Mid 1951 to Mid 1952* (September 4, 1951).

———. *Semiannual Report to the Director of Central Intelligence* (July–December 2000). https://www.cia.gov/library/readingroom/docs/DOC_0001311476.pdf.

———. *Soviet Acquisition of Western Technology and Its National Security Implications* (February 23, 1982). https://www.cia.gov/library/readingroom/docs/CIA-RDP83M00914R002000070021-4.pdf.

———. *Summary Report on Technology Transfer to Communist Countries and the Intelligence Community's Role and Effectiveness* (1981). https://www.cia.gov/library/readingroom/docs/CIA-RDP85T00176R000900020001-5.pdf.

———. *The Technology Acquisition Efforts of the Soviet Intelligence Services* (1982). https://www.cia.gov/library/readingroom/docs/CIA-RDP82M00786R000104810001-5.pdf.

Cha, Ariana Eunjung. "Even Spies Embrace China's Free Market." *Washington Post,* February 15, 2008.

Chappellet-Lanier, Tajha. "After Protest, Open Source Software Company Chef Will Let ICE Contract Expire." *Fedscoop,* September 23, 2019.

Civil Applications Committee (CAC) Blue Ribbon Study. *Independent Study Group Final Report* (2005). https://nsarchive2.gwu.edu//NSAEBB/NSAEBB229/40.pdf.

Clancy, Charles, and Emily Frye. "Is It Time to Designate Social Media as 'Critical Infrastructure'?" *The Hill,* July 27, 2020. https://thehill.com/opinion/cybersecurity/509154-is-it-time-to-designate-social-media-as-critical-infrastructure.

Cloud, David S., and Shashank Bengali. "Facebook Touts Free Speech: In Vietnam, It's Aiding in Censorship." *Los Angeles Times,* October 22, 2020. https://www.latimes.com/world-nation/story/2020-10-22/facebook-censorship-suppress-dissent-vietnam.

Coates, James, and Rogers Worthington. "How Spy Ring Went Shopping and Almost Stole the US Store." *Chicago Tribune,* October 23, 1983. https://www.cia.gov/library/readingroom/docs/CIA-RDP90-00552R000302530042-4.pdf.

Commission on CIA Activities Within the United States. *Report to the President* (Washington, DC, 1975). https://www.fordlibrarymuseum.gov/library/document/0005/1561495.pdf.

Commission to Review Department of Defense Security Policies and Practices. *Keeping the Nation's Secrets: A Report to the Secretary of Defense* (Washington, DC, 1985). https://www.cia.gov/library/readingroom/docs/CIA-RDP96B01172R000100090004-9.pdf.

Conger, Kate. "Google Removes 'Don't Be Evil' Clause from Its Code of Conduct." May 18, 2018. https://gizmodo.com/google-removes-nearly-all-mentions-of-dont-be-evil-from-1826153393.

Conger, Kate, and Daisuke Wakabayashi. "Google Employees Protest Secret Work on Censored Search Engine for China." *New York Times,* August 16, 2018. https://www.nytimes.com/2018/08/16/technology/google-employees-protest-search-censored-china.html?searchResultPosition=1.

Congressional Research Service. *CFIUS Reform Under FIRRMA* (Washington, DC, 2020). https://fas.org/sgp/crs/natsec/IF10952.pdf.

———. *Commercial Space: Federal Regulation, Oversight, and Utilization* (Washington, DC, 2018).

———. *The Committee on Foreign Investment in the United States (CFIUS)* (Washington, DC, 2020). https://crsreports.congress.gov/product/pdf/RL/RL33388.

———. *Critical Infrastructure: Emerging Trends and Policy Considerations for Congress* (Washington, DC, 2020).

———. *Critical Infrastructure: The National Asset Database* (Washington, DC, 2007). https://fas.org/sgp/crs/homesec/RL33648.pdf.

———. *Defense Primer: Information Operations* (Washington, DC, 2020). https://crsreports.congress.gov/product/pdf/IF/IF10771.

———. *Encryption Technology: Congressional Issues* (Washington, DC, 1998).

———. *The Export Administration Act: Evolution, Provisions, and Debate* (Washington, DC, 2003). https://fas.org/sgp/crs/RL31832.pdf.

———. *Federal Research and Development (R&D) Funding: FY 2020* (Washington, DC, 2020). https://crsreports.congress.gov/product/pdf/R/R45715.

———. *The Global Research and Development Landscape and Implications for the Department of Defense* (Washington, DC: Congressional Research Service, 2018).

———. *The International Emergency Economic Powers Act: Origins, Evolution, and Use* (Washington, DC, 2019).

———. *The International Emergency Economic Powers Act: Origins, Evolution, and Use* (Washington, DC, 2020). https://fas.org/sgp/crs/natsec/R45618.pdf.

———. *Internet Freedom in China: US Government Activity, Private Sector Initiatives, and Issues of Congressional Interest* (Washington, DC, 2018).

———. *The National Counterintelligence and Security Center (NCSC): An Overview* (Washington, DC, 2018). https://crsreports.congress.gov/product/pdf/IF/IF11006.

———. *Transfer of Defense Articles: Sale and Export of U.S.-Made Arms to Foreign Entities* (Washington, DC, 2020). https://crsreports.congress.gov/product/pdf/R/R46337.

———. *The U.S. Election Assistance Commission: Overview and Selected Issues for Congress* (Washington, DC, 2019). https://crsreports.congress.gov/product/pdf/R/R45770.

———. *US Export Control System and the Export Control Reform Initiative* (Washington, DC, 2020).

Corrigan, Jack. "DHS Contract Will Help Drones Automatically Spot Border Threats." May 10, 2018. https://www.nextgov.com/emerging-tech/2018/05/dhs-contract-will-help-drones-automatically-spot-border-threats/148088.

———. "DHS Is Exploring How Blockchain Can Stop Counterfeits and Forgeries." December 4, 2018. https://www.nextgov.com/emerging-tech/2018/12/dhs-exploring-how-blockchain-can-stop-counterfeits-and-forgeries/153273.

———. "DHS Startup Accelerator Awards Its First Final-Phase Contract." April 30, 2018. https://www.nextgov.com/emerging-tech/2018/04/dhs-startup-accelerator-awards-its-first-final-phase-contract/147864.

———. "NSA Cyber Chief Wants to Share Digital Threats Early and Often." September 5, 2019. https://www.nextgov.com/cybersecurity/2019/09/nsa-cyber-chief-wants-share-digital-threats-early-and-often/159673/.

———. "The Pentagon's Startup Outreach Office Is No Longer an Experiment." August 9, 2018. https://www.nextgov.com/cio-briefing/2018/08/pentagons-startup-outreach-office-no-longer-experiment/150408.

Cyberspace Solarium Commission. 2020. https://drive.google.com/file/d/1ryMCIL_dZ30QyjFqFkkf10MxIXJGT4yv/view.

Defense Personnel Security Research Center. *Espionage and Other Compromises of National Security* (Monterey, CA, 2009). https://fas.org/irp/eprint/esp-summ.pdf.

———. *Foreign Intelligence Threat Awareness Programs: A Review.* Prepared for the National Counterintelligence Policy Board (1998).

de la Merced, Michael J., and Peter Lattiman. "Appeals Court Limits Federal Law Used in Goldman Programmer Case." *New York Times,* April 12, 2012.

Denham, Hannah. "No Tech for ICE: Protesters Demand Amazon Cut Ties with Federal Immigration Enforcement." *Washington Post,* July 12, 2019.

Department of Commerce, Bureau of Export Administration. *Improvements Are Needed in Programs Designed to Protect Against the Transfer of Sensitive Technologies to Countries of Concern* (Washington, DC, 2000). https://www.oig.doc.gov/OIGPublications/IPE-12454.pdf.

———, Bureau of Industry and Security. *Deemed Export Controls May Not Stop the Transfer of Sensitive Technology to Foreign Nations in the U.S.* (Washington, DC, 2004).

———, Bureau of Industry and Security. *Fiscal Year 2019 President's Submission.* http://www.osec.doc.gov/bmi/budget/FY19CBJ/BIS_FY19_President%27s_Budget_FINAL.pdf.

Department of Defense. *National Security Implications of Climate-Related Risks and a Changing Climate* (Washington, DC, 2015). https://archive.defense.gov/pubs/150724-congressional-report-on-national-implications-of-climate-change.pdf.

———. *Report of the Defense Science Board Task Force on Basic Research* (Washington, DC: Office of the Undersecretary of Defense, for Acquisition, Technology and Logistics, 2012).

Department of Defense Industrial Security Review Committee. *Analysis of the Effectiveness of the Department of Defense Industrial Security Program and Recommendations for Program Improvement.* Report to the deputy undersecretary of defense for policy (Washington, DC, 1984). https://apps.dtic.mil/dtic/tr/fulltext/u2/a196076.pdf.

Department of Homeland Security. *Audit of Export Controls for Activities Related to China* (Washington, DC, 2006). https://www.governmentattic.org/2docs/4DHS-OIG_Reports_2005-2007.pdf.

———. *DHS Can Enhance Efforts to Protect Commercial Facilities from Terrorism and Physical Threats* (Washington, DC, 2020). https://www.oig.dhs.gov/sites/default/files/assets/2020-06/OIG-20-37-Jun20.pdf.

———. *DHS Made Limited Progress to Improve Information Sharing Under the Cybersecurity Act in Calendar Years 2017 and 2018* (Washington, DC, 2020). https://www.oig.dhs.gov/sites/default/files/assets/2020-09/OIG-20-74-Sep20.pdf.

———. *DHS Needs to Improve Cybersecurity Workforce Planning* (Washington, DC, 2019). https://www.oig.dhs.gov/sites/default/files/assets/2019-09/OIG-19-62-Sep19.pdf.

———. *Effectiveness of the Infrastructure Security Compliance Division's Management Practices to Implement the Chemical Facility Anti-Terrorism Standards Program* (Washington, DC, 2013). https://www.oig.dhs.gov/assets/Mgmt/2013/OIG_13-55_Mar13.pdf.

———. *Efforts to Identify Critical Infrastructure Assets and Systems* (Washington, DC, 2009). https://www.oig.dhs.gov/sites/default/files/assets/Mgmt/OIG_09-86_Jun09.pdf.

———. *Homeland Threat Assessment* (Washington, DC, 2020). https://www.dhs.gov/sites/default/files/publications/2020_10_06_homeland-threat-assessment.pdf.

———. *Implementing 9/11 Commission Recommendations: Progress Report* (Washington, DC, 2011). https://www.dhs.gov/xlibrary/assets/implementing-9-11-commission-report-progress-2011.pdf.

———. *Preventing and Defending Against Cyber Attacks* (Washington, DC, 2011). https://www.dhs.gov/xlibrary/assets/preventing-and-defending-against-cyber-attacks.pdf.

———. *Progress in Developing the National Asset Database* (Washington, DC, 2006).

———. *Protective Security Advisor Program Efforts to Build Effective Critical Infrastructure Partnerships: Oil and Gas Subsector* (Washington, DC, 2010). https://www.hsdl.org/?view&did=13056.

———. *Quadrennial Homeland Security Review 2010* (Washington, DC). https://www.dhs.gov/sites/default/files/publications/2010-qhsr-report.pdf.

———. *Quadrennial Homeland Security Review 2014* (Washington, DC). https://www.dhs.gov/sites/default/files/publications/2014-qhsr-final-508.pdf.

———. *Survey of the Information Analysis and Infrastructure Protection Directorate* (Washington, DC, 2004). https://www.oig.dhs.gov/sites/default/files/assets/Mgmt/OIG_SurveyIAIP_0204.pdf.

Department of Justice. *The Accomplishments of the U.S. Department of Justice, 2001–2009* (Washington, DC, 2010). https://www.justice.gov/sites/default/files/opa/legacy/2010/03/08/doj-accomplishments.pdf.

———. *Attorney General Guidelines for FBI Foreign Intelligence Collection and Foreign Counterintelligence Investigations* (Washington, DC, 1995). https://fas.org/irp/agency/doj/fbi/terrorismintel2.pdf.

———. *Audit of the Federal Bureau of Investigation's Implementation of Its Next Generation Cyber Initiative* (Washington, DC, 2015).

———. "Brooklyn Man Sentenced in Manhattan Federal Court to 15 Years in Prison for Providing Material Support to Al Qaeda." January 20, 2015. https://www.justice.gov/usao-sdny/pr/brooklyn-man-sentenced-manhattan-federal-court-15-years-prison-providing-material.

———. "California Resident Convicted of Conspiring to Illegally Export Fighter Jet Engines and an Unmanned Aerial Vehicle to China." June 9, 2016. https://www.justice.gov/usao-sdfl/pr/california-resident-convicted-conspiring-illegally-export-fighter-jet-engines-and-unmanned.

———. "Chinese Government Employee Charged in Manhattan Federal Court with Participating in Conspiracy to Fraudulently Obtain U.S. Visas." September 16, 2019. https://www.justice.gov/opa/pr/chinese-government-employee-charged-manhattan-federal-court-participating-conspiracy.

———. "Chinese Telecommunications Device Manufacturer and Its US Affiliate Indicted for Theft of Trade Secrets, Wire Fraud, and Obstruction of Justice." January 28, 2019. https://www.justice.gov/opa/pr/chinese-telecommunications-device-manufacturer-and-its-us-affiliate-indicted-theft-trade.

———. *The Federal Bureau of Investigation's Ability to Address the National Security Cyber Intrusion Threat* (Washington, DC, 2011).

———. *Follow-Up Audit of the Federal Bureau of Investigation's Efforts to Hire, Train, and Retain Intelligence Analysts* (Washington, DC, 2007). https://oig.justice.gov/reports/FBI/a0730/findings.htm.

———. "Former Dow Research Scientist Sentenced to 60 Months in Prison for Stealing Trade Secrets and Perjury." January 13, 2012. https://www.justice.gov/opa/pr/former-dow-research-scientist-sentenced-60-months-prison-stealing-trade-secrets-and-perjury.

———. "Former Emory University Professor and Chinese Thousand Talents Participant Convicted and Sentenced for Filing a False Tax Return." May 11, 2020. https://www.justice.gov/usao-ndga/pr/former-emory-university-professor-and-chinese-thousand-talents-participant-convicted.

———. "Former Russian Nuclear Energy Official Sentenced to 48 Months in Prison for Money Laundering Conspiracy Involving Foreign Corrupt Practices Act Violations." December 15, 2015. https://www.justice.gov/opa/pr/former-russian-nuclear-energy-official-sentenced-48-months-prison-money-laundering-conspiracy.

———. "Harvard University Professor Indicted on False Statement Charges." June 9, 2020. https://www.justice.gov/opa/pr/harvard-university-professor-indicted-false-statement-charges.

———. "Manhattan U.S. Attorney Announces Charges Against Seven Iranians for Conducting Coordinated Campaign of Cyber Attacks Against U.S. Financial Sector on Behalf of Islamic Revolutionary Guard Corps–Sponsored Entities." March 24, 2016. https://www.justice.gov/usao-sdny/pr/manhattan-us-attorney-announces-charges-against-seven-iranians-conducting-coordinated.

———. "NOAA National Weather Service Employee Indicted for Allegedly Downloading Restricted Government Files." October 20, 2014. https://www.justice.gov/usao-sdoh/pr/noaa-national-weather-service-employee-indicted-allegedly-downloading-restricted.

———. *Review of Four FISA Applications and Other Aspects of the FBI's Crossfire Hurricane Investigation* (Washington, DC, 2019). https://www.justice.gov/storage/120919-examination.pdf.

———. *A Review of the Federal Bureau of Investigation's Counterterrorism Program: Threat Assessment, Strategic Planning, and Resource Management* (Washington, DC, 2002). https://fas.org/irp/agency/doj/oig/fbi02sum.html.

———. *A Review of Various Actions by the Federal Bureau of Investigation and Department of Justice in Advance of the 2016 Election* (Washington, DC, 2018). https://www.oversight.gov/sites/default/files/oig-reports/o1804.pdf.

———. "Seven Iranians Working for Islamic Revolutionary Guard-Corps Affiliated Entities Charged for Conducting Coordinated Campaign of Cyber Attacks Against U.S. Financial Sector." March 24, 2016. https://www.justice.gov/opa/pr/seven-iranians-working-islamic-revolutionary-guard-corps-affiliated-entities-charged.

———. "Singaporean National Pleads Guilty to Acting in the United States as an Illegal Agent of Chinese Intelligence." July 24, 2020. https://www.justice.gov/opa/pr/singaporean-national-pleads-guilty-acting-united-states-illegal-agent-chinese-intelligence.

———. "Summary of Major U.S. Export Enforcement, Economic Espionage, Trade Secret, and Embargo-Related Criminal Cases." January 2008–present. https://www.hsdl.org/?view&did=825549.

———. "Three British Nationals Indicted on Charges of Conspiring to Use Weapons of Mass Destruction, Providing Material Support to Terrorist." April 12, 2005 https://www.justice.gov/archive/opa/pr/2005/April/05_crm_180.htm.

———. "Two Chinese Hackers Working with the Ministry of State Security Charged with Global Computer Intrusion Campaign Targeting Intellectual Property and Confidential Business Information Including COVID-19 Research." July 21, 2020. https://www.justice.gov/opa/pr/two-chinese-hackers-working-ministry-state-security-charged-global-computer-intrusion.

———. "Update to Sony Investigation." December 19, 2014. https://www.justice.gov/opa/pr/update-sony-investigation.

———. "U.S. Charges Five Chinese Military Hackers for Cyber Espionage Against U.S. Corporations and a Labor Organization for Commercial Advantage." May 19, 2014. https://www.justice.gov/opa/pr/us-charges-five-chinese-military-hackers-cyber-espionage-against-us-corporations-and-labor.

Department of State. *The Neutrality Acts, 1930s* (Washington, DC, undated) https://history.state.gov/milestones/1921-1936/neutrality-acts.

Department of the Treasury. *CFIUS Reform: The Foreign Investment and National Security Act of 2007* (Washington, DC, 2008). https://www.treasury.gov/resource-center/international/foreign-investment/Documents/Summary-FINSA.pdf.

———. *Summary of the Foreign Investment Risk Review Modernization Act of 2018* (Washington, DC, undated). https://www.treasury.gov/resource-center/international/Documents/Summary-of-FIRRMA.pdf.

Dockey, Stephen. "Chinese Company Will Sell Wind Farm Assets in CFIUS Settlement." *Wall Street Journal,* November 4, 2015.

The Economist. "Keep Your T-Bonds, We'll Take the Bank, Sovereign Wealth Funds." July 28, 2007.

———. "Spooks for Hire: America's Intelligence Agencies Find Creative Ways to Compete for Talent." March 3, 2018.

Ellis, Ryan. *Letters, Power Lines, and Other Dangerous Things: The Politics of Infrastructure Security* (Cambridge: Massachusetts Institute of Technology Press, 2020).

Federal Bureau of Investigation. *Annual Report 1966* (Washington, DC). https://ia801005.us.archive.org/29/items/FBIAnnualReport1966/FBI%20Annual%20Report%201966.pdf.

———. *Annual Report 1969* (Washington, DC). https://ia803109.us.archive.org/20/items/FBIAnnualReport1969/FBI%20Annual%20Report%201969.pdf.

———. *Annual Report 1971* (Washington, DC). https://ia803106.us.archive.org/23/items/FBIAnnualReport1971/FBI%20Annual%20Report%201971.pdf.

———. Bureau Bulletin no. 17: (D) Plant Surveys, 5-15-41. https://ia802702.us.archive.org/2/items/foia_FBI_Confidential_Informants-HQ-1a/FBI_Confidential_Informants-HQ-1a.pdf.

———. "Carnivore/DCS 1000 Report to Congress." February 24, 2003. https://www.epic.org/privacy/carnivore/2002_report.pdf.

———. *Chinese Communist Intelligence Activities in the United States* (Washington, DC, 1954). https://ia801908.us.archive.org/35/items/FBIPRCSpying/fbi-prc-spying.pdf.

———. *Chinese Talent Programs 2015* (Washington, DC). https://webcache.googleusercontent.com/search?q=cache:d58XN_T2cpQJ:https://compliance.fiu.edu/documents/SPIN%2520-%2520Chinese%2520Talent%2520Program.pdf+&cd=1&hl=en&ct=clnk&gl=us.

———. *Exposé of Soviet Espionage 1960* (Washington, DC). https://www.cia.gov/library/readingroom/docs/CIA-RDP65B00383R000200040033-2.pdf.

———. *The FBI's Counterterrorism Program Since September 2001* (Washington, DC, 2004).

———. *History of the S.I.S. Division.* Vol. 1. (Washington, DC, undated).

———. *History of the S.I.S. Division.* Vol. 2, *Accomplishment Argentina-Japan* (Washington, DC, undated).

———. *The Sabotage Plans and Potential of the Communist Party, USA* (Washington, DC, 1953). https://ia800809.us.archive.org/2/items/TheSabotagePlansAndPotentialOfTheCPUSA/The%20Sabotage%20Plans%20and%20Potential%20of%20the%20Communist%20Party%2C%20U.S.A%20%281953%29%20%5B1953-10%5D.pdf.

———. *Soviet Intelligence Targets in the United States, 1946–1953* (Washington, DC, 1953). https://www.governmentattic.org/2docs/FBI_Monograph_Soviet-Targets-US_1953.pdf.

———. *Soviet Military, Naval, and Air Representatives in the United States* (Washington, DC, 1955). https://ia800704.us.archive.org/12/items/SovietMilitaryNavalAndAir

RepresentativesInTheUnitedStates/Soviet%20Military%2C%20Naval%2C%20and%20Air%20Representatives%20in%20the%20United%20States.pdf.
———. *Terrorism 2002–2005* (Washington, DC, undated). https://www.fbi.gov/stats-services/publications/terrorism-2002-2005.
Federation of American Scientists. *Awareness of National Security Issues and Response* (1998). https://fas.org/irp/ops/ci/ansir.htm.
Feng, Ashley. "We Can't Tell If Chinese Firms Work for the Party." *Foreign Policy,* February 7, 2019. https://foreignpolicy.com/2019/02/07/we-cant-tell-if-chinese-firms-work-for-the-party.
Fialka, John. *War by Other Means* (New York: Norton, 1997).
Finklea, Kristin. *Renewed Crypto Wars* (Washington, DC: Congressional Research Service, 2016). https://fas.org/sgp/crs/misc/IN10440.pdf.
Fox, Chris, and Leo Kelion. "Coronavirus: Russian Spies Target Covid-19 Vaccine Research." *BBC News,* July 16, 2020. https://www.bbc.com/news/technology-53429506.
General Accounting Office. *Combating Terrorism: FBI's Use of Federal Funds for Counterterrorism-Related Activities (FYs 1995–1998)* (Washington, DC, 1998).
———. *Container Security: A Flexible Staffing Model and Minimum Equipment Requirements Would Improve Overseas Targeting and Inspection Efforts* (Washington, DC, 2005). https://www.gao.gov/new.items/d05557.pdf.
———. *Critical Infrastructure Protection: Challenges for Selected Agencies and Industry Sectors* (Washington, DC, 2003). https://www.gao.gov/assets/240/237449.pdf.
———. *Critical Infrastructure Protection: Federal Efforts Require a More Coordinated and Comprehensive Approach for Protecting Information Systems* (Washington, DC, 2002). https://www.gao.gov/new.items/d02474.pdf.
———. *Critical Infrastructure Protection: Significant Challenges in Developing Analysis, Warning, and Response Capabilities* (Washington, DC, 2001). https://www.gao.gov/assets/110/108944.pdf.
———. *Critical Infrastructure Protection: Significant Challenges in Developing National Capabilities* (Washington, DC, 2001). https://www.gao.gov/assets/160/157052.pdf.
———. *Economic Espionage: The Threat to US Industry* (Washington, DC, 1992). https://www.gao.gov/assets/110/104477.pdf.
———. *Export Controls: Actions Needed to Improve Enforcement* (Washington, DC, 1993). https://www.gao.gov/assets/160/154080.pdf.
———. *Foreign Investment* (Washington, DC, 1995). https://www.gao.gov/assets/230/221994.pdf.
———. *Information Superhighway: An Overview of Technology Challenges* (Washington, DC, 1995). https://www.govinfo.gov/content/pkg/GAOREPORTS-AIMD-95-23/pdf/GAOREPORTS-AIMD-95-23.pdf.
———. *US Munitions Export Controls Need Improvement* (Washington, DC, 1979). https://www.gao.gov/assets/130/126431.pdf.
———. *What Would Be the Impact of Raising or Repealing the Commercial Arms Sales Ceiling?* (Washington, DC, 1980). https://www.gao.gov/assets/130/128529.pdf.
Ghaffary, Shirin. "Google Employees Are Demanding an End to the Company's Work with Agencies Like CBP and ICE." August 14, 2019. https://www.vox.com/2019/8/14/20805562/human-rights-concerns-google-employees-petition-cbp-ice.
Golden, Daniel. *Spy Schools* (New York: Holt, 2017).
Government Accountability Office. *Critical Infrastructure Protection: Actions Need to Enhance DHS Oversight of Cybersecurity at High-Risk Chemical Facilities* (Washington, DC, 2020). https://www.gao.gov/assets/710/706972.pdf.
———. *Critical Infrastructure Protection: Department of Homeland Security Faces Challenges in Fulfilling Cybersecurity Responsibilities* (Washington, DC, 2005). https://www.gao.gov/new.items/d05434.pdf.

———. *Critical Infrastructure Protection: Key Private and Public Cyber Expectations Need to Be Consistently Addressed* (Washington, DC, 2010). https://www.gao.gov/assets/310/307222.pdf.

———. *Critical Infrastructure Protection: Observations on DHS Efforts to Implement and Manage Its Chemical Security Program* (Washington, DC, 2014). https://www.gao.gov/assets/670/663170.pdf.

———. *DHS Intelligence Analysis: Additional Actions Needed to Address Analytic Priorities and Workforce Challenges* (Washington, DC, 2014). https://www.gao.gov/assets/670/663794.pdf.

———. *DHS List of Priority Assets Needs to Be Validated and Reported to Congress* (Washington, DC, 2013). https://www.gao.gov/assets/660/653300.pdf.

———. *Improving Federal Recruiting and Hiring Efforts* (Washington, DC, 2019). https://www.gao.gov/assets/710/700657.pdf.

———. *Industrial Security: DOD Cannot Ensure Its Oversight of Contractors Under Foreign Influence Is Sufficient* (Washington, DC, 2005). https://www.gao.gov/assets/250/247113.html.

———. *Information Sharing: Federal Agencies Are Helping Fusion Centers Build and Protect Privacy, but Could Better Manage Results* (Washington, DC, 2010). https://www.gao.gov/assets/320/310268.pdf.

———. *Public Companies: Disclosure of Environmental, Social, and Governance Factors and Options to Enhance Them* (Washington, DC, 2020). https://www.gao.gov/assets/710/707949.pdf.

Greenhouse, Linda. "F.B.I. Defends Library Monitoring Program." *New York Times,* July 14, 1988.

Groll, Elias. "FBI Rolls Out Red Scare Film to Highlight Threat of Economic Espionage." *Foreign Policy,* July 23, 2015. https://foreignpolicy.com/2015/07/23/fbi-rolls-out-red-scare-film-to-highlight-threat-of-economic-espionage.

Heckman, Jory. "Launch of DHS Cyber Agency 'More of a Groundbreaking Than a Ribbon Cutting.'" *Federal News Radio,* November 16, 2018. https://federalnewsnetwork.com/cybersecurity/2018/11/launch-of-dhs-cyber-agency-more-of-a-groundbreaking-than-a-ribbon-cutting.

Hentoff, Nat. "The FBI in the Library." *Washington Post,* July 23, 1988. https://www.washingtonpost.com/archive/opinions/1988/07/23/the-fbi-in-the-library/f0ea90c7-4c52-46c0-a546-e567220fe0a1.

Hicks, Josh. "Homeland Security Is Laying Roots in Silicon Valley, and You Might Not Like Its Reasons." *Washington Post,* April 22, 2015.

Jackson, James K. *The Committee on Foreign Investment in the United States* (Washington, DC: Congressional Research Service, 2014).

———. *The Exon-Florio National Security Test for Foreign Investment* (Washington, DC, Congressional Research Service, 2013).

Jensen, Joan M. *Army Surveillance in America, 1775–1980* (New Haven: Yale University Press, 1991).

Johnson, David. "F.B.I. Is Proposing a Special Division for Hunting Spies." *New York Times,* June 26, 1999. https://www.nytimes.com/1999/06/26/world/fbi-is-proposing-a-special-division-for-hunting-spies.html.

Jones, Seth G., Catrina Doxsee, Nicholas Harrington, Grace Hwang, and James Suber. *The War Comes Home: The Evolution of Domestic Terrorism in the United States* (Washington, DC: Center for Strategic and International Studies, 2020). https://csis-website-prod.s3.amazonaws.com/s3fs-public/publication/201021_Jones_War_Comes_Home_v2.pdf.

Kania, Elsa B. *China's Threat to American Government and Private Sector Research and Innovation Leadership.* Testimony before the House Permanent Select Committee on Intelligence, July 19, 2018. https://docs.house.gov/meetings/IG/IG00/20180719/108561/HHRG-115-IG00-Wstate-KaniaE-20180719.pdf.

Kelly, John. "Why St. Elizabeth? More Musing on Washington's Hospital for the Mentally Ill." *Washington Post,* July 6, 2019. https://www.washingtonpost.com/local/why-st-elizabeth-more-musing-on-washingtons-hospital-for-the-mentally-ill/2019/07/05/f18c38be-9f78-11e9-b27f-ed2942f73d70_story.html.

Korn, Melissa. "Microsoft Brings U.S. and China Universities Together." *Wall Street Journal,* June 19, 2015.

Krige, John. "Regulating International Knowledge Exchange: The National Security State and the American Research University from the 1950s to Today." *Technology and Culture* 60 (January 2019): 252–277.

Kurland, Kevin J. *End Use Monitoring and Effective Export Compliance* (Washington, DC: Department of Commerce, 2016). https://www.bis.doc.gov/index.php/documents/pdfs/1593-end-user-verification-kurland/file.

Lamb, Gregory M. "Leaks Flow East—and West; US Industry and High-Tech Spies." *Christian Science Monitor,* December 28, 1982.

Levinson, Charles. "Comey: FBI 'Grappling with Hiring Policy Concerning Marijuana.'" *Wall Street Journal,* May 20, 2014. https://www.wsj.com/articles/BL-LB-48089.

Levy, Steven. "Battle of the Clipper Chip." *New York Times,* June 12, 1994.

Lichtblau, Eric. "Judge Tells Apple to Help Unlock iPhone Used by San Bernadino Gunman." *New York Times,* February 16, 2016. https://www.nytimes.com/2016/02/17/us/judge-tells-apple-to-help-unlock-san-bernardino-gunmans-iphone.html?searchResultPosition=5.

Lichtblau, Eric, and Katie Benner. "Apple Fights Order to Unlock San Bernadino Gunman's iPhone." *New York Times,* February 17, 2016. https://www.nytimes.com/2016/02/18/technology/apple-timothy-cook-fbi-san-bernardino.html?searchResultPosition=6.

Markoff, John. "Electronics Plan Aims to Balance Government Access with Privacy." *New York Times,* April 16, 1993.

———. "Industry Defies U.S. on Data Encryption." *New York Times,* January 14, 1994.

Marks, Joseph. "Only 6 Non-Federal Groups Share Cyber Threat Info with Homeland Security." June 27, 2018. https://www.nextgov.com/cybersecurity/2018/06/only-6-non-federal-groups-share-cyber-threat-info-homeland-security/149343.

McAllister, Bill. "FBI to Limit Probes of Library Users; Program to Detect Foreign Agents Is Altered to Guard Patron Privacy." *Washington Post,* November 15, 1988.

McFadden, Robert. "F.B.I. in New York Asks Librarians' Aid in Reporting on Spies." *New York Times,* September 18, 1987.

McGregor, Richard. "How the State Runs Business in China." *The Guardian,* July 25, 2019. https://www.theguardian.com/world/2019/jul/25/china-business-xi-jinping-communist-party-state-private-enterprise-huawei.

Metz, Cade, Erin Griffith, and Kate Conger. "What's a Palantir? The Tech Industry's Next Big IPO." *New York Times,* August 26, 2020. https://www.nytimes.com/2020/08/26/technology/palantir-ipo.html?searchResultPosition=4.

Mims, Christopher. "Google Outgrows Its Youthful Ideals." *Wall Street Journal,* August 17, 2018. https://www.wsj.com/articles/googles-risky-pragmatism-1534427069.

Morgan, Daniel. *Research and Development in the Department of Homeland Security* (Washington, DC: Congressional Research Service, 2003).

Nakashima, Ellen. "Chinese Breach Data of 4 Million Federal Workers." *Washington Post,* June 4, 2015. https://www.washingtonpost.com/world/national-security/chinese-hackers-breach-federal-governments-personnel-office/2015/06/04/889c0e52-0af7-11e5-95fd-d580f1c5d44e_story.html.

Nakashima, Ellen, and Barton Gellman. "As Encryption Spreads, U.S. Grapples with Clash Between Privacy, Security." *Washington Post,* April 10, 2015.

National Counterintelligence and Security Center. *National Counterintelligence Strategy of the United States of America, 2020–2022* (Washington, DC, undated). https://www.dni.gov/files/NCSC/documents/features/20200205-National_CI_Strategy_2020_2022.pdf.

National Counterintelligence Center. *Annual Report to Congress on Foreign Economic Collection and Industrial Espionage, 1995* (Washington, DC).

———. *Annual Report to Congress 2000* (Washington, DC). https://fas.org/irp/ops/ci/docs/fy00.htm.

National Counterintelligence Executive. *Annual Report to Congress on Foreign Economic Collection and Industrial Espionage, 2003* (Washington, DC). https://www.hsdl.org/?view&did=464996.

———. *Annual Report to Congress on Foreign Economic Collection and Industrial Espionage, 2004* (Washington, DC).

National Security Agency. *American Cryptology During the Cold War, 1945–1989.* Book 1, *The Struggle for Centralization, 1945–1960* (Ft. Meade, MD, 1995).

———. *American Cryptology During the Cold War, 1945–1989.* Book 3, *Retrenchment and Reform, 1972–1980* (1998). https://www.nsa.gov/Portals/70/documents/news-features/declassified-documents/cryptologic-histories/cold_war_iii.pdf.

Office of the National Counterintelligence Executive. *Annual Report to Congress on Foreign Economic Collection and Industrial Espionage, 2004* (2005).

———. *Foreign Spies Stealing US Economic Secrets in Cyberspace* (2011).

Overend, William. "FBI Also a Resident of S.F. Neighborhood: Soviet Consulate: Cow Hollow Intrigue." *Los Angeles Times,* July 28, 1985. https://www.latimes.com/archives/la-xpm-1985-07-28-mn-5383-story.html.

Paletta, Damian, Keith Johnson, and Sudeep Reddy. "Obama Blocks Chinese Firm from Wind-Farm Projects." *Wall Street Journal,* September 29, 2012.

Partnership for Public Service. *Best Places to Work: Agency Rankings* (Washington, DC, 2020).

———. *Cyber In-Security* (Washington, DC, 2015). https://ourpublicservice.org/wp-content/uploads/2015/04/5a6ae63596cc99f7039b9e409c70891a-1429280031.pdf.

Pedlow, Gregory W., and Donald E. Welzenbach. *The CIA and the U-2 Program, 1954–1974* (Central Intelligence Agency, 1998). https://www.cia.gov/library/center-for-the-study-of-intelligence/csi-publications/books-and-monographs/the-cia-and-the-u-2-program-1954-1974/u2.pdf.

Perlroth, Nicole. "Accused of Spying for China, Until She Wasn't." *New York Times,* May 9, 2015. https://www.nytimes.com/2015/05/10/business/accused-of-spying-for-china-until-she-wasnt.html.

———. "Two from China Are Charged in 2014 Anthem Data Breach." *New York Times,* May 9, 2019. https://www.nytimes.com/2019/05/09/technology/anthem-hack-indicted-breach.html.

Piller, Charles. "Federal Agency Created to Combat the Rise in Cyber-Crime Is Viewed with Distrust by Firms It Is Supposed to Protect." *Los Angeles Times,* March 5, 2000. https://www.latimes.com/archives/la-xpm-2000-mar-05-mn-5606-story.html.

Powers, Richard Gid. *Broken: The Troubled Past and Uncertain Future of the FBI* (New York: Free Press, 2004).

———. *G-Men* (Carbondale: Southern Illinois University Press, 1983).

Prados, John. *The Family Jewels* (Austin: University of Texas Press, 2013).

President's Foreign Intelligence Advisory Board. *Science at Its Best; Security at Its Worst: A Report on Security Problems at the US Department of Energy* (1999). https://www.energy.gov/sites/prod/files/cioprod/documents/pfiab-doe.pdf.

Rafalko, Frank, ed. *A Counterintelligence Reader: Post World War II to Closing the 20th Century* (Washington, DC: National Counterintelligence Executive, 2004).

Rando, Mark A. *The Department of Homeland Security Intelligence Enterprise: Operational Overview and Oversight Challenges for Congress* (Washington, DC: Congressional Research Service, 2010). https://fas.org/sgp/crs/homesec/R40602.pdf.

Rangan, K. Katsuri, Lisa Chase, and Sohel Karim. "The Truth About CSR." *Harvard Business Review,* January–February 2016. https://hbr.org/2015/01/the-truth-about-csr.

Reuters. "China's Kunlun Tech Agrees to U.S. Demand to Sell Grindr Gay Dating App." May 13, 2019. https://www.reuters.com/article/us-grindr-m-a-beijingkunlun/chinas-kunlun-tech-agrees-to-u-s-demand-to-sell-grindr-gay-dating-app-idUSKCN1SJ28N.

Riebling, Mark. *Wedge: The Secret War Between the FBI and CIA* (New York: Knopf, 1994).

Sanger, David E. "Grindr Is Owned by a Chinese Firm, and the US Is Trying to Force It to Sell." *New York Times,* March 28, 2019. https://www.nytimes.com/2019/03/28/us/politics/grindr-china-national-security.html.

———. "Russian Hackers Broke into Federal Agencies, U.S. Officials Suspect." *New York Times,* December 13, 2020. https://www.nytimes.com/2020/12/13/us/politics/russian-hackers-us-government-treasury-commerce.html?action=click&module=Top%20Stories&pgtype=Homepage.

Sanger, David E., and Nicole Perlroth. "FireEye, a Top Cybersecurity Firm, Says It Was Hacked by a Nation-State." *New York Times,* December 8, 2020. https://www.nytimes.com/2020/12/08/technology/fireeye-hacked-russians.html.

Savage, Charlie. "U.S. Tries to Make It Easier to Wiretap the Internet." *New York Times,* September 27, 2010.

Sciolino, Elaine. "Zakharov Charges a 'Setup' by F.B.I." *New York Times,* September 17, 1986.

Second Annual Report of the National Munitions Control Board for the Year Ending November 30, 1937. https://tinyurl.com/4s54xnjv.

Steinbrock, Dan. *The Challenges for America's Defense Innovation* (Washington, DC: Information Technology and Innovation Foundation, 2014).

Suro, Roberto. "New FBI Spy Unit Gets Reno's Approval." *Washington Post,* June 26, 1999. https://www.washingtonpost.com/archive/politics/1999/06/26/new-fbi-spy-unit-gets-renos-approval/63f6f03b-e6fe-485c-b4aa-493120b2713c.

Thornton, Mary. "Customs Fights KGB on High-Tech Thefts." *Washington Post,* February 5, 1986. https://www.cia.gov/library/readingroom/docs/CIA-RDP90-00965R000706710007-7.pdf.

Timberg, Craig. "Apple Will No Longer Unlock Most iPhones, iPads for Police, Even with Search Warrants." *Washington Post,* September 18, 2014. https://www.washingtonpost.com/business/technology/2014/09/17/2612af58-3ed2-11e4-b03f-de718edeb92f_story.html.

Treverton, Gregory F. *The Next Steps in Reshaping Intelligence* (Santa Monica: RAND, 2005).

Treverton, Gregory F., and Seth G. Jones. *Measuring National Power* (Santa Monica: RAND, 2005).

Tromblay, Darren E. *The FBI Abroad* (Boulder: Lynne Rienner, 2020).

———. *Protecting Partners or Preserving Fiefdoms?* (Washington, DC: Information Technology and Innovation Foundation, 2017). http://www2.itif.org/2017-counterintelligence-outreach-industry.pdf.

Tromblay, Darren, and Robert Spelbrink. *Securing U.S. Innovation: The Challenge of Preserving a Competitive Advantage in the Creation of Knowledge* (Lanham: Rowman and Littlefield, 2016).

Tucker, Patrick. "Pentagon Shakes Up Silicon Valley Outreach." May 11, 2016. https://www.defenseone.com/technology/2016/05/pentagon-shakes-silicon-valley-outreach/128198.

US-China Economic and Security Review Commission. *China's Pursuit of Next Frontier Tech: Computing, Robotics, and Biotechnology* (Washington, DC, 2017). https://www.uscc.gov/sites/default/files/transcripts/March%20Transcript.pdf.

———. *How Chinese Companies Facilitate Technology Transfer from the United States* (Washington, DC, 2019). https://www.uscc.gov/sites/default/files/Research/How%20Chinese%20Companies%20Facilitate%20Tech%20Transfer%20from%20the%20US.pdf.

———. *2009 Report to Congress* (Washington, DC).

———. *2019 Report to Congress* (Washington, DC). https://www.uscc.gov/sites/default/files/2019-11/2019%20Annual%20Report%20to%20Congress.pdf.

US Congress. *The Activities of the Federal Bureau of Investigation.* Pt 2. Before the Subcommittee on Crime of the Committee on the Judiciary, House of Representatives, 105th Congress (Washington, DC, 1997).

———. *Appropriations, Department of Justice, 1926.* Before a subcommittee of House Committee on Appropriations, 68th Congress (Washington, DC, 1924).

———. *The Chemical Facilities Anti-Terrorism Standards Program: Addressing Its Challenges and Finding a Way Forward.* Before the Committee on Homeland Security, 112th Congress (Washington, DC, 2012). https://www.govinfo.gov/content/pkg/CHRG-112hhrg76601/pdf/CHRG-112hhrg76601.pdf.

———. *The Chemical Facility Anti-Terrorism Standards Program.* Before the House Committee on Homeland Security, 116th Congress (Washington, DC, February 27, 2019). https://homeland.house.gov/imo/media/doc/Testimony-Wulf.pdf.

———. *Chemical Facility Anti-Terrorism Standards (CFATS) Program: A Progress Update.* Before the Committee on Energy and Commerce, House of Representatives, 113th Congress (Washington, DC, 2013). https://www.govinfo.gov/content/pkg/CHRG-113hhrg80377/pdf/CHRG-113hhrg80377.pdf.

———. *China's Pursuit of Emerging and Exponential Technologies.* Before the Committee on Armed Services, 115th Congress (Washington, DC, 2018). https://www.govinfo.gov/content/pkg/CHRG-115hhrg28966/pdf/CHRG-115hhrg28966.pdf.

———. *Commerce, Justice, Science, and Related Agencies Appropriations for 2010.* Pt. 1. Before a subcommittee of the Committee on Appropriations, House of Representatives, 111th Congress (Washington, DC, 2009).

———. *Commerce, Justice, Science, and Related Agencies Appropriations for 2014.* Pt. 2B. Before a subcommittee of the Committee on Appropriations, House of Representatives, 113th Congress (Washington, DC, 2013).

———. *Commerce, Justice, Science, and Related Agencies Appropriations for 2016.* Pt. 2B. Before a subcommittee of the Committee on Appropriations, House of Representatives, 114th Congress (Washington, DC, 2015).

———. *Commerce, Justice, Science, and Related Agencies Appropriations for 2017.* Before a subcommittee of the Committee on Appropriations, House of Representatives, 114th Congress (Washington, DC, 2016).

———. *Corporate and Industrial Espionage and Their Effects on American Competitiveness.* Before the Committee on International Relations, House of Representatives, 106th Congress (Washington, DC, 2000). https://www.govinfo.gov/content/pkg/CHRG-106hhrg68684/pdf/CHRG-106hhrg68684.pdf.

———. *Counterintelligence and National Security Information.* Before a subcommittee of the Committee on Government Operations, House of Representatives, 99th Congress (Washington, DC, 1985).

———. *Cyber Security: U.S. Vulnerability and Preparedness.* Before the Committee on Science, House of Representatives, 109th Congress (Washington, DC, 2005). https://www.govinfo.gov/content/pkg/CHRG-109hhrg23332/pdf/CHRG-109hhrg23332.pdf.

———. *Cybersecurity: Protecting America's Critical Infrastructure, Economy, and Consumers.* Before the Committee on Energy and Commerce, House of Representatives, 109th Congress (Washington, DC, 2006). https://www.govinfo.gov/content/pkg/CHRG-109hhrg31464/pdf/CHRG-109hhrg31464.pdf.

———. *Cybersecurity Recommendations for the Next Administration.* Before the Committee on Homeland Security, House of Representatives, 110th Congress (Washington, DC, 2008). https://www.govinfo.gov/content/pkg/CHRG-110hhrg48089/pdf/CHRG-110hhrg48089.pdf.

———. *Defending Our Democracy: Building Partnerships to Protect America's Elections.* Before the House Committee on Homeland Security (Washington, DC, 2019). https://homeland.house.gov/imo/media/doc/Testimony-Krebs.pdf.

———. *Department of Homeland Security Appropriations for 2005.* Pt. 1. Before a subcommittee of the Committee on Appropriations, House of Representatives, 108th Congress (Washington, DC, 2004).

———. *Department of Homeland Security Appropriations for 2012.* Pt. 4. Before a subcommittee of the Committee on Appropriations, House of Representatives, 112th Congress (Washington, DC, 2012).

———. *Department of Homeland Security Appropriations for 2013.* Pt. 5. Before the Committee on Appropriations, House of Representatives, 112th Congress (Washington, DC, 2012). https://www.govinfo.gov/content/pkg/CHRG-112hhrg77757/pdf/CHRG-112hhrg77757.pdf.

———. *Department of Homeland Security Appropriations for 2017.* Pt. 1C. Before the Committee on Appropriations, House of Representatives, 114th Congress (Washington, DC, 2016).

———. *Department of Justice Appropriation Bill for 1933.* Before a subcommittee of the House Committee on Appropriations, 72nd Congress (Washington, DC, 1932).

———. *Department of Justice Appropriation Bill for 1937.* Before the House Committee on Appropriations, 74th Congress (Washington, DC, 1936).

———. *Department of Justice Appropriation Bill for 1938.* Before a subcommittee of the Committee on Appropriations, House of Representatives, 75th Congress (Washington, DC, 1937).

———. *Department of Justice Appropriation Bill for 1941.* Before a subcommittee of the Committee on Appropriations, House of Representatives, 76th Congress (Washington, DC, 1940).

———. *Department of Justice Appropriation Bill for 1942.* Before a subcommittee of the Committee on Appropriations, House of Representatives, 77th Congress (Washington, DC, 1941).

———. *Department of Justice Appropriation Bill for 1943.* Before the Committee on Appropriations, House of Representatives, 76th Congress (Washington, DC, 1942).

———. *Departments of Commerce, Justice, and State, the Judiciary, and Related Agencies Appropriations for 1983.* Pt. 7. Before a subcommittee of the Committee on Appropriations, House of Representatives, 97th Congress (Washington, DC, 1982).

———. *Departments of Commerce, Justice, and State, the Judiciary, and Related Agencies Appropriations for 1984.* Pt. 6. Before a subcommittee of the Committee on Appropriations, House of Representatives, 98th Congress (Washington, DC, 1983).

———. *Departments of Commerce, Justice, and State, the Judiciary, and Related Agencies Appropriations for 1985.* Pt. 8. Before a subcommittee of the Committee on Appropriations, House of Representatives, 98th Congress (Washington, DC, 1984).

———. *Departments of Commerce, Justice, and State, the Judiciary, and Related Agencies Appropriations for 1990.* Pt. 2. Before a subcommittee of the Committee on Appropriations, House of Representatives, 101st Congress (Washington, DC, 1989).

———. *Departments of Commerce, Justice, and State, the Judiciary, and Related Agencies Appropriations for 1991.* Before a subcommittee of the Committee on Appropriations, House of Representatives, 101st Congress (Washington, DC, 1990).

———. *Departments of Commerce, Justice, and State, the Judiciary, and Related Agencies Appropriations for 1992.* Pt. 2. Before a subcommittee of the Committee on Appropriations, House of Representatives, 102nd Congress (Washington, DC, 1991).

———. *Departments of Commerce, Justice, and State, the Judiciary, and Related Agencies Appropriations for 1993.* Pt. 2B. Before the Committee on Appropriations, House of Representatives, 102nd Congress (Washington, DC, 1992).

———. *Departments of Commerce, Justice, and State, the Judiciary, and Related Agencies Appropriations for 1994.* Before the Committee on Appropriations, House of Representatives, 103rd Congress (Washington, DC, 1993).

———. *Departments of Commerce, Justice, and State, the Judiciary, and Related Agencies Appropriations for 1995.* Pt. 2A. Before a subcommittee of the Committee on Appropriations, House of Representatives, 103rd Congress (Washington, DC, 1994).

———. *Departments of Commerce, Justice, and State, the Judiciary, and Related Agencies Appropriations for 1996.* Pt. 2. Before a subcommittee of the Committee on Appropriations, House of Representatives, 104th Congress (Washington, DC, 1995).

———. *Departments of Commerce, Justice, and State, the Judiciary, and Related Agencies Appropriations for 1998.* Before a subcommittee of the Committee on Appropriations, House of Representatives, 105th Congress (Washington, DC, 1997).

———. *Departments of Commerce, Justice, and State, the Judiciary, and Related Agencies Appropriations for 1999.* Pt. 6. Before a subcommittee of the Committee on Appropriations, House of Representatives, 105th Congress (Washington, DC, 1998).

———. *Departments of Commerce, Justice, and State, the Judiciary, and Related Agencies Appropriations for 2000.* Pt. 6. Before the Committee on Appropriations, House of Representatives, 106th Congress (Washington, DC, 1999).

———. *Departments of Commerce, Justice, and State, the Judiciary, and Related Agencies Appropriations for 2001.* Pt. 2. Before a subcommittee of the Committee on Appropriations, House of Representatives, 106th Congress (Washington, DC, 2000).

———. *Departments of Commerce, Justice, and State, the Judiciary, and Related Agencies Appropriations for 2001.* Pt. 6. Before a subcommittee of the Committee on Appropriations, House of Representatives, 106th Congress (Washington, DC, 2000).

———. *Departments of Commerce, Justice, and State, the Judiciary, and Related Agencies Appropriations for 2002.* Pt. 6. Before the Committee on Appropriations, House of Representatives, 107th Congress (Washington, DC, 2001).

———. *Departments of Commerce, Justice, and State, the Judiciary, and Related Agencies Appropriations for 2004.* Pt. 10. Before a subcommittee of the Committee on Appropriations, House of Representatives, 108th Congress (Washington, DC, 2003).

———. *Departments of Commerce, Justice, and State, the Judiciary, and Related Agencies Appropriations for 2005.* Pt. 6. Before a subcommittee of the Committee on Appropriations, House of Representatives, 108th Congress (Washington, DC, 2004).

———. *Departments of Commerce, Justice, and State, the Judiciary, and Related Agencies Appropriations for 2005.* Pt. 10. Before a committee of the Committee on Appropriations, House of Representatives, 108th Congress (Washington, DC, 2004).

———. *Departments of State and Justice, the Judiciary, and Related Agencies Appropriations for 1957.* Before a subcommittee of the Committee on Appropriations, House of Representatives, 84th Congress (Washington, DC, 1956).

———. *Departments of State and Justice, the Judiciary, and Related Agencies Appropriations for 1959.* Before a subcommittee of the Committee on Appropriations, House of Representatives, 85th Congress (Washington, DC, 1958).

———. *Departments of State and Justice, the Judiciary, and Related Agencies Appropriations for 1962.* Before the Committee on Appropriations, House of Representatives, 86th Congress (Washington, DC, 1961).

———. *Departments of State, Justice, and Commerce Appropriations for 1955.* Before a subcommittee of the Committee on Appropriations, House of Representatives, 83rd Congress (Washington, DC, 1954).

———. *Departments of State, Justice, and Commerce, the Judiciary, and Related Agencies Appropriations for 1963.* Before the Committee on Appropriations, House of Representatives, 87th Congress (Washington, DC, 1962).

———. *Departments of State, Justice, and Commerce, the Judiciary, and Related Agencies Appropriations for 1964.* Before a subcommittee of the Committee on Appropriations, House of Representatives, 88th Congress (Washington, DC, 1963).

———. *Departments of State, Justice, and Commerce, the Judiciary, and Related Agencies Appropriations for 1965.* Before a subcommittee of the Committee on Appropriations, House of Representatives, 88th Congress (Washington, DC, 1964).

———. *Departments of State, Justice, and Commerce, the Judiciary, and Related Agencies Appropriations for 1966.* Before the Committee on Appropriations, House of Representatives, 89th Congress (Washington, DC, 1965).

———. *Departments of State, Justice, and Commerce, the Judiciary, and Related Agencies Appropriations for 1976*. Pt. 2. Before the Committee on Appropriations, House of Representatives, 94th Congress (Washington, DC, 1975).
———. *Departments of State, Justice, and Commerce, the Judiciary, and Related Agencies Appropriations for 1977*. Pt. 4. Before a subcommittee of the Committee on Appropriations, House of Representatives, 94th Congress (Washington, DC, 1976).
———. *Departments of State, Justice, and Commerce, the Judiciary, and Related Agencies Appropriations for 1978*. Pt. 5. Before the Committee on Appropriations House of Representatives, 95th Congress (Washington, DC, 1977).
———. *A DHS Intelligence Enterprise: Still Just a Vision or Reality?* Before the Committee on Homeland Security, House of Representatives, 111th Congress (Washington, DC, 2010).
———. *DHS's Vulnerability Assessments in Protecting Our Nation's Critical Infrastructure*. Before the Committee on Homeland Security, House of Representatives, 114th Congress (Washington, DC, 2016). https://www.govinfo.gov/content/pkg/CHRG-114hhrg25264/html/CHRG-114hhrg25264.htm.
———. *Domestic Intelligence Operations for Internal Security Purposes*. Pt. 1. Before the Committee on Internal Security, House of Representatives, 93rd Congress (Washington, DC, 1974).
———. *Economic Espionage*. Before the Subcommittee on Crime of the Committee on the Judiciary, House of Representatives, 104th Congress (Washington, DC, 1996).
———. *Economic Espionage: A Foreign Intelligence Threat to American Jobs and Homeland Security*. Before the Committee on Homeland Security, House of Representatives, 112th Congress (Washington, DC, 2012). https://www.govinfo.gov/content/pkg/CHRG-112hhrg79843/html/CHRG-112hhrg79843.htm.
———. *The Encryption Tightrope: America's Security and Privacy*. Before the Committee on the Judiciary, House of Representatives, 114th Congress (Washington, DC, 2016). https://www.govinfo.gov/content/pkg/CHRG-114hhrg98899/pdf/CHRG-114hhrg98899.pdf.
———. *Evaluating Internal Operation and Implementation of the Chemical Facility Anti-Terrorism Standards Program (CFATS) by the Department of Homeland Security*. Before the Committee on Energy and Commerce, House of Representatives, 112th Congress (Washington, DC, 2012). https://www.govinfo.gov/content/pkg/CHRG-112hhrg75573/pdf/CHRG-112hhrg75573.pdf.
———. *Examining the Mission, Structure, and Reorganization Effort of the National Protection and Programs Directorate*. Before the Committee on Homeland Security, House of Representatives, 114th Congress (Washington, DC, 2015).
———. *FBI Counterintelligence Visits to Libraries*. Before the Subcommittee on Civil and Constitutional Rights of the Committee on the Judiciary, House of Representatives, 100th Congress (Washington, DC, 1988).
———. *FBI Oversight and Authorization Request for Fiscal Year 1989*. Before the Subcommittee on Civil and Constitutional Rights of the Committee on the Judiciary, House of Representatives, 100th Congress (Washington, DC, 1988).
———. *FBI Oversight and Authorization Request for Fiscal Year 1992*. Before the Committee on the Judiciary, House of Representatives, 102nd Congress (Washington, DC, 1991).
———. *FBI Oversight and Authorization, Fiscal Year 1993*. Before the Committee on the Judiciary, House of Representatives, 102nd Congress (Washington, DC, 1992).
———. *Federal Bureau of Investigation: Emergency Supplemental Appropriations Bill for 1940*. Before a subcommittee of the Committee on Appropriations, House of Representatives, 76th Congress (Washington, DC, 1939).
———. *Fiscal Year 2010 Budget for the Office of Intelligence and Analysis of the Department of Homeland Security*. Before the Committee on Homeland Security, House of Representatives, 111th Congress (Washington, DC, 2009). https://www.govinfo.gov/content/pkg/CHRG-111hhrg51634/pdf/CHRG-111hhrg51634.pdf.

———. *Global Terrorism: Threats to the Homeland.* Pt. 2. Before the Committee on Homeland Security, House of Representatives, 116th Congress (Washington, DC, 2020). https://www.govinfo.gov/content/pkg/CHRG-116hhrg40463/pdf/CHRG-116hhrg40463.pdf.

———. *Going Dark: Lawful Electronic Surveillance in the Face of New Technologies.* Before the Committee on the Judiciary, House of Representatives, 112th Congress (Washington, DC, 2011). https://www.govinfo.gov/content/pkg/CHRG-112hhrg64581/pdf/CHRG-112hhrg64581.pdf.

———. *Hearing on Public-Private Initiatives to Secure the Supply Chain.* Before the Committee on Homeland Security, House of Representatives, 116th Cong. (Washington, DC, 2019). https://homeland.house.gov/imo/media/doc/Testimony-Kolasky1.pdf.

———. *Homeland Security Intelligence: Its Relevance and Limitations.* Before the Committee on Homeland Security, House of Representatives, 111th Congress (Washington, DC, 2009). https://www.govinfo.gov/content/pkg/CHRG-111hhrg49943/pdf/CHRG-111hhrg49943.pdf.

———. *Homeland Security Investigations: Examining DHS's Efforts to Protect American Jobs and Secure the Homeland.* Before the Subcommittee on Oversight, Investigations, and Management of the Committee on Homeland Security, House of Representatives, 112th Congress (Washington, DC, 2011). https://www.govinfo.gov/content/pkg/CHRG-112hhrg72254/pdf/CHRG-112hhrg72254.pdf.

———. *How Is America Safer? A Progress Report on the Department of Homeland Security.* Before the Select Committee on Homeland Security, House of Representatives, 108th Congress (Washington, DC, 2003).

———. *H.R. 2868: The "Chemical Facility Anti-Terrorism Act of 2009."* Before the Committee on Homeland Security, House of Representatives, 111th Congress (Washington, DC, 2009). https://www.govinfo.gov/content/pkg/CHRG-111hhrg51493/pdf/CHRG-111hhrg51493.pdf.

———. *H.R. 4007: The Chemical Facility Anti-Terrorism Standards Authorization and Accountability Act of 2014.* Before the Committee on Homeland Security, House of Representatives, 113th Congress (Washington, DC, 2014). https://www.govinfo.gov/content/pkg/CHRG-113hhrg88171/pdf/CHRG-113hhrg88171.pdf.

———. *H.R. 6588: National Intelligence Act of 1980.* Before the Permanent Select Committee on Intelligence, House of Representatives, 96th Congress, 2nd session (Washington, DC, 1980).

———. *Investigation and Study of the Administration, Operation, and Enforcement of the Export Control Act of 1949 and Related Acts.* Report of the Select Committee on Export Control, House of Representatives, 87th Congress (Washington, DC, 1962. https://books.google.com/books?id=iNQz97wiQbgC&pg=RA105-PA59&lpg=RA105-PA59&dq=secretary+of+commerce+revoked+soviet+license+1961&source=bl&ots=hZNjDZORfi&sig=ACfU3U1AaiQ2_7nTvwGC0s2EWNCd_Ag8Ow&hl=en&sa=X&ved=2ahUKEwiE1OPKoqPqAhV3mHIEHeGAAdEQ6AEwCXoECA0QAQ#v=onepage&q=secretary%20of%20commerce%20revoked%20soviet%20license%201961&f=false.

———. *Investigative Report on the U.S. National Security Issues Posed by Chinese Telecommunications Companies Huawei and ZTE.* Report by Chairman Mike Rogers and Ranking Member C. A. Dutch Ruppersberger of the Permanent Select Committee on Intelligence, House of Representatives, 112th Congress (Washington, DC, 2012).

———. *Is the Office of Intelligence and Analysis Adequately Connected to the Broader Homeland Communities?* Before the Subcommittee on Intelligence, Information Sharing, and Terrorism Risk Assessment of the Committee on Homeland Security, House of Representatives, 111th Congress (Washington, DC, 2010).

———. *PRC Acquisition of U.S. Technology.* Select Committee of the United States House of Representatives (Washington, DC, 1999). https://www.govinfo.gov/content/pkg/GPO-CRPT-105hrpt851/pdf/GPO-CRPT-105hrpt851-1-5.pdf.

———. *The President's Fiscal Year 2009 Budget Request for the Department of Homeland Security.* Before the Committee on Homeland Security, House of Representatives, 110th Congress (Washington, DC, 2008). https://www.govinfo.gov/content/pkg/CHRG-110hhrg44512/pdf/CHRG-110hhrg44512.pdf.

———. *The Progress of the DHS Chief Intelligence Officer.* Before the Subcommittee on Intelligence, Information Sharing, and Terrorism Risk Assessment of the Committee on Homeland Security, House of Representatives, 109th Congress (Washington, DC, 2006).

———. *Protecting American Interests Abroad: U.S. Citizens, Businesses, and Nongovernmental Organizations.* Before the Committee on Government Reform, House of Representatives, 107th Congress (Washington, DC, 2001). https://www.govinfo.gov/content/pkg/CHRG-107hhrg75955/pdf/CHRG-107hhrg75955.pdf.

———. *Science, the Departments of State, Justice, and Commerce, and Related Agencies Appropriations for 2007.* Pt. 10. Before the Committee on Appropriations, House of Representatives, 109th Congress (Washington, DC, 2006).

———. *Securing Our Nation's Chemical Facilities: Building on the Progress of the CFATS Program.* Before the Committee on Homeland Security, House of Representatives, 116th Congress (Washington, DC, 2019). https://www.govinfo.gov/content/pkg/CHRG-116hhrg35379/pdf/CHRG-116hhrg35379.pdf.

———. *Turning Spy Satellites on the Homeland: The Privacy and Civil Liberties Implications of the National Applications Office.* Before the Committee on Homeland Security, House of Representatives, 110th Congress (Washington, DC, 2007). https://www.govinfo.gov/content/pkg/CHRG-110hhrg48963/pdf/CHRG-110hhrg48963.pdf.

———. *United States Counterintelligence and Security Concerns, 1986.* Report of the House Permanent Select Committee on Intelligence (Washington, DC, 1987). https://www.cia.gov/library/readingroom/docs/CIA-RDP91B00390R000200160014-6.pdf.

———. *West Fertilizer, Off the Grid: The Problem of Unidentified Chemical Facilities.* Before the Committee on Homeland Security, House of Representatives, 113th Congress (Washington, DC, 2013). https://www.govinfo.gov/content/pkg/CHRG-113hhrg86244/pdf/CHRG-113hhrg86244.pdf.

———. *Worldwide Threats to the Homeland.* Before the Committee on Homeland Security, House of Representatives, 113th Congress (Washington, DC, 2014). https://www.govinfo.gov/content/pkg/CHRG-113hhrg93367/pdf/CHRG-113hhrg93367.pdf.

———. *Worldwide Threats to the Homeland: ISIS and the New Wave of Terror.* Before the Committee on Homeland Security, House of Representatives, 114th Congress (Washington, DC, 2016). https://www.govinfo.gov/content/pkg/CHRG-114hhrg25265/pdf/CHRG-114hhrg25265.pdf.

US Senate. *Biological Weapons: The Threat Posed by Terrorists.* Before the Subcommittee on Technology, Terrorism, and Government Information of the Committee on the Judiciary, 105th Congress (Washington, DC, 1998).

———. *Commerce, Justice, Science, and Related Agencies Appropriations, FY 2016.* Before the Appropriations Committee, 114th Congress (Washington, DC, 2016). https://www.govinfo.gov/content/pkg/CHRG-114shrg93106/pdf/CHRG-114shrg93106.pdf.

———. *Confronting the Terrorist Threat to the Homeland: Six Years After 9/11.* Before the Committee on Homeland Security and Governmental Affairs, 110th Congress (Washington, DC, 2007). https://www.govinfo.gov/content/pkg/CHRG-110shrg38842/pdf/CHRG-110shrg38842.pdf.

———. *Counterterrorism.* Before the Committee on the Judiciary, 107th Congress (Washington, DC, 2002).

———. *Counterterrorism: Evaluating the 5 Year Plan.* Before a subcommittee of the Committee on Appropriations, 105th Congress (Washington, DC, 1998).

———. *Counterterrorism, Counterintelligence, and the Challenges of "Going Dark."* Before the Select Committee on Intelligence, 114th Congress (Washington, DC,

2015). https://www.govinfo.gov/content/pkg/CHRG-114shrg27896/pdf/CHRG-114shrg27896.pdf.
———. *Critical Information Infrastructure Protection: The Threat Is Real.* Before the Committee on the Judiciary, 106th Congress (Washington, DC, 1999).
———. *Critical Infrastructure Protection: Toward a New Policy Directive.* Before the Committee on the Judiciary, 105th Congress (Washington, DC, 1998).
———. *Critical Infrastructure Protection: Who's in Charge?* Before the Committee on Governmental Affairs, 107th Congress (Washington, DC, 2001).
———. *Current and Future Worldwide Threats to the National Security of the United States.* Before the Committee on Armed Services, 109th Congress (Washington, DC, 2006). https://www.govinfo.gov/content/pkg/CHRG-109shrg32745/pdf/CHRG-109shrg32745.pdf.
———. *Current and Future Worldwide Threats to the National Security of the United States.* Before the Committee on Armed Services, 112th Congress (Washington, DC, 2012). https://www.govinfo.gov/content/pkg/CHRG-112shrg79855/pdf/CHRG-112shrg79855.pdf.
———. *Current and Projected National Security Threats to the United States.* Before the Select Committee on Intelligence, 105th Congress (Washington, DC, 1998). https://www.govinfo.gov/content/pkg/CHRG-105shrg51954/pdf/CHRG-105shrg51954.pdf.
———. *Current and Projected National Security Threats to the United States.* Before the Select Committee on Intelligence, 106th Congress (Washington, DC, 2000). https://www.govinfo.gov/content/pkg/CHRG-106shrg65329/pdf/CHRG-106shrg65329.pdf.
———. *Current and Projected National Security Threats to the United States.* Before the Select Committee on Intelligence, 107th Congress (Washington, DC, 2002). https://www.govinfo.gov/content/pkg/CHRG-107shrg82338/pdf/CHRG-107shrg82338.pdf.
———. *Current and Projected National Security Threats to the United States.* Before the Select Committee on Intelligence, 108th Congress (Washington, DC, 2003). https://www.govinfo.gov/content/pkg/CHRG-108shrg89797/pdf/CHRG-108shrg89797.pdf.
———. *Current and Projected National Security Threats to the United States.* Before the Select Committee on Intelligence, 109th Congress (Washington, DC, 2005). https://www.govinfo.gov/content/pkg/CHRG-109shrg22379/pdf/CHRG-109shrg22379.pdf.
———. *Current and Projected National Security Threats to the United States.* Before the Select Committee on Intelligence, 110th Congress (Washington, DC, 2007). https://www.govinfo.gov/content/pkg/CHRG-110shrg48098/pdf/CHRG-110shrg48098.pdf.
———. *Current and Projected National Security Threats to the United States.* Before the Select Committee on Intelligence, 112th Congress (Washington, DC, 2012). https://www.govinfo.gov/content/pkg/CHRG-112shrg74790/pdf/CHRG-112shrg74790.pdf.
———. *Current and Projected National Security Threats to the United States.* Before the Select Committee on Intelligence, 114th Congress (Washington, DC, 2016). https://www.govinfo.gov/content/pkg/CHRG-114shrg20544/pdf/CHRG-114shrg20544.pdf.
———. *Cyber Attack: Improving Prevention and Prosecution.* Before the Committee on the Judiciary, 106th Congress (Washington, DC, 2000).
———. *Cyber Attacks: The National Protection Plan and Its Privacy Implications.* Before the Committee on the Judiciary, 106th Congress (Washington, DC, 2000). https://www.govinfo.gov/content/pkg/CHRG-106shrg68776/pdf/CHRG-106shrg68776.pdf.
———. *Cyber Enabled Information Operations.* Before the Committee on Armed Services, 115th Congress (Washington, DC, 2017).
———. *Cyber Security: Responding to the Threat of Cyber Crime and Terrorism.* Before the Committee on the Judiciary, 112th Congress (Washington, DC, 2011).
———. *Cybersecurity, Terrorism, and Beyond: Addressing Evolving Threats to the Homeland.* Before the Committee on Homeland Security and Governmental Affairs, 113th Congress (Washington, DC, 2014). https://www.govinfo.gov/content/pkg/CHRG-113shrg92903/pdf/CHRG-113shrg92903.pdf.

———. *DEA and FBI.* Pt. 3. Before the Committee on the Judiciary, 100th Congress (Washington, DC, 1987).
———. *Department of Defense Authorization for Appropriations for Fiscal Year 2000 and the Future Years Defense Program.* Pt. 5. Before the Committee on Armed Services, 106th Congress (Washington, DC, 1999).
———. *Department of Homeland Security Appropriations for Fiscal Year 2004.* Before the Committee on Appropriations, 108th Congress (Washington, DC, 2003). https://www.govinfo.gov/content/pkg/CHRG-108shrg2910448/html/CHRG-108shrg2910448.htm.
———. *Department of Homeland Security Appropriations for Fiscal Year 2007.* Pt. 2. Before the Committee on Appropriations, 109th Congress (Washington, DC, 2006).
———. *Department of Homeland Security Status Report: Assessing Challenges and Measuring Progress.* Before the Committee on Homeland Security and Governmental Affairs, 110th Congress (Washington, DC, 2007).
———. *Department of Homeland Security's Budget Submission for Fiscal Year 2005.* Before the Committee on Governmental Affairs, 108th Congress (Washington, DC, 2004).
———. *Department of Homeland Security's Information Analysis and Infrastructure Protection Budget Proposal for Fiscal Year 2005.* Before the Select Committee on Homeland Security, 108th Congress (Washington, DC, 2004).
———. *Departments of State, Justice, and Commerce, the Judiciary, and Related Agencies Appropriations for Fiscal Year 1972.* Before a subcommittee of the Committee on Appropriations, 92nd Congress (Washington, DC, 1971).
———. *Departments of State, Justice, and Commerce, the Judiciary, and Related Agencies, Appropriations for Fiscal Year 1975.* Pt. 1. Before a subcommittee of the Committee on Appropriations, 93rd Congress (Washington, DC, 1974).
———. *Departments of State, Justice, and Commerce, the Judiciary, and Related Agencies Appropriations for Fiscal Year 1978.* Pt. 1. Before a subcommittee of the Committee on Appropriations, 95th Congress (Washington, DC, 1977).
———. *Departments of State, Justice, and Commerce, the Judiciary, and Related Agencies Appropriations for Fiscal Year 1978.* Pt. 6. Before a subcommittee of the Committee on Appropriations, 95th Congress (Washington, DC, 1977).
———. *Departments of State, Justice, and Commerce, the Judiciary, and Related Agencies Appropriations for Fiscal Year 1980.* Before a subcommittee of the Committee on Appropriations, 96th Congress (Washington, DC, 1979).
———. *Departments of State, Justice, Commerce, and the Judiciary Appropriations for 1951.* Pt. 1. Before the Committee on Appropriations, 81st Congress (Washington, DC, 1950).
———. *Departments of State, Justice, the Judiciary, and Related Agencies Appropriations, 1962.* Before the Committee on Appropriations, 87th Congress (Washington, DC, 1961).
———. *Economic Espionage and Trade Secret Theft: Are Our Laws Adequate for Today's Threats?* Before the Committee on the Judiciary, 113th Congress (Washington, DC, 2014). https://www.govinfo.gov/content/pkg/CHRG-113shrg96009/pdf/CHRG-113shrg96009.pdf.
———. *FBI Budget and Oversight for Fiscal Year 1987.* Before the Subcommittee on Security and Terrorism of the Committee on the Judiciary, 99th Congress (Washington, DC, 1986).
———. *FBI Oversight.* Before the Subcommittee on Security and Terrorism of the Committee on the Judiciary, 97th Congress (Washington, DC, 1982).
———. *FBI Oversight.* Before the Committee on the Judiciary, 109th Congress (Washington, DC, 2006).
———. *FBI Oversight and Authorization.* Before the Subcommittee on Security and Terrorism of the Committee on the Judiciary, 98th Congress (Washington, DC, 1983).

———. *FBI Oversight and Budget Authorization.* Before the Subcommittee on Security and Terrorism of the Committee on the Judiciary, 98th Congress (Washington, DC, 1984).
———. *FBI Oversight and Budget Authorization for Fiscal Year 1986.* Before the Subcommittee on Security and Terrorism of the Committee on the Judiciary, 99th Congress (Washington, DC, 1985).
———. *FBI Statutory Charter.* Pt. 1. Before the Committee on the Judiciary, 95th Congress (Washington, DC, 1979).
———. *Federal Bureau of Investigation Oversight.* Before the Committee on the Judiciary, 109th Congress (Washington, DC, 2005).
———. *Fifteen Years After 9/11: Threats to the Homeland.* Before the Committee on Homeland Security and Governmental Affairs, 114th Congress (Washington, DC, 2016). https://www.govinfo.gov/content/pkg/CHRG-114shrg25160/pdf/CHRG-114shrg25160.pdf.
———. *Final Report of the Select Committee to Study Governmental Operations with Respect to Intelligence Activities.* Book 3, *Supplementary Detailed Staff Reports on Intelligence Activities and the Rights of Americans.* 94th Congress (Washington, DC, 1976). https://www.intelligence.senate.gov/sites/default/files/94755_III.pdf.
———. *Final Report of the Select Committee to Study Governmental Operations with Respect to Intelligence Activities.* Book 6, *Supplementary Reports on Intelligence Activities.* 94th Congress (Washington, DC, 1976).
———. *Foreign Intelligence Surveillance Act of 1978.* Before the US Senate Select Committee on Intelligence, 95th Congress (Washington, DC, 1978).
———. *Foreign Terrorists in America: Five Years After the World Trade Center.* Before the Subcommittee on Technology, Terrorism, and Government Information of the Committee on the Judiciary, 105th Congress (Washington, DC, 1998).
———. *The Future of Homeland Security.* Before the Committee on Homeland Security and Governmental Affairs, 112th Congress (Washington, DC, 2012). https://www.govinfo.gov/content/pkg/CHRG-112shrg76059/pdf/CHRG-112shrg76059.pdf.
———. *The Homeland Security Department's Budget Submission for Fiscal Year 2011.* Before the Committee on Homeland Security and Governmental Affairs, 111th Congress (Washington, DC, 2010).
———. *The Homeland Security Department's Budget Submission for Fiscal Year 2012.* Before the Committee on Homeland Security and Governmental Affairs, 112th Congress (Washington, DC, 2011).
———. *Improving Our Ability to Fight Cybercrime: Oversight of the National Infrastructure Protection Center.* Before the Committee on the Judiciary, 107th Congress (Washington, DC, 2001).
———. *Meeting the Espionage Challenge: A Review of United States Counterintelligence and Security Programs.* Report of the Select Committee on Intelligence (Washington, DC, 1986). https://www.cia.gov/library/readingroom/docs/CIA-RDP90-00530R000300620021-3.pdf.
———. *Oversight of the Federal Bureau of Investigation.* Before the Committee on the Judiciary, 105th Congress (Washington, DC, 1997).
———. *Oversight of the Federal Bureau of Investigation.* Before the Committee on the Judiciary, 113th Congress (Washington, DC, 2013).
———. *Reforming the FBI in the 21st Century.* Before the Committee on the Judiciary, 107th Congress (Washington, DC, 2002).
———. *S. 2726 to Improve U.S. Counterintelligence Measures.* Before the Select Committee on Intelligence, 101st Congress (Washington, DC, 1990).
———. *Securing Our Infrastructure: Private/Public Information Sharing.* Before the Committee on Governmental Affairs, 107th Congress (Washington, DC, 2002).
———. *Special Report of the Select Committee on Intelligence, January 4, 1995–October 3, 1996.* 105th Congress (Washington, DC, 1996). https://www.govinfo.gov/content/pkg/CRPT-105srpt1/pdf/CRPT-105srpt1.pdf.

———. *Ten Years After 9/11: 2011.* Before the Committee on Homeland Security and Governmental Affairs, 112th Congress (Washington, DC, 2012).

———. *Threats to the Homeland.* Before the Committee on Homeland Security and Governmental Affairs, 113th Congress (Washington, DC, 2013). https://www.govinfo.gov/content/pkg/CHRG-113shrg86635/pdf/CHRG-113shrg86635.pdf.

———. *Threats to the Homeland.* Before the Committee on Homeland Security and Governmental Affairs, 115th Congress (Washington, DC, 2017). https://www.govinfo.gov/content/pkg/CHRG-115shrg29657/pdf/CHRG-115shrg29657.pdf.

———. *Threats to the US Research Enterprise: China's Talent Recruitment Plans.* Staff report (Washington, DC, 2019). https://www.hsgac.senate.gov/imo/media/doc/2019-11-18%20PSI%20Staff%20Report%20-%20China's%20Talent%20Recruitment%20Plans.pdf.

———. *Transfer of United States High Technology to the Soviet Union and Soviet Bloc Nations.* Before the Committee on Governmental Affairs, 97th Congress (Washington, DC, 1982). https://www.cia.gov/library/readingroom/docs/CIA-RDP85M00364R001001520004-2.pdf.

Wakabayashi, Daisuke, and Scott Shane. "Google Will Not Renew Pentagon Contract That Upset Employees." *New York Times,* June 1, 2018. https://www.nytimes.com/2018/06/01/technology/google-pentagon-project-maven.html.

Watkins, Ali. "Russia Escalates Spy Games After Years of US Neglect." *Politico,* June 1, 2017. https://www.politico.com/story/2017/06/01/russia-spies-espionage-trump-239003.

Weiner, Tim. "Lies and Rigged 'Star Wars' Test Fooled the Kremlin, and Congress." *New York Times,* August 18, 1993. https://www.nytimes.com/1993/08/18/us/lies-and-rigged-star-wars-test-fooled-the-kremlin-and-congress.html.

Weiss, Gus. "The Farewell Dossier: Duping the Soviets." 2007. https://www.cia.gov/library/center-for-the-study-of-intelligence/csi-publications/csi-studies/studies/96unclass/farewell.htm.

Wejrauch, Bruce. "Operation Exodus: The United States Government's Program to Intercept Illegal Exports of High Technology." *Computer Law Journal* 7, no. 2 (Fall 1986). https://repository.jmls.edu/cgi/viewcontent.cgi?article=1478&context=jitpl.

Whitney, Robert F. "2 Seized in Mexico in Sabotage Blasts of Towers in U.S." *New York Times,* June 19, 1961. https://timesmachine.nytimes.com/timesmachine/1961/06/19/97674007.pdf?pdf_redirect=true&ip=0.

Wingfield, Nick, Ted Bridis, and Neil King Jr. "Earthlink Says It Won't Install Device for FBI." *Wall Street Journal,* July 14, 2000.

Wirtz, James J. "The Cyber Pearl Harbor Redux: Helpful Analogy or Cyber Hype." *Intelligence and National Security* 33, no. 5 (2018). https://www.tandfonline.com/doi/abs/10.1080/02684527.2018.1460087?journalCode=fint20.

Yadron, Danny. "Iranian Hackers Infiltrated New York Dam in 2013." *Wall Street Journal,* December 20, 2015. https://www.wsj.com/articles/iranian-hackers-infiltrated-new-york-dam-in-2013-1450662559.

Yuan, Li, and Daisuke Wakabayashi. "Google Seeking a Return to China, Is Said to Be Building a Censored Search Engine." *New York Times,* August 1, 2018. https://www.nytimes.com/2018/08/01/technology/china-google-censored-search-engine.html.

Zonderman, Jon. "Policing High Tech Exports." *New York Times,* November 27, 1983. https://www.nytimes.com/1983/11/27/magazine/policing-high-tech-exports.html?searchResultPosition=5.

Index

About the Book

AS A PROVIDER OF VITAL INFRASTRUCTURE AND TECHNOLOGY, THE private sector has become an essential contributor to US national security—and the target of hackers and terrorists. Darren Tromblay traces the evolution of an often fraught public-private partnership to explore how the complex web of intelligence agencies has struggled to protect critical economic and industrial interests.

Darren E. Tromblay has served as an intelligence analyst with the US government since 2005.